ENCYCLOPEDIA OF THE MIDDLE EAST PEACE PROCESS AND THE MIDDLE EAST/NORTH AFRICA ECONOMIC COMMUNITY

ENCYCLOPEDIA OF
THE MIDDLE EAST PEACE PROCESS
AND
THE MIDDLE EAST/NORTH AFRICA
ECONOMIC COMMUNITY

Jerry M. Rosenberg
Professor of International Business
Director, Center for Middle East/North Africa Business Studies
Rutgers University
Graduate School of Management

JAI PRESS INC.

Greenwich, Connecticut *London, England*

Library of Congress Cataloging-in-Publication Data

Rosenberg, Jerry Martin.
Encyclopedia of the Middle East peace process and the Middle East/North Africa economic community / Jerry M. Rosenberg.
p. cm.
ISBN 0-7623-0350-6
1. Arab-Israeli conflict, 1993- —Encyclopedias. 2. Middle East—Economic relations—Encyclopedias. I. Title.
DS119.76.R67 1997
956'.003—dc21 97-39512
CIP

55 Old Post Road No. 2
Greenwich, Connecticut 06830

JAI PRESS LTD.
38 Tavistock Street
Covent Garden
London WC2E 7PB
England

ISBN: 0-7623-0350-6

Library of Congress Catalog Card Number: 97-39512

Manufactured in the United States of America

Affectionately Dedicated
To Those Who Have
Provided Me With Love And Support In
The Multiple Stages Of This Work

My family

Ellen
Liz and Jon
Lauren and Bob
Bess and Ella

INTRODUCTION

There is a lengthy history to the region that is referred to as the Middle East, at times including nations of North Africa, but more often countries bordering and near to Israel. More recently, definitions have attempted to make the distinction between those states east of Egypt to be named as part of the Middle East, while those on the western slope of Egypt as North Africa.

Events in the troubled area around Israel are measured in the sands of century-old conflicts and tensions. Peace barely survives as the world searches for the glue that will hold the entire process together. Arguably, the singular best hope to stabilize the area remains cross-border activities, especially economic.

The Madrid Conference of 1991, followed by Oslo I and Oslo II, Treaties with Jordan and Israel, the Palestinian Liberation Authority and Israel, annual economic gatherings in Casablanca, Amman, and Cairo, all reveal a mood that is unparalleled in the saga of the region.

This encyclopedia was compiled to both add to the literature and understanding of historic events leading up to the present efforts at a regional peace process and to the projected evolution of a Middle East/North Africa Economic Community, with the gradual free movement of capital, goods, and services for participating nations.

The intentional structure of an encyclopedia enables the student, teacher, researcher, government, and business representative to rapidly locate sources of needed information.

This encyclopedia covers specific facts, events, issues, ideologies, controversies, and actions needed to increase understanding in the pursuit of peace, cooperation, and productivity. Much of the material was drawn from numerous official sources, with special emphasis on interpreting key aspects.

I am especially thankful to the many individuals, authors, and international experts who have devoted their time to discussing the varying elements of this work. In particular, counsel from Dr. Mostafa Terrab,

Executive Secretariat of the Rabat-based Middle East/North Africa Economic Summits, was most valued and supportive. In addition, over three years, during interviews and meetings with dozens of specialists and distinguished participants in the process for peace and economics within the region, I have been provided with the most current insights and materials, making this work possible. The list includes the following, presented alphabetically, who are identified in their position at the time we met or when they spoke:

Ismat Abdelmajid, Secretary-General of the Arab League, Egypt
Samir Abdullah, Director, Department of Economic Policies and Project Selection, PECDAR, Palestinian Authority
Mayor Admon, Economic Advisor, Ministry of Finance of Israel
Abdulaziz Al-Dakhil, Minister of Planning and Minister of State for Cabinet Affairs of Kuwait
Abdulatif Al-Homad, Director-General and Chairman of the Board of Directors, Arab Fund for Economic and Social Development, Kuwait
Nasser Bin Hamad M. Al-Khalifa, Ambassador Extraordinary and Plenipotentiary, Permanent Representative of the State of Qatar to the United Nations
Abdullah Al-Nuaimi, Minister of Foreign Affairs of the United Arab Emirates
Refael Benvenisti, Senior Adviser to the Minister of Finance, Israel
Mohamed Bin Mubarak Al-Khalifa, Minister of Foreign Affairs of Bahrain
Jassim Al-Mannai, Chairman of the Board and Director-General, Arab Monetary Fund, Abu Dhabi
Mohammad bin Amain Al-Saleh, Director Planning, Ministry of Electricity and Water, Oman
Sheikh Hamad Bin Jassim Bin Jabr Al-Thani, Minister of Foreign Affairs of Qatar
Yasir Arafat, Chairman of the Palestine Liberation Authority
Hana Ashwari, Palestinian negotiator
Bassem Awadallah, Advisor to the Prime Minister of Jordan
Andre Azoulay, Senior Advisor to King Hassan, Morocco
Ehud Barak, Candidate for Israel's Labor Party leadership
Yossi Beilin, Minister of Economy and Planning of Israel and Candidate for Labor Party leadership
Khalil Benharbit, Director of the Domaines Agricoles, Morocco
Refael Benvenisti, Senior Advisor, Ministry of Finance, Israel
Youssef Boutros-Ghali, Minister of State at the Council of Ministers for International Cooperation, Egypt
David Brodet, Director-General, Ministry of Finance, Israel

Sayed El Bous, Advisor to the Minister of Economy and External Trade, Egypt
Oded Eran, Deputy Director General, Economic Affairs, Israel, Israeli Ambassador to Jordan
Murtadha bin Mohammad bin Fadhil, Head, Oman Development Bank
Nabil Fahmy, Political Advisor to the Minister of Foreign Affairs, Egypt
Abdelatif Filali, Prime Minister, Minister of Foreign Affairs and Cooperation of Morocco
Jacob Frenkel, Governor, Bank of Israel
Mohamed Ghannouchi, Minister of International Cooperation and Foreign Investment of Tunisia
Onur Gokce, Ambassador of Turkey to Israel
Mahdi Adul Hadi, Palestinian PASSIA Research Institute
Mohammed Halaiqah, Director-General, Jordan Export Development Corporation, Jordan
Badr Hamam, Assistant Minister for Arab Affairs and the Middle East, Ministry of Foreign Affairs, Egypt
Micha Harish, Israel Minister of Trade and Industry
His Majesty King Hassan II of Morocco
Hassan Bin Talal, HRH Crown Prince of Jordan
Jawad Naji Hirzallah, Under Secretary, Ministry of Economy and Trade and Industry, Palestinian Authority
His Majesty King Hussein of Jordan
Osama bin Jafar Faquh, Minister of Trade and Industry of Saudi Arabia
Basel Jardaneh, Minister of Finance of Jordan
Yousuf Kamal, Undersecretary, Ministry of Finance, Economics and Trade of Qatar
Alon Liel, Director-General, Ministry of Economy and Planning, Israel
Hamdi Loza, Head, International Economy Affairs Division, Ministry of Foreign Affairs, Egypt
Essmat Abdul Majid, Secretary General of the Arab League, Cairo
Nawaf Masalha, Israeli Deputy Minister of Health
Ephraim Ben-Matityahu, Consul for Academic Affairs and Ministry of Foreign Affairs, Israel
Dan Meridor, Finance Minister of Israel
Amre Mahmoud Moussa, Minister of Foreign Affairs, Egypt
Hani Mulki, President, Royal Scientific Society, Jordan
Ora Namir, Israel Minister of Labor and Social Welfare
Mohamad Z. Nashashibi, Minister of Finance of the Palestinian Authority
Benjamin Netanyahu, Prime Minister of Israel
Shimon Peres, Prime Minister/Foreign Minister of Israel
Zohar Peri, Director of Foreign Trade Administration, Israel Ahmed Qurie, Economic Affairs, Palestinian Liberation Authority

Yitzhak Rabin, Israel Prime Minister/Defense Minister
Itamar Rabinovich, Former Israeli Ambassador to the United States and Ettinger Professor of Contemporary Middle East History, Tel Aviv University
Ali H. Abu Ragheb, Minister of Trade and Industry of Jordan
Raouf Saad, Deputy Assistant Foreign Minister for Regional Economic Cooperation, Egypt
Uri Savir, Director General, Israel Ministry-Foreign Affairs, President of the Peres Institute for Peace
Nabil Sha'ath, Minister of Planning and International Cooperation of the Palestinian Authority
Natan Sharansky, Minister of Trade and Industry, Israel
Mohammed Smadi, Director-General, Ministry of Industry and Trade, Jordan
Yosi Snir, Director General, Israel Minister of Trade
Ahmed Snoussi, Permanent UN Representative of Morocco
Mark Sofer, Director, Department for Middle East Economic Affairs, Ministry of Foreign Affairs, Israel
Abd Al-Aziz As-Sulaim Sulaiman, Minister of Commerce of Saudi Arabia
Makboul Bin Ali Bin Sultan, Minister of Commerce and Industry of Oman
Mostafa Terrab, Secretary-General, Executive Secretariat, Middle East/ North Africa Economic Summit, Morocco
Ahmed Tibi, Special Advisor to Palestinian Chairman Arafat
Ezer Weizman, President of Israel
Al-Habib bin Yahya, Minister of Foreign Affairs of Tunisia
Joseph Vardi, Advisor to the Israeli Government in Regional Economic Operations

To my wife Ellen, a 37-year companion and long-time supporter of my varied projects and interests, who has traveled with me and provided great insight and wisdom, I remain forever grateful. To Liz and Jon, to Lauren and Bob, granddaughters Bess and Ella, my immediate family, I continue to be thankful for their enduring encouragement, and most thankful for their love.

This encyclopedia is a companion work to my earlier book *The Peace Dividend: Creating a Middle East/North Africa Community,* commissioned and published by the Secretariat of the Middle East/North Africa Economic Summits (Rabat), and then distributed at the Cairo Economic Conference in November 1996.

August 1997

A

AAA: Agreed Adjacent Area. See Hebron (Appendix).

AAAID: See Arab Authority for Agricultural Investment and Development.

AAB: See Arab African Bank.

Abdullah Ibn Hussein (King): Jordan's King Abdullah is assassinated in Jerusalem on July 20, 1951 in the presence of his grandson who succeeded to the Jordanian throne a year later. See chronology (July 20, 1951), Hussein ibn Talal (His Majesty).

Abdullah, Ali (Colonel): Born 1942. Seeking to mend fences with Saudi Arabia after five years of isolation in the wake of Yemen's support for Iraq's occupation of Kuwait. Became president and commander-in-chief of the armed forces in July 1978 after participation in the 1974 coup landed him the top security post in Taiz province. See Yemen.

Abdullah bridge: See Cairo Agreement.

ABEDA: See Arab Bank for Economic Development in Africa.

Abou Amar: See Arafat, Yasir.

Abu Amar: See Arafat, Yasir.

Abu Dhabi: See United Arab Emirates.

Abu Mazen Plan: See Beilin-Abu Mazen Plan.

abuses: See Treaty of the Middle East/North Africa Community (Article 12).

ACC: See Arab Cooperation Council.

Achille Lauro: See Palestine Liberation Front, terrorism, United States.

ACM: See Arab Common Market.

Aden: See Yemen.

Ad Hoc Liaison Committee of Donors: See Agreement on Security Arrangements for Hebron.

ADIBOR: Abu Dhabi Interbank Offered Rate.

adu Allah: Used by some Islamists to describe the enemies of God.

advertising: See television.

AED: The international standard code for the currency of the United Arab Emirates—the UAE dirham.

AFESD: See Arab Fund for Economic and Social Development.

AG: See Arabian Gulf.

Agenda For Peace: See Israel-Jordan Common Agenda.

Agenda 21: See Multilateral Working Group on the Environment.

Agreed Minutes (Egypt and Israel): See Egyptian-Israel Peace Treaty.

Agreed Minutes (Gaza Strip and Jericho): See Declaration of Principles on Interim Self-Government Arrangements.

Agreement of Understanding by Israel and Lebanon: On April 26, 1996 Israel, Syria, and Lebanon agreed on a written set of understandings to stop 16 days of Hezbollah rocketing and shelling in northern Israel and southern Lebanon. The Israeli attacks (Operation Grapes of Wrath) had killed more than 150 Lebanese, forced hundred of thousands of people in both countries to flee their homes and caused millions of dollars of damage.

U.S. Secretary of State Christopher played a dominant role in this Agreement.

The agreement did not end fighting in the zone of southern Lebanon that Israel occupied as a buffer, but aimed to contain it, preventing the attacks on civilians that flared into the current conflict. Under the unsigned Agreement of Understanding reached between Israel and Lebanon, in consultation with Syria:

The United States understands that after discussions with the governments of Israel and Lebanon, and in consultation with Syria, Lebanon, and Israel will ensure the following:

1. Armed groups in Lebanon will not carry out attacks by Katyusha rockets or by any kind of weapons into Israel.
2. Israel and those cooperating with it will not fire any kind of weapon at civilians or civilian targets in Lebanon.
3. Beyond this, the two parties commit to ensuring that under no circumstances will civilians be the target of attack and that civilian populated areas and industrial and electrical installations will not be used as launching groups for attacks.
4. Without violating this understanding, nothing herein shall preclude any party from exercising the right of self-defense.

A Monitoring Group is established consisting of the United States, France, Syria, Lebanon, and Israel. Its task will be to monitor the application of the understanding stated above. Complaints will be submitted to the Monitoring Group. In the event of a claimed violation of the understanding, the party submitting the complaint will do so within 24 hours. Procedures for dealing with the complaints will be set by the Monitoring Group.

The United States will also organize a Consultative Group, to consist of France, the European Union, Russia, and other interested parties, for the purpose of assisting in the reconstruction needs of Lebanon.

It is recognized that the understanding to bring the current crisis between Lebanon and Israel to an end cannot substitute for a permanent solution. The United States understands the importance of achieving a comprehensive peace in the region. Toward this end, the United States proposes the resumption of negotiations between Syria and Israel and between Lebanon and Israel at a time to be agreed upon, with the objective of reaching comprehensive peace. The United States understands that it is desirable that these negotiations be conducted in a climate of stability and tranquility.

The understanding will be announced simultaneously at 1800 hours, April 26, 1996, in all countries concerned. The time set for implementation is 0400 hours, April 27, 1996. See Israel, Lebanon, Syria.

Agreement on Preparatory Transfer of Powers and Responsibilities: See Preparatory Transfer of Powers and Responsibilities.

Agreement on Security Arrangements for Hebron (1994): Signed by Major-General Amnon Lipkin-Shahak for Israel and Dr. Nabeel Sha'ath for the PLO on March 31, 1994.

Agreement

In the aftermath of the horrendous massacre in Hebron, and in response to the heightened needs of the Palestinians for security throughout the West Bank and the Gaza Strip and particularly in Hebron and in accordance with Security Council Resolution 904, delegations of Israel and the PLO met and agreed to take measures set out in this agreement. As soon as this agreement is signed, its implementation will begin and the Gaza-Jericho negotiations will be resumed, as set out below:

A. Temporary International Presence in the City of Hebron

1. In response to the unique situation created in Hebron in the aftermath of the massacre, a temporary international presence will be established in the city of Hebron ("TIPH"). As detailed in paragraph 3 below, the TIPH will assist in promoting stability and in monitoring and reporting the efforts to restore normal life in the city of Hebron, thus creating a feeling of security among Palestinians in the city of Hebron.
2. The two sides shall request donor countries to provide 160 persons, citizens of Norway, Denmark, and Italy, as TIPH personnel, consisting of field observers, office staff, and support personnel, as agreed between the two sides. Changes in the composition of the TIPH may be made from among the donor countries with the consent of both sides. Consistent with its stated tasks, the TIPH personnel shall have no military or police functions.
3. The tasks of TIPH personnel will be:
   a. to provide by their presence a feeling of security to the Palestinians of Hebron.
   b. to help promote stability and an appropriate environment conducive to the enhancement of the well-being of the Palestinians of Hebron and their economic development;
   c. to monitor the efforts to restore the safety of Palestinians and events affecting it and the return to normal life in the city of Hebron; and
   d. to provide reports as set out in paragraph 5 below.
4. In order to facilitate the carrying out of TIPH tasks, a building will be chosen in the city of Hebron as a seat for the TIPH.

5. The TIPH will report to the following:
   a. on specific events—to a Joint Hebron Committee ("JHC"), comprised of two representatives from each side. The senior Palestinian representative will be the Mayor of Hebron and the senior Israeli representative will be the head of the Civil Administration in the District of Hebron. A representative of the TIPH will be invited on a biweekly basis to participate in the JHC meeting in order to report on the TIPH activities.
   b. periodically—to the Joint Israeli-Palestinian Liaison Committee established pursuant to the DOP.

   In addition, the TIPH will provide periodic reports to the Chair of the Ad Hoc Liaison Committee of the Donors.
6. The members of the TIPH shall wear distinctive uniforms with a special emblem, as agreed by the two sides, and their vehicles shall be marked with the same emblem. TIPH members may carry pistols for self-defense purposes.
7. The TIPH will enjoy freedom of movement for the performance of its tasks within the city of Hebron. Such freedom of movement shall not be restricted, except for reasons of imperative military necessity, and then only as an exceptional and temporary measure.
8. The TIPH will establish the modalities of its presence and activity with the agreement of the two sides, with due regard being given to its aforementioned tasks.
9. The expenses of the TIPH will be borne by the donor countries.
10. The TIPH will commence its operation immediately after the signing of this agreement and continue to function for a period of three months. With the consent of the two sides, the TIPH may extend the period or change the scope of its operation, as agreed.

B. The Gaza-Jericho Negotiations

1. The Gaza-Jericho negotiations shall be resumed on Thursday, March 31, 1994. These negotiations shall be accelerated with the objective of making up for lost time.
2. Israel agrees to shorten the withdrawal schedule and accelerate the withdrawal, being guided by the target dates set in the DOP.
3. Immediately after the conclusion of the Gaza-Jericho Agreement, early empowerment negotiations will commence, and the two sides will explore possible expansion of the scope of these negotiations beyond the five spheres.
4. The two sides will intensify the negotiations on the interim arrangements consistent with the DOP and guided by its target date.
5. The two sides reiterate their commitment to commence permanent status negotiations as soon as possible, but not later than the begin-

ning of the third year of the interim period, as provided for in Article V of the DOP.

6. Gradual movement into Gaza and Jericho of Palestinian policemen will start one week after the resumption of the Gaza-Jericho negotiations, in order to commence preparations for assuming powers and responsibilities, as agreed by the two sides.

See Declaration of Principles, Gaza-Jericho negotiations.

Compare Agreement on the Temporary International Presence in the City of Hebron (1997).

Agreement on Technical Barriers to Trade of the Final Act of the Uruguay Round of Trade Negotiations: See Protocol on Economic Relations (Article III).

Agreement on the Establishment of a Free Trade Area between the Government of Israel and the Government of the United States of America: See Free Trade Agreement (United States and Israel).

Agreement on the Facilitations of Trade (1981): See Arab (economy).

Agreement on the Gaza Strip and the Jericho Area: See Gaza Strip and the Jericho Area Agreement.

Agreement on the Temporary International Presence in the City of Hebron (1997): Pursuant to Article VII of Annex I of the Israeli-Palestinian Interim Agreement on the West Bank and the Gaza Strip, dated September 28, 1995 ("the Interim Agreement"), which deals with the redeployment of Israeli military forces in the city of Hebron and provides that there will be a Temporary International Presence in Hebron, Israel and the PLO agree as follows:

1. A Temporary International Presence in Hebron (hereinafter "TIPH") is hereby established. It will supersede the existing TIPH operating under the Agreement on the Temporary International Presence in the city of Hebron, dated May 9, 1996. It will be stationed in and operate within the city of Hebron as delineated on map No. 9 attached to the Interim Agreement. In all its activities, the TIPH will relate to Hebron as one city. As detailed in paragraph 5 below, the TIPH will assist in monitoring and reporting the efforts to maintain normal life in the city of Hebron, thus creating a feeling of security among Palestinians in the city of Hebron. The organizational structure, operational guidelines, logistics, support, and privileges and immunities of the TIPH shall be in accordance with the modalities on the establishment of the TIPH concluded by the participating countries with the agreement of the two sides.

2. The two sides shall request Norway, Italy, Denmark, Sweden, Switzerland, and Turkey to provide 180 persons as TIPH personnel. Norway shall be requested to coordinate the establishment and activity of the TIPH. The number of on-duty observers at any given time and changes in the composition of the TIPH shall be with the consent of both sides.
3. TIPH personnel shall have no military or police functions, nor will they interfere in disputes, incidents, or the activities of Israeli security forces or the Palestinian Police.
4. In order to facilitate the carrying out of TIPH tasks, a building will be chosen in Hebron as a seat for the TIPH.
5. The tasks of the TIPH personnel will be:
   a. to promote by their presence a feeling of security to the Palestinians of Hebron;
   b. to help promote stability and an appropriate environment conducive to the enhancement of the well-being of the Palestinians of Hebron and their economic development;
   c. to observe the enhancement of peace and prosperity among Palestinians;
   d. to assist in the promotion and execution of projects initiated by the donor countries;
   e. to encourage economic development and growth in Hebron;
   f. to provide reports as set out in paragraph 7 below; and
   g. to coordinate its activities with the Israeli and Palestinian authorities in accordance with paragraph 7 below.
6. The development projects referred to in paragraph 5.d above will be identified by the Hebron municipality and the relevant Palestinian ministries in cooperation with the member countries of the TIPH.
7. a. Representatives of the TIPH shall be situated at the District Coordination Office ("DCO") in Har Manoah/Jabel Manoah and may also be present at the DCO sub-office in the city of Hebron, in order to coordinate TIPH activity with both sides. The TIPH representatives shall be equipped with the necessary means of communication to enable direct and immediate contact with the TIPH headquarters and field observers.
   b. A Joint Hebron Committee shall be established comprising the Israeli Military Commander and the Palestinian Police Commander of the Hebron District, the Israel and Palestinian heads of the Hebron District Civil Liaison Office, and a representative of the TIPH. This committee shall receive reports from the TIPH on specific events and shall deal with any issues arising from the presence and activity of the TIPH which cannot be

dealt with by the DCO. The committee shall meet on a weekly basis, or on the request of any of its members.

c. The Monitoring and Steering Committee established pursuant to the Interim Agreement will receive periodic reports from the TIPH. Each side of this committee will designate one of its members to form a group together with representatives of the TIPH in order to discuss matters of policy on a biweekly basis or on the request of the TIPH.

8. While on duty, the members of the TIPH shall wear distinctive uniforms with a special emblem, as agreed by the two sides. Their vehicles shall be marked with the same emblem. TIPH members may carry pistols for self-defense purposes.
9. TIPH personnel shall enjoy freedom of movement for the performance of their tasks with the city of Hebron, as set out in the modalities referred to in paragraph 1.
10. The TIPH and its members shall, in conformity with the modalities referred to in paragraph 1, be accorded such privileges and immunities as are necessary for the fulfillment of their tasks, and for the independent exercise of their functions, including immunity of the TIPH premises and the right to import operational supplies and equipment exempt from import taxes and duties. All imported equipment, materials, and goods exempted from import taxes and duties will be re-exported or donated, at the conclusion of the mission of the TIPH, according to applicable customs procedures agreed upon between the two sides.
11. The presence and activities of the TIPH personnel will be in accordance with Annex I, Article VII and other relevant provisions of the Interim Agreement and the Protocol Concerning the Redeployment in Hebron.
12. The expense of the TIPH will be borne by the participating countries.
13. This agreement shall enter into force on February 1, 1997 and shall supersede the Agreement on the Temporary International Presence in the City of Hebron, dated May 9, 1996.
14. The TIPH may commence its operation immediately upon the approval of its modalities of operation by the two sides, and shall continue to function for a period of three months renewable for an additional period of three months unless otherwise agreed between the two sides. With the consent of the two sides, the TIPH may extend the period or change scope of its operation, as agreed.

Date: January 21, 1997

Compare Agreement on Security Arrangements for Hebron (1994).

agriculture:

Agriculture in the Middle East/North Africa

	% of Land used for Agric.	% Annual Change in Agri. Land	% of Agric. Land Irrigated	Cereal Imports (m.tons)	Population Density	Popul. (mill)
Algeria	16.3	0	0.9	5,185	10	25.6
Egypt	2.6	0.2	100	8,580	51	53.6
Iran	36.0	0	9.8	6,250	33	59.9
Iraq	21.6	0	18.5	2,834	40	18.7
Israel	28.5	−0.2	47.8	1,802	218	5.
Jordan	13.1	0.2	4.0	1,491	34	4.1
Kuwait	7.7	0	0.7	427	115	2.1
Lebanon	30.4	0	27.7	356	343	2.8
Libya	8.8	0	1.5	2,290	2	4.7
Morocco	67.5	1.1	4.2	1,578	55	25.7
Oman	4.9	0	3.9	338	7	1.6
Saudi Arabia	40.0	0	0.5	5,273	7	15.4
Sudan	46.5	0	1.7	586	10	25.9
Syria	73.3	−2.2	4.7	2,091	63	12.8
Tunisia	49.2	−0.2	3.5	1,439	48	8.2
UAE	2.6	0.5	2.3	576	6	1.6
Yemen	33.2	0	1.4	1,047	21	12.1

See Agriculture (Center of the Middle East/North Africa Community), Amman Middle East/North Africa Economic Summit, Cairo Middle East/North Africa Economic Conference, Israel-Jordan Peace Treaty, pesticides, Protocol on Economic Relations (Article VIII), water.

Agriculture (Center of the Middle East/North Africa Community): Increased cooperation in agricultural projects is primary to any progress in the community, such as routing water sources to arid regions, joint use of irrigation methods, and utilizing existing know-how and experience in this field; research and development of agricultural products and fertilizers; storage of agricultural products, and so on.

Regional agricultural cooperation is not only necessary but feasible throughout the community. To the millions, it is in their best interest to cultivate crops and produce what takes advantage of the area's unique climatic conditions. Cooperative efforts can occur in:

- agricultural free trade zones
- farms for production and technology transfer
- plant prevention and pest control
- prevention and treatment of livestock diseases
- joint natural disaster hotline
- training of personnel
- joint R&D in husbandry of specially bred livestock

- reduction of post harvest losses of yields through storage and preservation methods
- marine agriculture
- gene banks
- new crops

See Amman Middle East/North Africa Economic Summit, Cairo Middle East/North Africa Economic Conference, Middle East/North Africa Community, Water and Desertification (Center of the Middle East/North Africa Community).

AIB: See Arab International Bank.

AIC: See Arab Investment Company.

aid: See Treaty of the Middle East/North Africa Community (Article 14).

air pollution: See pollution.

air transport: See Amman Middle East/North Africa Economic Summit, aviation, Palestinian Airlines, Transportation (Center of the Middle East/North Africa Community).

al-adu: referred by many Islamists for Israelis, the enemy.

Al-Aqsa mosque: See Jordan.

Alawaites: See minorities.

Albright, Madeleine K.: By mid-June 1997 the U.S. administration gathered to rethink its strategy for the region of the Middle East—how to get the peace negotiations restarted. President Clinton finally adapted a proposal first made by Prime Minister of Israel Netanyahu to speed up on the timetable for talks between Israel and the Palestinians, and address the most nettlesome issues of a final settlement. Officials of the Administration agreed that the central premise of the Oslo peace accords—that gradual negotiations would build trust—no longer held. Administration officials concluded that the time had come to speak out and pressure all sides to move directly to formulate a final peace agreement.

U.S. Secretary of State Albright (1997-) spoke out in early August urging the need to revitalize the peace discussions. She indicated that she would take her first trip to the Middle East as the President's representative by the end of the summer 1997. In addition, she is expected to participate in the Doha Middle East/North Africa Economic Summit.

See Doha Middle East/North Africa Economic Summit.

alcoholism: See Hamas Covenant.

ALF: See Arab Liberation Front.

al-Fatah: The main military wing of the Palestine Liberation Organization. See Fatah.

al-Filastini: See PLO Charter.

Algeria: A guerrilla war was waged by the Algerian FLN (Front de Liberation Nationale) against the French military and administrative authorities from November 1954 until March 1962. Independence from France was proclaimed on July 3, 1962.

Algeria has an economy with great but as-yet unrealized potential. The country boasts a population of 27 million people. It is blessed with substantial natural gas and oil reserves. Algeria's location also offers investors the potential to take advantage of low wage rates, a relatively well-developed infrastructure, as well as cheap and readily accessible energy supplies to export to the European Community and the rest of North Africa.

The transition to a market economy has been slow. In part, this is due to Algeria's hydrocarbon wealth which provided USD $9-12 billion a year in export earnings and 55 percent of government revenues, and enabled successive governments to put off painful macroeconomic reforms. The fall in oil prices in 1993 coupled with Algeria's high external debt payments obliged the government of Algeria (GOA) to conclude an accord with the IMF in May 1994 so that the country could reschedule USD $5.3 billion in external debt payments owed to Algeria's official Paris Club creditors through May 1995. As part of its agreement with the IMF, Algeria already has undertaken a number of important reforms including a 50 percent devaluation of the dinar.

As a result of the Paris Club rescheduling as well as new funds from the IMF, World Bank, France, the European Economic Community, and others, merchandise imports are projected to increase from USD $7.8 billion in 1993 to USD $9.4 billion in 1994. Cereal imports are projected to grow substantially because of what was expected to be another drought-induced poor harvest in 1994. Another major import priority is likely to be raw materials and capital goods for the depressed housing sector. In addition to the above, best prospects for U.S. exports include oil and gas, telecommunications, water resources and other agricultural equipment, computers, pharmaceuticals, and food processing and mining equipment.

Despite a new investment code that offers a number of attractive incentives, the investment climate in 1996 remains plagued by dangerous security conditions. Personnel and facilities of foreign companies have been the targets of terrorist violence since September 1993. The U.S. Department of State has warned U.S. citizens not to travel to Algeria and those that are there to depart unless they have effective protection. All U.S. companies in Algeria have adopted vigorous security programs.

Given the current security situation, most U.S. companies have reduced American staff to a minimum, but are maintaining Algerian staff in order to continue to export to the Algerian market and to position themselves for potential investments should the security environment improve. Algeria is seeking to diversify its foreign suppliers and sources of finance. The United States has been a major beneficiary.

With its constitutional referendum in November 1996 the Algerian government outlawed Islamic parties, stripped power from the Parliament, and allowed the president to rule by decree.

On June 5, 1997 parties representing Algeria's military-backed authorities swept to victory in parliamental elections, but a low turnout and accusations of fraud undermined the government's assertion that the vote represented a big step toward democracy. Nevertheless, the vote was the first multiparty election since one in 1992 ushered in an undeclared war between the military authorities and Muslim guerrilas.

See Ben Bella, Mohammed Ahmed; Front de Liberation Nationale, Maghreb Agreement, Union du Maghreb Arabe.

Algerian dinar: The currency of Algeria.

Al Haram Al Irahimi: See Hebron.

aliyah: See World Zionist Organization.

al-jumahir: Used by some Arab nationalists when referring to the masses at a public meeting.

al-kifah al-musallah: Used by some Arab nationalists for a conflict—the armed struggle.

Allenby Bridge: See Cairo Agreement.

Alliance of Palestinian Forces: A front of ten Palestinian rejectionist groups opposed to the Arab-Israeli peace process and the Israel-PLO accord.

Allon, Yigal: See Allon Plan, Beilin-Eitan agreement.

Allon Plan: The Jordan Valley was settled according to this plan. Yigal Allon foresaw difficulties in retaining the densely populated territories that came into Israel control after the 1967 war. He recognized the strategic importance of retaining the Jordan Valley, the Northern Shore of the Dead Sea, and the Golan Heights under Israeli rule. His plan called for the creation of settlements in these areas, while making provisions for the return of those areas in Judea, Samaria, and the Gaza Strip that had a large Arab population. See Allon-Plus Plan, Beilin-Eitan Agreement.

Allon-Plus Plan: In June 1997 Prime Minister Netanyahu presented his concept of Israel's future borders to his inner cabinet, saying that Israel should retain the greater Jerusalem area, the Jordan Valley, a security belt east of the pre-1967 armistice line, and Gush Etzion.

Dubbed the Allon-Plus Plan, it is an enlarged version of the formula proposed by Yigal Allon after the Six Day War. The Prime Minister suggested that its provisions be presented to the Palestinians in final status talks. The plan would enlarge the pre-1967 Jerusalem Corridor, thereby placing all the capital's new satellite communities under Israeli sovereignty.

The plan was rejected by Chairman Arafat as a violation of the Oslo accords.

Al-Maghtas: See Cairo Agreement.

al-Majali, Abdessalam: Jordanian Prime Minister. See Washington Declaration.

al-mihaq: See PLO Charter.

Alpher, Joseph: See Alpher Plan, Moderate Territorial Compromise.

Alpher Plan: A Palestine-Israel border plan developed by Joseph Alpher, a former member of Israeli intelligence who is director of the Israel/Middle East office of the American Jewish Committee in Jerusalem. It calls for Israel to annex 11 percent of the disputed territories, leaving the majority of Jewish settlers under Israeli sovereignty; Israel would deploy a temporary force near the West Bank's border with Jordan, giving the Israelis the ability to rapidly deploy in case of attack from the east. Israel would control all of Jerusalem.

Compare Beilin-Abu Mazen Plan, Palestinian National Authority Plan, Sharon Plan, Third Way Plan.

Al-Qadhafi, Muammar: See Libya.

Al-Sabah Dynasty: The ruling Kuwait family, descended from its eighteenth-century founder. There are 1,200 members of the dynasty. See Kuwait.

al-Sadat: See Sadat, Anwar.

al-Shuhada Street: See Hebron (American Plan on Al-Shuhada Street).

al-Watani: See PLO Charter.

American Plan on Al-Shuhada Street: See Hebron (American Plan on Al-Shuhada Street).

Amir, Yigal: See Rabin, Yitzhak.

Amman Declaration: Read at the Middle East/North Africa Economic Summit in Amman, Jordan.

See Amman Middle East/North Africa Economic Summit, Cairo Declaration.

Amman Economic Summit: See Amman Middle East/North Africa Economic Summit.

Amman Middle East/North Africa Economic Summit: On October 29-31, 1995 the second Middle East/North Africa Economic Summit was held in Amman, Jordan under the patronage of His Majesty King Hussein bin Talal. The summit, cosponsored by President Clinton of the United States and President Yeltsin of the Russian Federation, with the support of the European Union, Canada, and Japan, brought together government and business leaders from the Middle East and North Africa, Europe, the Americas, and Asia.

Summit participants thanked His Majesty King Hussein for his able leadership and for the extraordinary efforts by the Hashemite Kingdom of Jordan to make this summit a success. The participants also expressed their appreciation for the partnership of the World Economic Forum, which assisted so ably in organizing this event.

The goals of the summit were to facilitate the expansion of private-sector investment in the region, to cement a public-private partnership which will ensure that end, and to work to enhance regional cooperation and development.

In this spirit, business leaders from the Middle East, North Africa, and other regions were able to conclude a number of significant commercial and business transactions at the summit that will help augment the productive capacity of the region and contribute to its broad-based economic development. These ventures involved projects in the fields of tourism, telecommunications, and transportation. Reflecting this public-private partnership, a number of these ventures will benefit from government guarantees, technical assistance, and other support from the international community.

Government representatives conducted a series of negotiations over the past year on institutional arrangements as called for in the Casablanca Declaration, which would help underpin the peace process. In this respect, the following agreement has been reached:

- A Bank for Economic Cooperation and Development in the Middle East and North Africa (originally labeled a Middle East Regional Development Bank) will be established in Cairo with an initialization capital of $5 billion. About 60 percent of the equity will be owned by development banks from OECD countries and 40 percent by banks within the region. The bank—as described in its draft articles—will be structured to promote development of the private sector, support regional infrastructure projects, and provide a forum to promote regional economic cooperation. The task force will finalize its negotiations by December 31, 1995 and will continue to explore proposals for the creation of a project presentation and financial intermediation facility. Those wishing to join the bank will begin their national ratification processes thereafter. Others who wish to leave open the option of joining the bank at a later date, in light of the evolution of institutional arrangements and other developments, may do so. The economic summit will review these issues at its next meeting.
- The establishment of a Regional Tourism Board, the Middle East-Mediterranean Travel and Tourism Association, to facilitate tourism and promote the region as a unique and attractive tourist destination. The board will include both public and private representatives.
- The establishment of a Regional Business Council to promote cooperation and trade among the private sectors of the countries of the region.
- The formal inauguration of the Economic Summit Executive Secretariat, which is located in Rabat and works to advance the public-private partnership, promoting contacts, sharing data, and fostering private sector investment in the region. The participants expressed

their appreciation to the Moroccan government for its contribution to this effort, and confirmed their support for its ongoing activities.

As a complement to the regional institutions called for at Casablanca, the Steering Group of the Multilateral Peace Negotiations has decided to establish the REDWG Monitoring Committee Secretariat as a permanent regional economic institution to be based in Amman. All participating parties have agreed that this institution will promote and strengthen regional economic cooperation in the Middle East and North Africa. The regional parties strongly recommend that the secretariat's activities will cover the range of sectors within the REDWG Monitoring Committee's work, that is, infrastructure, tourism, trade, finance, and areas within the Copenhagen Plan of Action. The core parties in close consultation with the European Union and other members of the Monitoring Committee undertake to finalize the appropriate document on the structure and operational functions of this institution, which will be submitted to the next meeting of the REDWG plenary, with a view to the commencement of the institution's activities in the first half of 1996. This REDWG plenary will consider the matter, take appropriate action, and report to the upcoming meeting of the Multilateral Steering Group.

The participants at the summit expressed their strong support for continued progress in the peace process begun at Madrid four years ago, and the importance of achieving a comprehensive peace. Participants took particular note of advances made in the past year. They welcomed the signing of the Israeli-Palestinian Interim Agreement in the West Bank and the Gaza Strip, and took favorable note of the significant progress made in implementing the Treaty of Peace between Israel and Jordan. They welcomed the decision to organize in Paris in December 1995 the Ministerial Conference on Economic Assistance for the Palestinians. The summit also took note of the positive contribution made toward peace by multilateral working groups. While welcoming an increasingly positive atmosphere of openness in the region, the summit recognized that the circle of peace needs to be widened. Attenders expressed the hope that peace agreements between Israel and Syria and Israel and Lebanon would be concluded as soon as possible. The summit welcomed significant steps taken by regional parties to the Taba Declaration and by the GCC with regard to lifting the boycott on Israel, and expressed its support for additional efforts to end the boycott.

Those at the summit declared their intent to implement as soon as possible the understandings reached in Amman. With respect to commercial activities, the business representatives reaffirmed their intention to follow through on the commercial ventures reached here and to explore new opportunities to expand trade and investment in the region. On the part of government, the officials attending the summit declared their intention to support the activities of the private sector, most particularly by getting the new institutions established in Amman up and running as soon as possible. They also welcomed the measures taken by regional parties to open their economies and join the global economy.

His Majesty King Hussein conducted the necessary consultations with the distinguished representatives of the two brotherly states, as well as with other interested parties. He gladly announced that Qatar had graciously conceded its offer to host the next summit in favor of Egypt, who will host it. And it had been agreed by all, including Jordan, the present host, as well as Egypt and others, that Qatar will be the venue of the Middle East/North Africa Economic Summit in 1997.

(Six days following his participation in the Amman Economic Summit, Israel's Prime Minister Yitzhak Rabin was assassinated).

See Council on Foreign Relations, joint ventures; Rabin, Yitzhak.

Compare Cairo Middle East/North Africa Economic Conference, Casablanca Middle East/North Africa Economic Summit: Hassan, Bin Talal (HRH Crown Prince).

AMF: See Arab Monetary Fund.

Amir, Yigal: See Rabin, Yitzhak.

AMU: See Arab Maghreb Union.

animal health controls: See Health and Housing (Center of the Middle East/North Africa Community), Regional Animal Health Cooperation Program.

ANM: See Arab Nationalist Movement.

antiterrorism accord: See terrorism, United States.

APM: See Arab Peace Movement.

approximation of laws: See Treaty of the Middle East/North Africa Community (Article 15).

Aqaba: See Camp David Accords, Israel-Jordan Peace Treaty.

Aqaba-Eilat developments: See Jordanian-Israeli bilateral accords.

aquifer: See Agriculture (Center of the Middle East/North Africa Community, Amman Middle East/North Africa Economic Summit, Cairo Middle East/North Africa Economic Conference, Water and Desertification (Center of the Middle East/North Africa Community).

Arab (economy): The Arab world has 230 million people, all speaking a common language, with a combined annual GDP of $500 billion. This renders the region, in output, roughly equal to Spain (1994 GDP $483 billion), except that the Spaniards enjoy around three times the purchasing power and 10 times the per capita income.

In addition, the regional economy is less diverse. Most Middle East production for export consists mainly of petroleum or agricultural products. Manufactured consumer goods are usually assembled behind protective walls for domestic markets. As a result, trade among Arab nations was $22.8 billion in 1994, according to the Arab League. This amounted to less than 10 percent of their total trade. Including Turkey and Iran, the figure declines to 7 percent, a number that has actually fallen four points since 1990.

Seventy-five percent of the region's exports are fuels and minerals, while 75 percent of its imports are manufactured goods.

Although there are signs of change, economic constraints have served to divide the Arab world. High customs duties, heavy government regulation, import substitution strategies, poor regional infrastructure, price controls, inadequate patent protection, prevalent government monopolies, and high transaction costs have convinced global entrepreneurs that profits presently will be minimal in the region. As a result, the region draws just one percent of the world's total FDI and exports have fallen to nearly half of their 1980 levels.

A draft implementation of the Arab Free Trade Zone was signed in February 1997 that will honor commitments from the Agreement on the Facilitations of Trade in 1981. The draft will commit members to a 10-year phase-out of tariffs on industrial goods and a quicker but as yet unscheduled elimination of agricultural duties. It should bring an immediate reduction of at least 10 percent across the board.

On the issue of foreign investment, while over $315 billion was invested by U.S. and European institutional investors in the emerging markets of the world during 1996, only around $3.1 billion went to the Middle East/North Africa region; a very small amount when compared to the market capitalization of all nine Arab stock exchanges which, by March 1997, had exceeded $125 billion.

The obstacles to investment in the Arab economies range from restrictions on the acquisition of real assets to general nationalistic resistance and an exaggerated fear of Western domination.

Arab (inter-trade): An inter-Arab trade potential awaits being tapped, offering considerable hope for stimulating jobs and growth should Arab economies be able to produce the sort of goods and services their neighbors want to buy.

Interregional trade in 1994 was no more than 8 percent of aggregate import-export flows. This is despite the considerable efforts of regional development banks and agencies to help Arab states fullfill the promise for such trade.

The Arab world's total trade remains dominated by the import of finished goods from the industrialized countries and by exports of hydrocarbons. In the Arab Maghreb Union (UMA), the five members—Libya, Algeria, Mauritania, Morocco, and Tunisia conduct 70 percent of their trade with the European Union.

The slow integration within the Arab world is changing. On January 1, 1995 there was a unification of certain economic and monetary policies for a regional free trade area in the Gulf states, the GCC (Gulf Cooperation Council), when the agreement to harmonize customs tariffs on 25 percent of goods went into effect.

The Jeddah-based Islamic Development Bank (IDB) has established the Islamic Corporation for the Insurance of Investment and Export Credit. This will provide guarantees and other insurance for exporters stuck with perennial obstacles in obtaining credit and securing payment for their goods. IDB has also established the Islamic International Company for Lease Finance, whose initial local operations will focus upon the promotion of "ijara" (lease finance) in the UAE, Morocco, and Kuwait.

A multilateral agency attacking low levels of intra-regional commerce is the Arab Trade Finance Program. The Abu Dhabi-based ATFP relaxed its credit terms in the second half of 1993 in an attempt to stimulate export activity. The Bahrain Development Bank can offer export and project finance services. Oman has created the first GCC state export insurer, the Export Guarantee and Financing Unit, which numbers the International Finance Corporation (IFC) among its shareholders and is a part of the Oman Development Bank. For those Arab exporters who can use their facilities these developments represent a much needed support for emerging exporters. All these examples and others not mentioned here are some of the shining indications of how far most Arab nations have progressed in only a few short years.

See Camp David Accords, intra-trade.

Arab African Bank (AAB): Established in 1964 and headquartered in Cairo, Egypt, banks of varying Arab nations maintain the bank whose function is to serve as a financial and economic link between Arab and African countries.

Arab-Arab cooperation: See Cairo Middle East/North Africa Economic Conference.

Arab Authority for Agricultural Investment and Development (AAAID): A regional Arab organization involving 15 member countries including five Gulf States, AAAID is concerned with investment in and development of agriculture and the related fields, with the view to enhance production of foods in its member countries. AAAID is headquartered in Khartoum, Sudan.

Arab Bank for Economic Development in Africa (ABEDA): Headquartered in Khartoum, Sudan and founded in 1973, its purpose is to spread Arab capital into non-Arab Africa for development projects.

Synonymous with Banque Arabe pour le Developpement Economique en Afrique (BADEA).

Arab boycott: See boycott.

Arab Common Market (ACM): A 1965 customs union between Egypt, Iraq, Sudan, Syria, the United Arab Emirates, and Yemen under the guidance of the Council of Arab Economic Unity.

Arab Cooperation Council (ACC): Established in 1989 by Egypt, Iraq, Jordan, and North Yemen; terminated in August 1990 following Iraq's invasion of Kuwait. Formed in Baghdad with Arab nations not members of the Gulf Cooperation Council that gave aid to Iraq during its war with Iran.

Arab East: Popular phrase for the Arab-speaking Middle East nations, excluding Arab West countries, and Sudan and Somalia. Includes Bahrain, Egypt, Lebanon, Occupied Arab Territories, Oman, Qatar, Saudi Arabia, Iraq, Jordan, Kuwait, Syria, the United Arab Emirates, and Yemen.

Compare Arab West.

Arab East Jerusalem: See East Jerusalem.

Arab Economic Council: See Council of Arab Economic Unity.

Arab Free Trade Zone: See Arab (economy).

Arab Fund for Economic and Social Development (AFESD): Formed in 1968 to finance joint Arab programs in Arab nations that may be of general Arab interest; headquartered in Kuwait.

Arabia: See League of Nations Mandate for Palestine.

Arabic Republic of Syria: See Syria.

Arab International Bank (AIB): Established in 1971, with headquarters in Cairo, Egypt, its purpose is to assist in the financing of development and of overseas trade, particularly for member nations and other Arab countries.

Arab Investment Company (AIC): Formed in 1974 as a joint stock company. Headquartered in Saudi Arabia, its functions include promoting the investment of Arab capital by undertaking investment projects itself and by operating in the financial markets.

Arab-Israeli joint venture: See joint ventures.

Arab-Israeli trade: See input sharing.

Arab-Israeli War (I): From May 14, 1948 to January 7, 1949.

Israel secured its existence as a sovereign state during this 18-month conflict. Immediately following the passage of UN Resolution 181 favoring partition of the Land of Israel, Arab forces opposed to the establishment of a Jewish state began attacking Jewish targets. Hostilities broke out on November 30, 1947 when Arabs fired on a Jewish public transportation bus killing five passengers. Israel declared its independence approximately a half year after the start of fighting on the termination of the British Mandate. At the outbreak of the war the Jewish community possessed a force composed of 2,100 Palmach members, 9,500 Haganah members, and 4,000 from the Etzel and the Lehi undergrounds. An additional 15,000 recruits were drafted during the war. The number of soldiers and weapons among the Arab forces greatly outnumbered those of the Jews. However, the Arab forces were disorganized, divided in leadership, and composed of soldiers with little military experience. The Arab-Israeli War was Israel's longest and most costly war in terms of human life—11 percent of Israelis were killed or wounded as a result of the fighting.

Synonymous with Palestinian War, War of Independence, War of Liberation.

Arab-Israeli War (II): See Suez War.

Arab-Israeli War (III): Synonymous with June 1967 War, Six Day War. See Six Day War.

Arab-Israeli War (IV): From October 6 to 25, 1973.

Synonymous with October 1973 War, Ramadan War, Yom Kippur War. See Yom Kippur War.

Arab-Jewish Agreement at Versailles: An agreement signed on January 3, 1919 at Versailles, France. His Royal Highness, the Emir Feisal (Feisal ibn-Hussein), representing and acting on behalf of the Arab Kingdom of Hedjaz, and Dr. Chaim Weizman, representing and acting on behalf of the Zionist Organization, mindful of the radical kinship and ancient bonds existing between the Arabs and the Jewish people, and realizing that the surest means of working out the consummation of their national aspirations is through the closest possible collaboration in the development of the Arab State and Palestine, and being desirous further of confirming the good understanding that exists between them, have agreed upon the following articles:

Article I—The Arab State and Palestine in all their relations and undertakings shall be controlled by the most cordial goodwill and understanding, and to this end Arab and Jewish duly accredited agents shall be established and maintained in the respective territories.

Article II—Immediately following the completion of the deliberations of the Peace Conference, the definite boundaries between the Arab State and Palestine shall be determined by a commission to be agreed upon by the parties hereto.

Article III—In the establishment of the Constitution and Administration of Palestine all such measures shall be adopted as will afford the fullest guarantees for carrying into effect the British government's declaration of the 2nd of November, 1917.

Article IV—All necessary measures shall be taken to encourage and stimulate immigration of Jews into Palestine on a large scale, and as quickly as possible to settle Jewish immigrants upon the land through closer settlement, and intensive cultivation of the soil. In taking such measures the Arab peasant and tenant farmers shall be protected in their rights, and shall be assisted in forwarding their economic development.

Article V—No recognition nor law shall be made prohibiting or interfering in any way with the free exercise of religion; and further the free exercise and enjoyment of religious profession and worship without discrimination or preference shall forever be allowed. No religious test shall ever be required for the exercise of civil or political rights.

Article VI—The Mohammedan Holy Places shall be under Mohammedan control.

Arab Kingdom of Hedjaz: See Arab-Jewish Agreement at Versailles.

Arab League: Founded in 1945 with headquarters in Egypt. Provides for mutual support, cooperation in social and economic programs, and coordination in mutual defense. Members include Algeria, Bahrain, Djibouti, Egypt, Iraq, Jordan, Kuwait, Lebanon, Libya, Mauritania, Morocco, Oman, Palestine Liberation Organization, Qatar, Saudi Arabia, Somalia, Sudan, Syria, Tunisia, and Yemen.

In 1978 its headquarters were moved from Cairo to Tunis. The league serves both as a body favoring economic unity and as the mouthpiece of moderate Arab opinion.

Arab League Summits, with head-of-state participation have been held periodically since the first in Cairo in January 1964, which dealt with the "struggle against the robbery of the waters of Jordan by Israel." Specific to the Peace Process, the most recent summits are:

- June 1988 (Casablanca)—provided future funding to the PLO to continue the Palestinian uprising in the Israeli-occupied area.
- May 1989 (Casablanca)—readmitted Egypt to the Arab League.
- May 1990 (Bagdad)—condemned the large flow of Soviet Jews emigrating to Israel.
- August 1990 (Cairo)-12 members of the 20 participants condemned Iraq for its invasion and annexation of Kuwait.
- June 1996 (Cairo)—met to discuss implications of a Likud Party victory in Israel with Netanyahu as its new prime minister. See Arab Summit (1995).

On March 31, 1997 foreign ministers of the Arab League recommended freezing economic and diplomatic ties with Israel.

Synonymous with League of Arab States.

See Arab Summit, boycott, Camp David Accords, Casablanca Middle East/North Africa Economic Summit, chronology (March 22, 1945), Council of Arab Economic Unity.

Arab League Summits: See Arab League.

Arab Legion: See Jordan.

Arab Liberation Front (ALF): Iraqi-backed radical PLO faction. Opposes Arafat's peace initiative.

Arab Maghreb Union (AMU): See Arab (inter-trade), MCM, Middle East/North Africa (MENA).
Synonymous with Mahgreb Common Market.
Compare Mashreq.

Arab Monetary Fund (AMF): An international financial organization, created in 1977 and headquartered in Abu Dhabi. Twenty countries, plus the Palestine Liberation Organization are represented. It maintains close ties to the International Monetary Fund, the World Bank, and the Bank for International Settlements.

Arab Nationalist Movement (ANM): Founded in the late 1940s as student movement with Palestinian nationalist branch. Ideological predecessor of Popular Front for the Liberation of Palestine.

Arab Peace Movement (APM): Created on January 30, 1997 with the encouragement and support of the Danish government.
See International Alliance for Peace in the Middle East.

Arab Republic of Egypt: See Egypt, Egyptian-Israeli Peace Treaty.

Arab Summit (1996): Held in Cairo on June 22-23, 1996, the first since the Gulf War when Iraq invaded Kuwait in 1990, in part to react to the election of Prime Minister of Israel Benjamin Netanyahu. Leaders from 21 participating Arab states cautioned Israel not to back away from its peace promises but stopped short of calling for any slowdown in economic normalization with Israel.

The Summit's communique called on Israel to withdraw from all occupied Palestinian lands, to allow Palestinians to make East Jerusalem the capital of their independent state, and to pull out of the Golan Heights. Should Israel's right-wing government back away from the principle of land for peace, the statement said, that would "compel all the Arab states to reconsider steps taken in the context of the peace process vis-à-vis Israel. The Israeli government will bear sole and full responsibility for this situation."

The summit's final pronouncement leaves the door open for Arab tourism and industrial projects with Israel.

In addition, the final statement reflected the Arab world's lingering divisions. Iraq, alone among the 22 members of the Arab League, was not invited to the meeting. Jordan won a blanket condemnation of terrorism, but the communique made clear that term didn't apply to Muslim guerrillas trying to out Israel from southern Lebanon. The assembled also called on Iran to "prevent sabotage aimed at Bahrain," asked Turkey to reconsider its military pact with Israel but didn't strongly condemn the agreement as Syria had urged.

See Netanyahu, Benjamin.

Arab Trade Finance Program (ATFP): See Arab (inter-trade).

Arab West: Popular phrase used to identify all nations of the Arab North Africa—Algeria, Libya, Morocco, Tunisia, and Mauritania. The Libyan Desert is the dividing line to the Arab East.

Compare Arab East.

Arab world: Doubling every 25 years, the population of the Arab countries presently approaches 285 million people, with a wealthy and poor class in each nation.

In the early 1990s the Arab countries imported $130 billion worth of goods. Over 70 percent of these imports were from the industrialized countries. Trade between the Arab countries accounted for only 5 percent of total imports. The cost of transporting products throughout the Middle East will be much lower than the cost of shipping from other distant regions of the globe.

The present trade structure of nations within the region shows that:

- most nations import more goods than they export. Trade deficits are compensated by positive balances in the service sector and by international financial transfers.
- the European Union constitutes the major trading partner, supplying 50 percent of the region's imports and purchasing between 30-60 percent of its exports.
- regional trade in goods is relatively underdeveloped, accounting for only 10-15 percent of exports.

See Arab (economy), Arab (inter-trade), Arab Summit, Camp David Accords, Middle East Peace Process, militant Islam.

Arafat, Yasir (Yasser): It was never in question that Arafat (born 1929, also known as Abou Amar, Abu Amar) (Ra'ees for Chairman) would be the PNA's first president. Well before the beginning of Palestinian self-rule in May 1994, he was chairman of the PLO and dominant long before that as founder of the organization's largest faction, Fatah. Only Arafat could have made the historic handshake in Washington, DC with Israel's Rabin in September 1993.

It is Arafat, rather than the stone-throwing youths of the intifadah or the refugees strewn across the camps of the Middle East, who now represents the public face of Palestinian nationalism, despite the vicious internal struggles that have sometimes threatened to topple him. Arafat pushed through the crucial Palestine National Council resolution in 1974 stating that the PLOs key objectives were to create a national authority in any liberated area of Palestine, a policy brought to fruition when he returned to run Gaza and Jericho in June 1994. Since the September 1993 accord, Arafat has been criticized by donors over lack of policy transparency and for poor management. He generally responds that the international community is attempting to dictate self-rule policy.

Arafat is under pressure. Long speeches littered with traditional Arab nationalist rhetoric may still play well among core supporters, but do not appeal to the Palestinian's new international supporters. Islamist militants in the Hamas movement have gained ground as criticism mounts in the Palestinian heartland that Arafat has conceded too much to Israel. In a movement that remains highly personalized (and where the leader seems loathe to delegate), Arafat will continue to be the focus of criticism. There are those who claim that Arafat was a leader more adept at organizing the freedom campaign from abroad than organizing an independent state.

Arafat won 88 percent of the votes in the first Palestinian election on January 20, 1996 as a leaders of an interim government and an 88-member legislative council.

On February 12, 1996 (See chronology) Yasir Arafat took the oath of office as the first leader of the Palestinian self-rule government. Under the self-rule agreement with Israel, Arafat will head a Palestinian government during an interim period until both sides agree on the final status of the West Bank and Gaza Strip.

By May 1996 it appeared that Yasir Arafat had been transformed to a proponent of peace. He was warmly welcomed to the White House, all based on his commitment to turn away from implacable opposition to Israel and move toward reconciliation (the covenant calling for the destruction of Israel had just been revoked). He had sustained a crackdown against partisans of violence in the West Bank and Gaza. Arafat and the Palestinian Authority represent one of the most encouraging prospects in Mideast peace.

After a long delay, Chairman Arafat met with the new Prime Minister of Israel, Benjamin Netanyahu, in early September 1996.

See Arab Liberation Front, Casablanca Middle East/North Africa Economic Summit, Declaration of Principles, Democratic Front for the Liberation of Palestine, East Jerusalem, European Union, Fatah, Fatah Revolutionary Council, Fatah Uprising, Force 17, Gaza Strip, Gaza Strip and Jericho Area Agreement, Har Homa, Hebron, Interim Agreement (on the West Bank and the Gaza Strip), Islamic National Salvation Party; Larsen, Terje; Letters of Mutual Recognition, National Salvation Front;

Netanyahu, B.; Oslo Agreement, Palestine Authority, Palestinian National Authority, Palestine Liberation Organization, perceptions, PLO Charter, Popular Front for the Liberation of Palestine, Popular Struggle Front; Rabin, Yitzhak; Revolutionary Palestinian Community Party, Saiqa, terrorism, Washington Declaration.

Arafat, Yasir (Press Statement of December 14, 1988): Allow me to explain my viewpoints before you. Our desire for peace is strategic and not a temporary tactic. We work for peace regardless of whatever may happen, whatever may happen.

Our state provides salvation for the Palestinians and peace for both the Palestinians and Israelis. The right to self-determination means the existence of the Palestinians and our existence does not destroy the existence of the Israelis, as their rulers claim.

In my speech yesterday, I referred to UN Resolution No. 181 as a basis for Palestinian independence. I also referred to our acceptance of Resolutions 242 and 338 as a basis for negotiations with Israel within the framework of the international conference.

Our PNC accepted these three resolutions at the Algiers session. Also in my speech yesterday it was clear that we mean our people's right to freedom and national independence in accordance with Resolution No. 191 as well as the right of all parties concerned with Middle East conflict to exist in peace and security, including—as I said—the State of Palestine, Israel, and other neighbors in accordance with Resolutions 242 and 338.

Regarding terrorism, yesterday I announced beyond doubt—and nevertheless I repeat for the sake of recording stands, I repeat for the sake of recording stands—that we totally and categorically reject all forms of terrorism, including individual, group, and state terrorism.

We explained our stand in Geneva and Algiers. Any talk to the effect that the Palestinians must offer more—do you remember this slogan—or that what was offered is insufficient or that the Palestinians are playing propaganda games or public relations maneuvers will be harmful and unfruitful. That is enough.

All outstanding issues should be discussed on the table and at the international conference. Let it be perfectly clear that neither Arafat nor anyone else can stop the uprising.

The uprising will stop only when practical and tangible steps are taken toward the attainment of its national goals and establishment of its Palestinian state.

Within this framework, I expect the EEC states to play a more effective role in consolidating peace in our region. They assume a political and moral responsibility and they can deal with this.

Finally, I announced before you and ask you to convey these words on my behalf: We want peace, we want peace, we are committed to peace, we are committed to peace, and we want to live in our Palestinian state and let others live. Thank you.

Delivered in Geneva, Switzerland.

Arafat, Yasir (Letter to Y. Rabin recognizing Israel's right to exist in peace, September 9, 1993): Mr. Prime Minister,

The signing of the Declaration of Principles marks a new era in the history of the Middle East. In firm conviction thereof, I would like to confirm the following PLO commitments:

The PLO recognizes the right of the State of Israel to exist in peace and security.

The PLO accepts UN Security Council Resolutions 242 and 338.

The PLO commits itself to the Middle East Peace Process, and to a peaceful resolution of the conflict between the two sides and declares that all outstanding issues relating to permanent status will be resolved through negotiations.

The PLO considers that the signing of the Declaration of Principles constitutes a historic event, inaugurating a new epoch of peaceful coexistence, free from violence and all other acts which endanger peace and stability. Accordingly, the PLO renounces the use of terrorism and other acts of violence and will assume responsibility over all PLO elements and personnel in order to assure their compliance, prevent violations, and discipline violators.

In view of the promise of a new era and the signing of the Declaration of Principles and based on Palestinian acceptance of Security Council Resolutions 242 and 338, the PLO affirms that those articles of the Palestinian Covenant which are inconsistent with the commitments of this letter are now inoperative and no longer valid. Consequently, the PLO undertakes to submit to the Palestinian National Council for formal approval the necessary changes in regard to the Palestinian Covenant.

Sincerely,

Yasser Arafat

Chairman, The Palestine Liberation Organization

See Declaration of Principles, Palestine National Council, PLO Charter; Rabin, Yitzhak (Letter to Arafat recognizing the PLO, September 9, 1993); Resolution 242, Resolution 338.

Arava: See Israel-Jordan Peace Treaty.

Arbitration Committee (Gaza Strip and Jericho): See Gaza Strip and the Jericho Area Agreement (Article XVIII), Declaration of Principles.

Arcru: A unit of account first used in 1974; based on the movement of 12 Arab currencies against the U.S. dollar.

ard al-Isra: The Land of Isra, which refers to Mohammed's night journey from Mecca to Jerusalem as called by Islamists.

ard al-Masra: The Land of Ascension, which is what many Islamists call Palestine.

arid regions: See water.

arms control: See Moscow Multilateral Middle East Conference, security and arms control.

Asad, (Assad), Hafez (Hafiz) (President): Born October 6, 1930. Prominent in the Ba'ath Party before and up to the October 1970 coup d'etat, Asad's reputation for toughness was established during an early military-air force career. He had a brief spell as prime minister and defense minister before being elected to the presidency in March 1971. Since then Asad has been forged in conflict with Israel and the West, at least until the collapse of his ally—the Soviet Union. His skillful repositioning of Syria with the West during the 1990-1991 Gulf conflict, which saw his Ba'athist rival Iraq's Saddam Hussain lose out, has yielded dividends and underlined Asad's reputation as one of the region's most cunning politicians, although his conservatism has done little to promote the opening of Syria's economy.

See Saudi Arabia, Syria.

Asiatic Turkey: See League of Nations Mandate for Palestine.

Asifa: The military branch of Fatah used as a cover for the first guerrilla operations against Israel in 1965. See Fatah.

Assad, Hafiz: See Asad, Hafez.

assist database: See Energizing Business Actors.

Aswan Dam: See Nasser, Gamal Abdal Abd al-.

ATFP (Arab Trade Finance Program): See Arab (inter-trade).

attitudes: See perceptions.

audio-visual: See Tourist (Center of the Middle East/North Africa Community).

Auja: See Cairo Agreement.

aviation: Regular air flights between Israel and Jordan began on April 1, 1996. An aviation agreement between the two countries allows El Al and Royal Jordanian up to five flights weekly.

Marka Airport in Amman will be the base for a new Jordanian airline that will carry passengers on short-haul flights around the region.

See Israel-Jordan Peace Treaty, Jordanian-Israeli bilateral accords, Palestinian Airlines, transport.

B

Bahrain: Iran made claim to the islands but was rejected by a United Nations commission in 1970. Bahrain declared independence from British protection in August 1971. The constitution, drafted by a partly elected constituent assembly, specified a National Assembly of 42, with 30 deputies to be elected on a limited franchise. The first assembly, elected in December 1973, was dissolved in August 1975 and the constitution suspended.

To fulfill the need for reform the emir appointed a 30-member advisory council in 1993. However, in December 1994 anti-regime demonstrations followed, with protestors demanding the restoration of the dissolved parliament.

Bahrain's ruler is a good friend to the West. With only 500,000 people Bahrain is also the administrative headquarters of America's Fifth Fleet, which provides strategic importance in relation to events in Iran and Iraq.

Although the government of Bahrain has controlling interests in many of the island's major industrial establishments, its overall approach to economic policy, especially those policies that affect demand for U.S. exports, can best be described as laissez-faire. Except for certain basic foodstuffs, the price of goods in Bahrain is determined by market forces, and the importation and distribution of foreign commodities and manufactured products is carried out by the private sector. Owing to its historical position as a regional trading center, Bahrain has a well-developed and highly competitive mercantile sector in which products from the entire world are represented. Import duties are primarily a revenue device for the government and are assessed at a 10 percent rate on most products. The Bahrain dinar (BD) is freely convertible, and there are no restrictions on the remittance of capital or profits. With the exception of the petroleum sector, Bahrain does not tax either corporate or individual earnings.

Over the last two decades the government of Bahrain has encouraged economic diversification by investing directly in such basic industries as aluminum smelting, petrochemicals, and ship repair, and by creating a regulatory framework which has fostered the development of Bahrain as a regional financial and commercial center. Despite diversification efforts, the oil and gas sector remains the cornerstone of the economy. Oil and gas revenues constitute over 60 percent of governmental revenues, and oil and related products account for about 80 percent of the island's exports.

Bahrain's total imports, excluding oil, weapons, and aircraft (not reported in Bahrain trade statistics), accounted to 2.5 billion U.S. dollars in 1993. The United States, as in other recent years, was the number one source of imported goods for Bahrain, selling goods worth $349.4 billion; 14.3 percent of total imports in 1993. By U.S. Department of Commerce figures, U.S. export of goods of all descriptions to Bahrain in 1993 were even higher, amounting to $653.1 million, a jump of 33.6 percent over 1992. U.S. sales to Bahrain benefit from generally warm U.S.-Bahraini relations and a local perception that U.S. firms offer high-quality and reliable prices, with the relative weakness of the dollar also helping to make American goods more price-competitive, particularly in relation to Japanese goods. Not included in the figures for trade in goods is a large U.S. advantage in the sale of services, particularly in the engineering, financial, and petroleum services.

As the headquarters of Gulf Air, Bahrain provides a major market for aircraft sales; Gulf Air spending for aircraft during 1993-1995 is expected to total over $1.7 billion, with a large portion of that going to aircraft from Boeing. Planned or proposed projects such as new electricity and desalination plants, a new container port, new hospitals, and a refinery upgrade offer major possibilities for sales in these sectors. On a long-term basis, U.S. products have done well in the categories of construction and building machinery and equipment, furniture, air-conditioning equipment, and vehicles.

Bahrain offers what many foreign businessmen judge to be the most free and most orderly business environment in the Gulf, or for that matter, in the Middle East-South Asia region.

In July 1996 Citibank opened Citi Islamic Investment Bank in Bahrain, providing special financial services required by Islamic law to avoid violating a ban on paying or charging interest. Instead the bank will give customers a share of the profits. They will also add a profit margin when providing funds to buy goods. Citibank has been providing Islamic banking services since the early 1980s but this is its first such stand-alone unit.

See Gulf Cooperation Council, Gulf nations, Islamic banking, Multilateral Working Group on the Environment.

Bahrain Development Bank: See Arab (inter-trade).

Bahraini dinar: The currency of Bahrain.

baksheesh (bakshish): Arab for an unethical, often illegal, gratuity to facilitate a business arrangement. In Middle Eastern nations it has historically been considered appropriate and customary.

Balfour Declaration: On November 2, 1917 the British Foreign Secretary Balfour declared British support for the establishment of a Jewish national home in Palestine provided that safeguards could be reached to protect the rights of the non-Jewish population. A declaration was confirmed by all the allied governments and formed the foundation for the League of Nations mandate for Palestine assigned to the British in 1920.

The policy statement to Lord Rothschild, a prominent British Zionist, in the form of a letter reads:

Foreign Office
2nd November 1917

Dear Lord Rothschild:

I have much pleasure in conveying to you on behalf of His Majesty's Government the following declaration of our sympathy with Jewish Zionist aspiration which has been submitted to, and approved by, the Cabinet. "His Majesty's Government view with favour the establishment in Palestine of a National Home for the Jewish people, and will use their best endeavours to facilitate the achievement of this object, it being clearly understood that nothing shall be done which may prejudice the civil and religious rights of existing non-Jewish communities in Palestine, or the rights and political status enjoyed by Jews in any other country." I should be grateful if you would bring this declaration to the knowledge of the Zionist Federation.

Yours Sincerely,
(Arthur James Balfour)

See chronology (November 2, 1917), Hamas Covenant, Israel (Declaration of Establishment), PLO Charter; Sadat, Anwar, San Remo Agreement.

Bani sahyun: Used by Islamists to describe the sons of Zion, a term used in the Koran.

Bank for Economic Cooperation and Development: See Amman Middle East/North Africa Economic Summit, Cairo Middle East/North Africa Economic Conference.

banking: See Bahrain, investment, Islamic banking.

Banking and Financial Services (Center of the Middle East/North Africa Community): Three financial liberties are called for in the community—the freedom to establish branches abroad, the freedom to sell across borders, and the freedom to move money. Financial services do form an important element in the economy of all participating nations as a source of employment and of net exports. They are important both in their own right (possibly 5 to 7% of the communities gross domestic product) and because of their role in oiling the wheels of the competitive market economy.

A general approach on financial services must be closely linked to the program of liberalization of capital movements, as a result of which residents of any community nation will have access to the financial systems of other community countries and all the financial products that are available there. Equally, there should be no restrictions on capital transfers and no discrimination in the form of fiscal measures. A program in financial services sector purports to break down national regulatory barriers that obstruct freedom of establishment and free trade in services which could continue even after exchange controls are fully removed. A unified banking system is needed where a bank can establish branches anywhere in the community and offer its services throughout its territories; an insurance market where insurance can be bought on the most competitive terms and provide community-wide cover; and a securities and capital market with enough capacity to meet the region's financing needs and capable of attracting investors from around the globe.

No economic or trade integration can succeed without the establishment of a cross-border banking system, that along with its traditional functions must set a means for moneylending and interest charges, a sensitive area within the Muslim world. Examples have been found where international and Arab banks are able to provide personal and commercial loans without violating religious tradition. Islamic scholars and bankers are creating an interest-free banking system that relies on lease agreements, mutual funds, and other methods to avoid paying interest.

An absolute precondition for an open community market in banking services is that money should be able to move freely across nations.

Member governments will be asked to respect each others' interpretation of those rules and to allow financial firms to sell in each other's market whatever they can sell in the home country, and under control from home. Mutual recognition of rules is equivalent to a far tougher trade demand—equivalent access—the right of a branch of an Israeli bank to do in Egypt what it is allowed to do in Israel. This will result in a single license allowing a bank to open in all community nations, or to provide services across frontiers, as a matter of right. Each branch in the community nations will be supervised by authorities in the head office nation.

All community countries regulate access to their banking system and supervise its operations, but controls differ. A bank based in one nation wishing to establish branches in the others currently needs authorization from different supervising authorities, each with their own conditions for granting authorization and for subsequent supervision.

At the heart of these proposals is the single banking license, enabling banks licensed in one country to establish branches and provide cross-border services throughout the region. Licenses for banks based outside the community will be based on the principle of reciprocity. The noncommunity bank will have similar rights within the community to those by community-based banks operating in that nonmember nation.

In addition, member nations shall have the right to set up insurance branches in other member nations.

If the community is to become a unified market, as opposed to a group of national markets, a community securities market system should be created to meet the needs of both investors and companies who go to the markets for capital and borrowings. Financial intermediaries authorized in one nation shall be able to operate throughout the community on the basis of a single license given in their home country.

Likewise, a single market in which goods, services, and persons circulate freely can only function efficiently if the related capital movements are unrestricted. In order to complete a unified market, restrictions on capital transfers must be abolished, and residents of any community nation must have free access to the financial systems and products of other community nations.

See Amman Middle East/North Africa Economic Summit, Cairo Middle East/North Africa Economic Conference, Casablanca Middle East/North Africa Economic Summit, Centers of the Middle East/North Africa Community, investment, Islamic banking, Middle East/North Africa Community.

Banque Arabe pour le Developpement Economique en Afrique: See Arab Bank for Economic Development in Africa.

Baqura Area: See Israel-Jordan Peace Treaty.

Barcelona Conference: See European Union.

barriers to trade: See free trade, logistic trade centers.

Basel Program: See World Zionist Organization.

Basle Concordat: See Protocol on Economic Relations (Article III).

batil: Used by many Islamists for nations that represent systems that falsify information.

Bedouins: See minorities.

Begin, Menacham: Born in Poland in 1913; became Israel's Prime Minister on June 21, 1977 holding the position for six years. Throughout his career he maintained a continuous struggle against the Palestine Liberation Organization and control over the predominantly Arab towns on the West Bank of the Jordan. Accepted the Nobel Peace Prize for 1978, along with President Sadat of Egypt for their work for an Israeli-Egyptian peace treaty. In August 1983 he resigned the prime ministership. He died in 1992.

See Camp David Accords, Egyptian-Israeli Peace Treaty; Peres, Shimon; Sadat, Anwar.

Begin-Sadat Center for Strategic Studies, The (BESA): Located at Israel's Bar-Ilan University, the center is dedicated to the memory of Israeli Prime Minister Menachem Begin and Egyptian President Anwar Sadat, who concluded the first Arab-Israel peace agreement. The center, a nonpartisan and independent institute, seeks to contribute to the advancement of Middle East peace and security by conducting policy-relevant research on strategic subjects, particularly as they relate to the national security and foreign policy of Israel.

Beijing Conference: See Water and Desertification (Center of the Middle East/North Africa Community), Water Data Banks Project.

Beilin-Abu Mazen Plan: In October 1995 a boundary plan separating Israel from Palestine was presented. No official map of this plan has been released, but it is believed to be based on a map published by Netivot Shalom, a religious peace movement. It is significant because it is a joint attempt at a compromise by high-ranking Israelis and Palestinians. Israel would annex 7 percent of the territories, mainly large settlement blocs where most Israeli settlers live. In exchange, Israel would recognize a Palestinian state and a Palestinian capital in expanded boundaries of Jerusalem.

See Beilin-Eitan agreement.

Compare Alpher Plan, Palestinian National Authority Plan, Sharon Plan, Third Way Plan.

Beilin-Eitan agreement: Published on January 22, 1997; named after its authors—Yossi Beilin, former minister in the Labor governments of Rabin and Peres and a driving force behind the secret negotiations in Oslo, and Michael Eitan, head of the ruling Likud's parliamentary faction. The Israeli Parliament passed this agreement by a majority of six in favor, three against, and two absentions.

The document is less forthcoming to the Palestinians version of the arrangement worked out between Beilin and the PA's Abu Mazen in October 1995. The model is less the Beilin-Abu Mazen Plan, but more the Allon Plan, which in 1967 proposed the annexation of some 40 percent of the West Bank, half the Gaza Strip, and the establishment of an "autonomous framework" for the territories' Palestinian inhabitants.

First and foremost, the document is less a negotiating text than the latest expression of Israel's national consensus on the need to remain in effective permanent control of the territories occupied almost 30 years ago.

The Beilin-Eitan agreement is best understood within the context of domestic Israeli politics, where the desire for consensus among the major parties in the wake of the Rabin assassination remains a basic feature of the Israeli political scene. Far more than a roadmap to the next stage of Israeli-Palestinian talks, Beilin-Eitan reflects the latest in Beilin's efforts to heal the breach between the Labor and Likud that was exacerbated by Rabin's murder.

Beilin had already moved the Peres government more explicitly than ever toward support of the permanent maintenance of all Israeli settlements under exclusive Israeli control, an objective Peres had sought since the outset of Oslo.

The Beilin-Eitan agreement declares that Israel must reaffirm the "strategic decision" endorsed by Rabin and Peres—the construction of "a relationship of good neighbors between Jews and Arabs in the Land of Israel."

The authors enumerate the three principles around which an Israeli national consensus is to be built. The first principle notes that it has become necessary for Israel—if it seeks to secure the advantages offered by its presence in the West Bank, East Jerusalem, and the Gaza Strip—

to "permit" the creation of a Palestinian "entity" of undetermined status. Whether this entity is called a state, as Beilin is prepared to concede, or an "enlarged autonomy," as Eitan prefers, the fact of extensive Israeli limits on Palestinian sovereign power remains constant.

The second principle is a demand for continuing Israeli strategic control over whatever territory is transferred to the Palestinian entity. This demand goes far beyond the recent negotiations over Israel's right of "hot pursuit," and it highlights the extent to which Israel views the rapprochement with the Palestinians as a vehicle for preserving, rather than reevaluating, its security interests in territories occupied in June 1967.

The third guiding principle concerns the future of settlements anywhere in the "Western Land of Israel," and it follows from the second principle. The authors declare than the continuing existence of all settlements must be part of an agreement with the Palestinians. Such an agreement must preserve settler rights not only to keep Israeli citizenship, but also to maintain their individual and communal ties to Israel. The preservation of these ties—personal, legal, and territorial—therefore becomes one of the basic security interests that Israel insists must be preserved.

These three elements of Israeli policy for the final status of the occupied territories resemble nothing so much as the principles first formulated years ago by Yigal Allon and later Moshe Dayan. Far from representing a break with the past—as did the Beilin-Abu Mazen agreement—the new agreement is securely within the historical consensus expressed by Israel's leadership during the last generation. The discussion of borders, including the designation of the entire Jordan Valley as a "special security zone" and the demand for continuing IDF supervision of borders, is evidence of this fact.

So, too, is the view of the settlements—even those that Israel is prepared to concede will lie outside the bounds of "full Israeli sovereignty"—as the rationale for far-reaching Israeli demands for military access and control throughout the territories. This view stands in stark contrast to the one put forward in the Beilin-Abu Mazen text, according to which 100 of 140 West Bank settlements, with a population of 40,000 would eventually come under Palestinian sovereignty. Were that agreement to be implemented, it would make possible a rethinking of Israel's traditional strategic requirements and impose real limitations on the authority exercised by the IDF over Palestinian areas.

The National Agreement notes that if no agreement is reached before the date for the third "further redeployment" in mid-1998, Israel will retain at least 50 percent of the West Bank as it enters final status discussions. Since Palestinian control in Areas A and B already approaches 30 percent of the West Bank, this means that over the next year and a half, Israel would surrender no more than 20 percent of the West Bank.

See Allon Plan, Allon-Plus Plan, Beilin-Abu Mazen Plan.

Synonymous with National Agreement Regarding the Negotiations on the Permanent Settlement with the Palestinians.

Beilin, Yossi: See Beilin-Abu Mazen Plan, Beilin-Eitan agreement, investing in peace, Oman.

Ben Ali, Zine el-Abdine: Took over in Tunisia, on November 7, 1987, in a coup that removed ailing president-for-life Habib Bourguiba, so averting a major political crisis and renewing Tunisia's then-tarnished image for stability. Ben Ali's policy is based on accelerated economic development with a gradual opening to political pluralism. He has concentrated power around the presidency and state agencies, earning criticism from the opposition. The president's picture is visible everywhere, smiling out on a country that has seen no major disturbances since he took power. Reelected in March 1994, pulling 99.9 percent of votes cast.

See Tunisia.

Ben Bella, Mohammed Ahmed: Born 1916; President of Algeria from 1963-1965; overthrown in a military coup by Colonel Boumedienne in 1965 and kept in prison until July 1979. Returned to Algeria from Paris in 1990.

See Algeria.

benefits (of Middle East/North Africa regionalization):

a. with size comes strength
b. the quest for unity unites others
c. there will be quantum leaps in technology and research
d. regional integration helps build global markets
e. devastation turns to profit
f. regional economic agreements can counter other regional trade accords.

See Middle East/North Africa Community.

Ben-Gurion: Born in Poland in 1886; became the first Prime Minister of Israel when the State of Israel was proclaimed on May 14, 1948. He died in 1973.

See Israel (Declaration of Establishment).

Berber: See minorities.

Berberophones: See sovereignty.

BHD: The international standard code for the currency of Bahrain—the Bahraini dinar.

BESA: See Begin-Sadat Center for Strategic Studies, The.

Bibor: The abbreviation for Bahrain interbank offered rate.

BIG (Business Interaction Group): See Middle East and North Africa Business Interaction Group.

bilateral negotiations: Meant to resolve the conflicts of the past, the first-ever direct talks between Israel and Syria, Lebanon, Jordan, and the Palestinians opened in Madrid on November 3, 1991, immediately following the Madrid Conference. These talks were, in fact, four separate sets of bilateral negotiations. While the talks with the three Arab states were aimed at achieving peace treaties, the negotiations between Israel and the Palestinians were based on a two-stage formula. In the first stage the Israeli-Palestinian talks were focusing on achieving agreement regarding five-year interim self-government arrangements. In the second phase negotiations concentrated on the permanent status issues.

See Gaza Strip and the Jericho Area Agreement, Israel-Jordan Common Agenda, Israel-Jordan Peace Treaty, Madrid Conference, multilateral negotiations, Trilateral Israel-Jordan-U.S. Economic Committee.

bilateral peace negotiations: See bilateral negotiations, Israel-Jordan Common Agenda, Jordanian-Israeli bilateral accords.

bilateral relations (Japan and Israel): See Japan.

birds: The Middle East and North Africa region is a giant funnel twice a year when many millions of birds migrate back and forth between nesting grounds to wintering sites in Africa. The stork, white pelican, lesser spotted eagle, and the Levant sparrowhawk are among the many species. Many of these birds fly overland because they cannot migrate over large expanses of water.

MENA should take responsibility for creating an international institution to study bird migration.

Black September: Terrorist group within Fatah which operated from 1970-1974. Organized attacks against Israeli athletes at Munich Olympics, U.S. diplomats in Sudan, and Jordanian officials.
See Fatah.

Blatt, Gregory: See World Economic Forum.

BOI: See Protocol on Economic Relations (Article III).

border conflicts: Across the breadth of the Middle East and North Africa at least 14 border disputes remain in 1997, such as:

1. Morocco and the West Sahara: disputes over land rights
2. Israel: Palestinians want to take full control of the West Bank
3. Israel-Lebanon: Israel's self-declared "security zone"
4. Israel-Syria: Golan Heights
5. Egypt-Sudan: boundary between the two countries
6. Syria-Turkey: access to Tigris and Euphrates rivers
7. Turkey-Iraq: water resources from Tigris and Euphrates rivers
8. Syria-Iraq: access to waters of the Euphrates
9. Iraq-Kuwait: Iraq disputes new border line set by the U.N.
10. Kuwait-Saudi Arabia: the small islands of Qaru and Umm al Maradim
11. Saudi Arabia-Yemen: a 1,300-mile section of the border
12. Bahrain-Qatar: the Huwar Islands
13. Iran-United Arab Emirates: the Abu Musa, Tunbs, and Sir Bani Yas Islands
14. Yemen-Eritrea: the Hanish Islands.

See Israel-Jordan Common Agenda.

borders: See Alpher Plan, Beilin-Eitan agreement, Beilin-Abu Mazen Plan, border conflicts, fences, Palestinian National Authority Plan, separation, Sharon Plan, Third Way Plan.

Boumedienne, Houari (Colonel): President of Algeria from 1965-1978. Broke with President Sadat of Egypt over the Camp David Accords. His dream of a North African socialist federation of Arab states never materialized.
See Ben Bella, Mohamed Ahmed.

boundaries: See borders.

Bourguiba, Habib ibn Ali: President of Tunis from 1957-1987.
See Tunisia (Tunis).

Boutros-Ghali, Boutros: See water.

boycott: The Arab economic boycott of Israel was the longest lived economic boycott in modern history. It has caused tens of billions of dollars in estimated economic damage to Israel since its inception. Lifting the boycott will have an effect in short order on both direct and indirect investments. Companies from around the world will be able to invest in Israel without the concern of being blacklisted. Firms that establish manufacturing operations in Israel will be able to subcontract a portion of their production to the West Bank or Gaza, finish the products in Israel is they wish, and ship them duty-free, home to the United States, or elsewhere.

On July 21, 1995 a Parliamentary panel endorsed legislation that would formally end Jordan's part in the Arab economic boycott of Israel. When passed, Jordan became the second Arab country to bow out of the 48-year-old Arab League boycott on direct and indirect trade with Israel.

The primary boycott can only be lifted by a unanimous decision of the Arab League heads of state. The secondary boycott is an attempt to prevent any trade between Arab states and non-Israeli firms which maintain "prohibited" economic contacts with Israel. Companies and individuals are automatically blacklisted if they have directors or management who are deemed to belong to a Zionist organization or support Israel and/or Zionism.

The tertiary or voluntary boycott seeks to prohibit non-Arab companies, out of fear of the boycott and of Arab reactions, from doing business with firms that are already blacklisted by the boycott. The quaternary boycott is the boycott of people with "Zionist" tendencies and has anti-Semitic overtones.

Egypt, which signed a peace treaty with Israel in 1979, no longer participates. Morocco and Tunisia established low-level economic ties with Israel in 1994, and the Persian Gulf states enforce only the part of the boycott banning direct trade with Israel.

Foreign ministers of the Arab League ended a two-day meeting on March 31, 1997 deciding to freeze relations with Israel and to reactivate the Arab economic boycott of products from Israel.

Although it did not threaten Israel's treaties with Egypt and Jordan, or its agreements with the Palestinians, the greater concern was with future relations with Qatar, Oman, Tunisia, and Morocco.

See Arab League, Camp David Accords, Casablanca Declaration, Casablanca Middle East/North Africa Economic Summit, chronology (1946), Gulf Cooperation Council, Jordan, "Singapore of the Middle East," Taba Declaration.

Boycott of Israel Law: See Jordan.

Brith Shalom: A Jewish association established in 1926 that advocated peace and brotherhood between Jews and Arabs living in the Land of Israel. The organization counted businessmen, philosophers, and scholars among its members. Its goal was to reach an understanding and a common way of life between Jews and Arabs. This was to be achieved in a framework of full political equality between the two peoples. The association remained small, but it was important primarily because of its renowned leaders.

broadcasting: See Tourism (Center of the Middle East/North Africa Community).

brucellosis: See Regional Animal Health Cooperation Program.

Brussels Definition of Valuation: See Protocol on Economic Relations (Article III).

Business Council (Regional): See Casablanca Declaration.

Business Interaction Group (BIG): See Middle East and North Africa Business Interaction Group.

C

CAC: See Cairo Agreement, Civil Affairs Cooperation and Cooperation Committee.

CAEU: See Council of Arab Economic Unity.

Cairo Agreement: The first implementation agreement, known as the Cairo Agreement, covers three contentious issues: control of border crossings, control of overall security for Gaza, and the size of the Jericho Area. On these issues Israel received the security guarantees it sought over border crossings and Gaza roads, while the Palestinians secured their use of symbols of nationhood, such as a Palestinian flag, documentation, and reduced Israeli visibility at border crossings. The Cairo Agreement was signed by PLO and Israeli officials on February 9, 1994. It reads:

The two sides agreed on the attached text to be included in the Gaza-Jericho Agreement.

The Jericho Area

1. The size of the Jericho Area will be as depicted on the agreed map attached to this Agreement.
2. In addition, while not part of the Jericho Area:
   a. Pending the entry into force of the Interim Agreement, the holy site of Nebi Mousa will be under the auspices of the Palestinian Authority for religious purposes.
   b. During religious events that take place three times a year and other special occasions that will be coordinated with the Israeli authorities, Palestinians will have the right to religious pilgrimages to the al-Maghtas under the Palestinian flag.
   c. Palestinian private projects, as well as joint ventures in accordance with the Declaration of Principles, will be located as agreed on the shore of the Dead Sea.

d. Safe passage will be provided from the Jericho Area to Nebi Mousa, al-Maghatas and the project and ventures as agreed in paragraph c above on the shore of the Dead Sea for the above-mentioned purposes. Details regarding the safe passage arrangements will be included in the Gaza-Jericho Agreement.

3. Roads within Jericho city will be under Palestinian control. Joint patrols on the main roads will be operated, led by the Palestinian vehicle. The issue of Auja and its roads will be negotiated in the immediate future in Taba.
4. Religious affairs in the "Shalom Al Israel" Synagogue in Jericho will be under the auspices of the Israeli authorities.

The Gaza Strip

1. In accordance with the DOP, during the interim period the Gush Katif and Erez settlement areas, as well as the other settlements in the Gaza Strip, and the Israeli military installation area along the Egyptian border in the Gaza Strip, as indicated on the attached map, will be under Israeli authority. In the areas delineated in yellow on the attached map and without derogating from Palestinian Authority, responsibility will be shared as follows: The Israeli authorities will have the overriding responsibility and powers for security and the Palestinian Authority will have the responsibility and powers for civil affairs, subject to the Gaza-Jericho Agreement. In addition, with regard to those areas delineated in yellow, cooperation and coordination in security matters, including joint patrols, as agreed, will be implemented. Possible changes in the area designated yellow in the Southern Security Zone will be dealt with in Taba.
2. Without derogating from Palestinian Authority and in accordance with the DOP:
   (a) On the three lateral roads connecting the Israeli settlements in the Gaza Strip to Israel, namely: The Kissufim-Gush Katif Road; the Sufa-Gush Katif Road; and the Nahal Oz-Karni Nezzarim Road, including the adjacent sides upon which the security of traffic along these roads is dependent, the Israeli authorities will have all necessary responsibilities and powers in order to conduct independent security activity, including Israeli patrols.
   (b) Joint Israel-Palestinian patrols will operate along these roads and the adjacent sides. Such joint patrols will be led by the Israeli vehicle.
   (c) Where the Israeli authorities carry out engagement steps, they will do so with a view to transferring, at the earliest opportunity,

the continued handling of the incidents falling within Palestinian responsibility to the Palestinian Police.

(d) Overpasses will be constructed on intersections between the lateral roads and the main north-south road.

(e) These arrangements will be reviewed by the JSCCC after one year from the date of completion of the withdrawal of Israeli forces from the Gaza Strip.

3. Zoning questions will be dealt with in Taba.

Other Issues

The early empowerment agreement will be negotiated in Taba, after the completion of the Gaza-Jericho Agreement. The Interim Agreement, including modalities for elections and redeployment of forces in the West Bank, will be negotiated in Washington, DC.

Article-Passages

1. General

a. While Israel remains responsible during the interim period for external security, including along the Egyptian border and the Jordanian line, border crossings shall take place according to the arrangements included in this Article. These arrangements aim at creating a mechanism that facilitates the entry and exit of people and goods, reflecting the new reality created by the Israeli-Palestinian Declaration of Principles, while providing full security for both sides.

b. The arrangements included in this Article shall apply to the following border crossings:
   (1) The Allenby Bridge crossing; and
   (2) The Rafah crossing.

c. The same arrangements will be applied by the parties, with the necessary adjustments, to agreed seaports, airports, or other international crossings, such as Abduallah and Damya bridges.

d. The two sides are determined to do their utmost to maintain the dignity of persons passing through the border crossings. To this end, the mechanism created will rely heavily on brief and modern procedures.

e. In each border crossing there will be one terminal, consisting of two wings. The first wing will serve Palestinian residents of the Gaza Strip and West Bank and visitors to these areas (hereinafter "the Palestinian Wing"). The second wing will serve Israelis and others (hereinafter "the Israeli Wing"). There will be a closed Israeli checking area and a closed Palestinian checking area, as set out below.

f. Special arrangements will apply to VIPs crossing through the Palestinian Wing. The liaison bureau to be established pursuant to paragraph 5 below (hereinafter "the Liaison Bureau") will define the scope and the nature of these special arrangements.

2. <u>Control and Management of the Passages</u>

a. or the purpose of this Article, "passage" is defined to mean the area from the crossing barrier at the Egyptian border or the Allenby Bridge, passing through and including the terminal and;
   (1) with regard to the Allenby Bridge crossing, from the terminal up to the Jericho Area; and
   (2) with regard to the Rafah crossing, from the terminal up to the outer limit of the Israel military location along the Egyptian border.

b.
   (1) Israel will have the responsibility for security throughout the passage, including the terminal.
   (2) An Israeli director-general will have the responsibility for the management and security of the terminal.
   (3) The director-general will have two deputies who will report to him:
      (a) an Israeli deputy who will be the manager of the Israeli Wing. Israel will have the exclusive responsibility for the management of the Israeli Wing; and
      (b) a Palestinian deputy, appointed by the Palestinian Authority, who will be the manager of the Palestinian Wing.
   (4) Each deputy will have an assistant for security and an assistant for administration. The assignments of the Palestinian deputies for security and administration will be agreed upon by the two sides in Taba.
   (5) There will be maximum coordination between the two sides. Both sides will maintain cooperation and coordination on matters of mutual concern.
   (6) The director-general will continue to use Palestinian contractors to provide bus services and other administrative and logistical services.
   (7) Palestinian policemen present at the terminal will be armed with handguns. Their deployment will be decided upon in Taba. Other Palestinian officials present at the terminal will be unarmed.
   (8) The details of management and security and Liaison Bureau issues will be dealt with in Taba.

(9) The two sides will work together in Taba in order to seek ways for additional arrangements in the Rafah terminal.
(10) Both parties will review these procedures in a year's time.

c. Except for the arrangements included in this Article the current procedures and arrangements applicable outside the terminal shall continue to apply throughout the passage.

d.

(1) Once incoming passengers have crossed the terminal they will proceed to the Jericho Area or the Gaza Strip, as appropriate, without any interference from Israeli authorities (safe passage).
(2) Outgoing passengers may proceed to the terminal without any interference from Israeli authorities after joint verification that such passengers hold the necessary documentation for exiting the area to Jordan or Egypt, as set out in this Agreement.

3. <u>Arrangement for Entry from Egypt and Jordan Through the Palestinian Wing</u>

a. At the entrance to the Palestinian Wing there will be a Palestinian policeman and a raised Palestinian flag.

b. Before entering the Palestinian Wing passengers will identify their personal luggage and it will be placed on a conveyor belt. Each side will be able to inspect such luggage inside its own checking area, using its own personnel and, if necessary, may open the luggage for inspection in the presence of the owner and a Palestinian policeman.

c. Persons entering the Palestinian Wing will pass through a magnetic gate. An Israeli policeman and a Palestinian policeman will be posted on each side of this gate. In the event of suspicion, each side will be entitled to require a physical inspection to be conducted in inspection booths to be located adjacent to the gate. Passengers will be inspected by a Palestinian policeman in the presence of an Israeli policeman. Accompanying personal belonging may also be inspected at this point.

d. Having completed the above phase, persons entering the Palestinian Wing will pass through one of three lanes for the purpose of identification and document control, as follows:

(1) The first lane will be used by a Palestinian residents of the Gaza Strip and Jericho Area. These passengers will pass via a Palestinian counter, where their documents and identity will be checked. Their documents will be checked by an Israeli officer who will also check their identity indirectly in an invisible manner.

(2) The second lane will serve other Palestinian residents of the West Bank. These passengers will first pass via a Palestinian counter, where their documents and identity will be checked. Then they will continue via an Israeli counter, where their documents and identity will be checked. The two counters will be separated by tinted glass and a revolving door.

(3) The third lane will serve visitors to the Gaza Strip and West Bank. An identical procedure as in paragraph 3.d(2) above will apply to such visitors, except that they will first pass via the Israeli counter, and then continue via the Palestinian counter.

e. In the event of suspicion regarding a passenger in any of the three lanes described in paragraph d above, each side may question such passenger in its closed checking area. Suspicion justifying questioning in the closed checking area may be one of the following:

(1) the passenger was involved, directly or indirectly, in criminal or planned criminal activity, in terrorist or planned terrorist activity, and is not a beneficiary of the amnesty provisions of this Agreement.

(2) the passenger conceals arms, explosives, or related equipment;

(3) the passenger holds forged or nonvalid documentation or the details included in the documentation are inconsistent with those included in the population registry (in case of a resident) or in the database (in case of a visitor), except that questions relating to such inconsistency will initially be raised at the counter and the passenger will be questioned in the closed checking area only if the suspicion has not been removed; or

(4) the passenger acts in an obviously suspicious behavior during the passage via the terminal. If, at the conclusion of this questioning, the suspicion has not been removed, such passenger may be apprehended, after the other side has been notified. In case of a Palestinian suspect being apprehended by the Israeli side, a Palestinian policeman will be asked to meet with the suspect. Following notification to the Liaison Bureau, any further treatment of the apprehended person will be in accordance with Annex III (Protocol Concerning Legal Arrangements in Criminal Matters).

f. In the Palestinian Wing each side will have the authority to deny the entry of persons who are not residents of the Gaza Strip and West Bank. For the purpose of this Agreement, "residents of the Gaza Strip and West Bank" shall mean persons who, on the date of entry into force of this Agreement, are registered as residents of these areas in the population registry maintained by the military government of the Gaza Strip and West Bank, as well as persons who have subsequently obtained permanent residency in these areas with the approval of Israel, as set out in this Agreement.

g. Following the above procedure, the passengers will collect their luggage and proceed to the customs area. (The procedures will be agreed upon in Paris.)

h. The Palestinian side will provide passengers whose entry is approved with an entry permit stamped by the Palestinian side and attached to their documents.

At the conclusion of the direct and indirect checking of the documents and identity of passengers passing via the first lane and stamping their entry permits, the Palestinian officer will provide the passenger with a white card issued by the Israeli officer. A Palestinian official posted at the exit of the Palestinian Wing will verify that the passenger holds such a white card and will collect the cards with indirect and invisible Israeli checking.

For passengers going through the second and third lanes, the Israeli officer will provide the passengers with a blue card, after checking their documents and identity, and verifying their entry permits. An Israeli and a Palestinian official posted at the exit of the Palestinian Wing will verify and collect the cards. White and blue cards collected will be checked by Israeli and Palestinian officials.

In cases where either side denies the entry of a nonresident passenger, that passenger will be escorted out of the terminal and sent back to Jordan or Egypt, as appropriate, after notifying the other side.

4. <u>Arrangements for Exit to Egypt and Jordan Through the Palestinian Wing</u>

Passengers exiting to Egypt or Jordan through the Palestinian Wing will enter the terminal without their luggage. Thereafter, the same procedures described in paragraph 3 above will apply to them, except that the order of passing via the Israeli and Palestinian counters will be reversed.

5. <u>Liaison Bureau</u>

a. There will be a liaison bureau at each crossing point in order to deal with matters arising regarding passengers through the Palestinian Wing, issues requiring coordination, and differences regarding the implementation of these arrangements. Without derogating from Israel's responsibility for security, the bureau will also deal with incidents.

b. This bureau will be comprised of an equal number of representatives from each side and will be located at a specified location inside each terminal.

c. This bureau will be subordinate to the CAC and to the relevant RCCO.

6. Miscellaneous

a. Special arrangements will be agreed upon by the two sides regarding the passage of goods, buses, trucks, and privately owned vehicles. Pending this agreement, the current arrangements will continue to apply.
b. Israel will attempt to complete the structural alterations on the Rafah and Allenby Bridge terminals not later than the date of the completion of the withdrawal of Israeli forces from the Gaza Strip and Jericho Area. If these structural alterations are not completed by that time, the arrangements described in this Article shall apply, except for those arrangements that cannot be implemented without the structural alterations.
c. In order to cross through the crossing points in and out of the Gaza Strip and Jericho Area, residents of these areas will use documents as detailed in the Annex (developed by the civilian committee). Pending the entry into force of the Interim Agreement, other West Bank residents will continue to use the existing documents issued by the military government and its civil administration.
d. Visitors to the Gaza Strip and Jericho Area will be permitted to remain in these areas for a period of up to three months granted by the Palestinian Authority and approved by Israel. The Palestinian Authority may extend this three-month period for an additional period of up to three months and will inform Israel about the extension. Any further extensions require the approval of Israel. The Palestinian request for a four-month period and an additional four months will be negotiated in Taba in the immediate future.
e. The Palestinian Authority will ensure that visitors referred to in paragraph d above will not overstay the duration of their entry permit and authorized extensions.

See Agreement on Security Arrangements for Hebron, chronology (February 9, 1994), Declaration of Principles, Taba Agreement.

Cairo Declaration: On November 12-14, 1996 the Middle East and North Africa Economic Conference was held in Cairo, Egypt under the presidency of His Excellency Hosni Mubarak. The conference cosponsored by the United States and the Russian Federation, with the support of Canada, the European Union, and Japan, brought together senior government and private-sector leaders from the Middle East and North Africa, as well as from other parts of the world.

Conference participants thank President Mubarak and the government of Egypt for hosting this event, and for the excellent organization and generous hospitality provided. The participants expressed their appreciation for Egypt's leadership in the quest for a comprehensive, just, and lasting peace in the Middle East and expressed their unwavering commitment on the basis of the terms of reference of the Madrid Peace Conference, established by U.N. Security Council resolutions 242 and 338. In this vein, they affirmed their determination to build upon the agreements reached among the parties and underlined the utmost importance of the faithful and expeditious implementation of those agreements by all parties in particular on the Israeli-Palestinian track. They further recommitted themselves to broadening and deepening peace and achieving further progress on all outstanding issues on all the Arab-Israeli negotiating tracks of the peace process. They urged all parties to pursue measures and policies which would help build confidence between the people of the region.

The theme for the Cairo Economic Conference was "Building for the Future, Creating an Investor-Friendly Environment." The conference provided an opportunity to encourage international and regional investment in the Middle East and North Africa. The region's economic, commercial, and trade potential was highlighted, which is being greatly enhanced by important economic reform programs currently being undertaken by many states in the region. These reforms which include privatization, structural reform, and removing trade barriers have provided for a more business-friendly economic climate throughout the region.

International private-sector representatives were given the opportunity to investigate in detail the increased economic and commercial opportunities in the region. Individual countries presented their investment and development programs, and cross-border opportunities were highlighted as well. Constructive and fruitful discussions were held on topics of particular relevance to both the countries of the region and the international business community.

The participants stressed the crucial importance of the development of the Palestinian economy. They noted with concern that the already weak Palestinian economy is suffering from restrictions and closures that hinder the daily movement of Palestinian labor and trade. They recognized the need of all parties in the region to live in peace, prosperity, and security, the improvement of which will enhance the economic viability of the region as a whole. They reiterated that removing restrictive measures and closures will prevent the decline of, and contribute positively to the performance of, the Palestinian economy, as well as the political atmosphere surrounding the peace process in its entirety.

The status of the economic institutions called for by the MENA conferences previously held at Casablanca and Amman was reviewed during the conference. The significant progress made on establishing the Middle East-Mediterranean Travel and Tourism Association in Tunis was welcomed by the participants. They underscored the importance of the Bank for Economic Cooperation and Development in the Middle East and North Africa in Cairo and its potential contribution to the promotion of capital flow to the region, to building infrastructure projects, and to the development of the private sector in the region. The conclusion of the drafting of the agreement establishing the bank was welcomed. Countries were encouraged to sign the agreement and complete their funding and ratification procedures promptly in order to enable the bank to begin operations in 1997. Work on establishing a Regional Business Council was also reviewed, and the relevant parties recommitted themselves to moving this important initiative forward.

The executive secretariat of the MENA conferences in Rabat has continued to develop successfully its programs and activities in fostering public/private partnership in the region between conferences.

The activities of the REDWG Monitoring Committee Secretariat established in Amman pursuant to the Amman Declaration, and formally inaugurated and institutionalized in May 1996, were reviewed. The activities of REDWG and the work of its Monitoring Committee in areas covered by the Copenhagen Action Plan, that is, infrastructure, tourism, trade, and finance were also reviewed. Participants expressed their appreciation for the work done by the committee.

Government and private-sector participants at the conference reaffirmed their commitment to continue to work as partners for peace and prosperity in the Middle East and North Africa. They underscored the importance of the economic underpinnings of peace. Nevertheless, they reaffirmed the urgency of achieving concrete progress in the political dimension of the Middle East peace process.

The participants expressed their appreciation for the unique role played by the World Economic Forum, whose tireless efforts were critical to the success of the conference. They also expressed their gratitude to the Council on Foreign Relations in New York, for its important contribution to a number of conference sessions.

The participants decided to meet again in late 1997 in Doha, Qatar for the fourth Middle East/North Africa Economic Conference.

Cairo, November 14, 1996

See Amman Middle East/North Africa Economic Summit, Cairo Middle East/North Africa Economic Conference, Casablanca Middle East/North Africa Economic Summit, Copenhagen Action Plan.

Compare Casablanca Declaration.

Cairo Middle East/North Africa Economic Conference: Held in Cairo, Egypt from November 12-14, 1996 and organized by the World Economic Forum, with the assistance of the U.S.-based Council on Foreign Relations; downgraded from a summit (with heads of state) to an economic conference. It attracted more than 1,500 representatives from the international business community, as well as senior government officials, representing 87 nations.

One of the key messages that Egypt purported to send through the MENA conference is that regional cooperation does not revolve around Israel; there is plenty of room for consolidating inter-Arab ties and Arab relations with other countries.

Egypt almost scrapped the meeting after the May election of Israeli Prime Minister Netanyahu, whose policies had slowed the region's peace process. But, President Mubarak decided that the conference would be a perfect forum for his country to present its increasingly market-led economy and its International Monetary Fund-backed economic restructuring program.

The common theme in the opening address of President Mubarak was that economic dividends of peace would not be forthcoming without progress in the Arab-Israeli peace negotiations.

Departing from the previous summits, MENA III saw political pressure being brought to bear on Israel, rather than on the participating Arab states.

"If you lose peace, everyone will be the loser. If we put the peace process back on track, everyone will gain," Prime Minister of Egypt Ganzouri said in the closing session.

The day following the conference the Egyptian government announced that during the conference business people and governments reached deals worth an astonishing $6.8 billion. A major announcement by U.S. Secretary Christopher (his last visit as Secretary to the Middle East) was made for the Middle East Development Bank's charter. The bank is expected to begin operation by the end of 1997, aiming to be a catalyst in supporting private-sector projects, promoting regional projects and dialogue, and focusing on the region's growing infrastructure needs.

One of the major regional projects that was announced during the conference was a memorandum of understanding between the Egyptian General Petroleum Corporation, Amaco Egypt Oil Company, a subsidiary of Amaco Corporation, and Turkey's Botas Petroleum Pipeline Corporation. This $2.4 billion deal is due to export Egyptian gas to Turkey, up to 10 billion cubic meters by the year 2000.

Doha, Qatar, it was announced, will host the fourth MENA economic conference in November 1997.

Synonymous with MENA III.

Compare Amman Middle East/North Africa Economic Summit, Casablanca Middle East/North Africa Economic Summit.

Camp David Accords: From September 5 to 17, 1978 conversations were held between Israeli Prime Minister Begin and Egyptian President Sadat at Camp David, Maryland. U.S. President Jimmy Carter served as intermediary. Agreement provided for normal diplomatic and economic relations between the two countries. Also, a demilitarized security zone along the frontier was established, along with gradual Israeli withdrawal from Sinai, and a five-year transitional period on the West Bank allowing for the development of some Palestinian autonomous authority.

The total cost to the United States of military and economic aid to Israel and Egypt, promised in order to secure the Camp David Accords, was put at $10 billion, with two-thirds going to Israel.

A peace treaty was finally signed in Washington on March 26, 1979. In the first phase of Israeli withdrawal from the Sinai Peninsula Israel vacated two-thirds of the peninsula from El Arish on the Mediterranean coast to Ras Muhammad on the Red Sea on January 23, 1980. The two sides exchanged ambassadors on February 26, 1980. The final Israeli withdrawal occurred on April 26, 1982.

Egypt was expelled from the Arab League while the accord was opposed by the PLO. Nevertheless, Israel returned the Sinai to Egypt on April 25, 1982.

Both the FRAMEWORK FOR PEACE IN THE MIDDLE EAST and the FRAMEWORK FOR THE CONCLUSION OF A PEACE TREATY BETWEEN EGYPT AND ISRAEL were signed at Camp David, outside of Washington, DC, on September 17, 1978, by Anwar Sadat for the government of the Arab Republic of Egypt, Menachim Begin for the government of Israel, and witnessed by Jimmy Carter, President of the United States of America.

I. Framework for Peace in the Middle East Agreed at Camp David

Muhammad Anwar al-Sadat, President of the Arab Republic of Egypt, and Menachem Begin, Prime Minister of Israel, met with Jimmy Carter, President of the United States of America, at Camp David from September 5 to September 17, 1978, and have agreed on the following framework for peace in the Middle East. They invite other parties to the Arab-Israeli conflict to adhere to it.

Preamble

The search for peace in the Middle East must be guided by the following:

- The agreed basis for a peaceful settlement of the conflict between Israel and its neighbors is United Nations Security Council Resolution 242.
- After four wars during 30 years, despite intensive human efforts, the Middle East, which is the cradle of civilization and the birthplace of three great religions, does not yet enjoy the blessings of peace. The people of the Middle East yearn for peace so that the vast human and natural resources of the region can be turned to the pursuits of peace and so that this area can become a model for coexistence and cooperation among nations.
- The historic initiative of President Sadat in visiting Jerusalem and the reception accorded to him by the Parliament, government, and people of Israel, and the reciprocal visit of Prime Minister Begin to Ismailia, the peace proposals made by both leaders, as well as the warm reception of these missions by the peoples of both countries, have created an unprecedented opportunity for peace which must not be lost if this generation and future generations are to be spared the tragedies of war.
- The provisions of the Charter of the United Nations and the other accepted norms of international law and legitimacy now provide accepted standards for the conduct of relations among all states.
- To achieve a relationship of peace, in the spirit of Article 2 of the United Nations Charter, future negotiations between Israel and any neighbor prepared to negotiate peace and security with it, are necessary for the purpose of carrying out all the provisions and principles of Resolutions 242 and 338.
- Peace requires respect for the sovereignty, territorial integrity, and political independence of every state in the area and their right to live in peace within secure and recognized boundaries free from threats or acts of force. Progress toward that goal can accelerate movement toward a new era of reconciliation in the Middle East marked by cooperation in promoting economic development, in maintaining stability, and in assuring security.
- Security is enhanced by a relationship of peace and by cooperation between nations that enjoy normal relations. In addition, under the terms of peace treaties, the parties can, on the basis of reciprocity,

agree to special security arrangements such as demilitarized zones, limited armaments areas, early-warning stations, the presence of international forces, liaison, agreed measures for monitoring, and other arrangements that they agree are useful.

Framework

Taking these factors into account, the parties are determined to reach a just, comprehensive, and durable settlement of the Middle East conflict through the conclusion of peace treaties based on Security Council Resolutions 242 and 338 in all their parts. Their purpose is to achieve peace and good neighborly relations. They recognize that, for peace to endure, it must involve all those who have been most deeply affected by the conflict. They therefore agree that this framework as appropriate is intended by them to constitute a basis for peace not only between Egypt and Israel, but also between Israel and each of its neighbors which is prepared to negotiate peace with Israel on this basis. With that objective in mind, they have agreed to proceed as follows:

A. West Bank and Gaza

1. Egypt, Israel, Jordan, and the representatives of the Palestinian people should participate in negotiations on the resolution of the Palestinian problem in all aspects. To achieve that objective, negotiations relating to the West Bank and Gaza should proceed in three stages:

(a) Egypt and Israel agree that, in order to ensure a peaceful and orderly transfer of authority, and taking into account the security concerns of all the parties, there should be transitional arrangements for the West Bank and Gaza for a period not exceeding five years. In order to provide full autonomy to the inhabitants, under these arrangements the Israeli military government and its civilian administration will be withdrawn as soon as a self-governing authority has been freely elected by the inhabitants of these areas to replace the existing military government. To negotiate the details of a transitional arrangement, the government of Jordan will be invited to join the negotiations on the basis of this framework. These new arrangements should give due consideration both the principle of self-government by the inhabitants of these territories and to the legitimate security concerns of the parties involved.

(b) Egypt, Israel, and Jordan will agree on the modalities for establishing the elected self-governing authority in the West Bank and Gaza. The delegations of Egypt and Jordan may include Palestinians from the West Bank and Gaza or other Palestinians as mutually agreed. The parties will negotiate an agreement that will define the powers and responsibilities of the self-governing authority to be exercised in the West Bank and Gaza. A withdrawal of Israeli armed forces will take place and there will be a redeployment of the remaining Israeli forces into specified security locations. The agreement will also include arrangements for assuring internal and external security and public order. A strong local police force will be established, which may include Jordanian citizens. In addition, Israeli and Jordanian forces will participate in joint patrols and in the manning of control posts to assure the security of the borders.

(c) When the self-governing authority (administrative council) in the West Bank and Gaza is established and inaugurated, the transitional period of five years will begin. As soon as possible, but not later than the third year after the beginning of the transitional period, negotiations will take place to determine the final status of the West Bank and Gaza and its relationship with its neighbors, and to conclude a peace treaty between Israel and Jordan by the end of the transitional period. These negotiations will be conducted among Egypt, Israel, Jordan, and the elected representatives of the inhabitants of the West Bank and Gaza. Two separate but related committees will be convened; one committee, consisting of representatives of the four parties which will negotiate and agree on the final status of the West Bank and Gaza, and its relationship with its neighbors, and the second committee, consisting of representatives of Israel and representatives of Jordan to be joined by the elected representatives of the inhabitants of the West Bank and Gaza, to negotiate the peace treaty between Israel and Jordan, taking into account the agreement reached on the final status of the West Bank and Gaza. The negotiations shall be based on all the provisions and principles of U.N. Security Council Resolution 242. The negotiations will resolve, among other matters, the location of the boundaries and the nature of the security arrangements. The solution from the negotiations must also recognize the legitimate rights of the Palestinian people and their just requirement. In this way, the Palestinians will participate in the determination of their own future through:

(1) The negotiations among Egypt, Israel, Jordan, and the representatives of the inhabitants of the West Bank and Gaza to agree on the final status of the West Bank and Gaza and other outstanding issues by the end of the transitional period.
(2) Submitting their agreement to a vote by the elected representatives of the inhabitants of the West Bank and Gaza.
(3) Providing for the elected representatives of the inhabitants of the West Bank and Gaza to decide how they shall govern themselves consistent with the provisions of their agreement.
(4) Participating as stated above in the work of the committee negotiating the peace treaty between Israel and Jordan.

2. All necessary measures will be taken and provisions made to assure the security of Israel and its neighbors during the transitional period and beyond. To assist in providing such security, a strong local police force will be constituted by the self-governing authority. It will be composed of inhabitants of the West Bank and Gaza. The police will maintain continuing liaison on internal security matters with the designated Israeli, Jordanian, and Egyptian officers.

3. During the transitional period representatives of Egypt, Israel, Jordan, and the self-governing authority will constitute a continuing committee to decide by agreement on the modalities of admission of persons displaced from the West Bank and Gaza in 1967, together with necessary measures to prevent disruption and disorder. Other matters of common concern may also be dealt with by this committee.

4. Egypt and Israel will work with each other and with other interested parties to establish agreed procedures for a prompt, just, and permanent implementation of the resolution of the refugee problem.

B. Egypt-Israel

1. Egypt and Israel undertake not to resort to the threat or the use of force to settle disputes. Any disputes shall be settled by peaceful means in accordance with the provisions of Article 33 of the Charter of the United Nations.

2. In order to achieve peace between them, the parties agree to negotiate in good faith with a goal of concluding within three months from the signing of this Framework a peace treaty between them, while inviting the other parties to the conflict to proceed simultaneously to negotiate and conclude similar peace treaties with a view to achieving a comprehensive peace in the area. The Framework for the Conclusion of a Peace Treaty Between Egypt and Israel will govern the peace negotiations between them. The parties will agree on the modalities and the timetable for the implementation of their obligations under the treaty.

C. Associated Principles

1. Egypt and Israel state that the principles and provisions described below should apply to peace treaties between Israel and each of its neighbors—Egypt, Jordan, Syria, and Lebanon.

2. Signatories shall establish among themselves relationships normal to states at peace with one another. To this end, they should undertake to abide by all the provisions of the Charter of the United Nations. Steps to be taken in this respect include:

(a) full recognition;
(b) abolishing economic boycotts;

(c) guaranteeing that under their jurisdiction the citizens of the other parties shall enjoy the protection of the due process of law.

3. Signatories should explore possibilities for economic development in the context of final peace treaties, with the objective of contributing to the atmosphere of peace, cooperation, and friendship which is their common goal.

4. Claims Commissions may be established for the mutual settlement of all financial claims.

5. The United States shall be invited to participate in the talks on matters related to the modalities of the implementation of the agreements and working out the timetable for the carrying out of the obligations of the parties.

6. The United Nations Security Council shall be requested to endorse the peace treaties and ensure that their provisions shall not be violated. The permanent member of the Security Council shall be requested to underwrite the peace treaties and ensure respect for their provisions. They shall also be requested to conform their policies and actions with the undertakings contained in this Framework.

II. Framework for the Conclusion of a Peace Treaty between Egypt and Israel

In order to achieve peace between them, Israel and Egypt agree to negotiate in good faith with a goal of concluding within three months of the signing of this Framework a peace treaty between them.

It is agreed that:

- The site of the negotiations will be under a United Nations flag at a location or locations to be mutually agreed.
- All of the principles of U.N. Resolution 242 will apply in this resolution of the dispute between Israel and Egypt.
- Unless otherwise mutually agreed, terms of the peace treaty will be implemented between two or three years after the peace treaty is signed.

The following matters are agreed between the parties:

(a) the full exercise of Egyptian sovereignty up to the internationally recognized border between Egypt and mandated Palestine;

(b) the withdrawal of Israeli armed forces from the Sinai;

(c) the use of airfields left by the Israelis near El Arish, Rafah, Ras en Naqb, and Sharm el Sheikh for civilian purposes only, including possible commercial use by all nations;

(d) the right of free passage by ships of Israel through the Gulf of Suez and the Suez Canal on the basis of the Constantinople Convention of 1888 applying to all nations; the Strait of Tiran and the Gulf of Aqaba are international waterways to be open to all nations for unimpeded and nonsuspendable freedom of navigation and overflight;

(e) the construction of a highway between the Sinai and Jordan near Elat with guaranteed free and peaceful passage by Egypt and Jordan; and

(f) the stationing of military forces listed below.

Stationing of Forces

A. No more than one division (mechanized or infantry) of Egyptian armed forces will be stationed within an area lying approximately 50 kilometers (km) east of the Gulf of Suez and the Suez Canal.

B. Only United Nations forces and civil police equipped with light weapons to perform normal police functions will be stationed within an area lying west of the international border and the Gulf of Aqaba, varying in width from 20 km to 40 km.

C. In the area within 3 km east of the international border there will be Israeli limited military forces not to exceed four infantry battalions and United Nations observers.

D. Border patrol units, not to exceed three battalions, will supplement the civil police in maintaining order in the area not included above. The exact demarcation of the above areas will be as decided during the peace negotiations. Early warning stations may exist to ensure compliance with the terms of the agreement. United Nations forces will be stationed:

(a) in part of the area in the Sinai lying within about 20 km of the Mediterranean Sea and adjacent to the international border, and

(b) in the Sharm el Sheikh area to ensure freedom of passage through the Strait of Tiran; and these forces will not be removed unless such removal is approved by the Security Council of the United Nations with a unanimous vote of the five permanent members. After a peace treaty is signed, and after the interim withdrawal is complete, normal relations will be established between Egypt and Israel, including: full recognition, including diplomatic, economic, and cultural relations; termination of economic boycotts and barriers to the free movement of goods and people; and mutual protection of citizens by the due process of law.

Interim Withdrawal

Between three months and nine months after the signing of the peace treaty all Israeli forces will withdraw east of a line extending from a point east of El Arish to Ras Muhammad, the exact location of this line to be determined by mutual agreement.

(More than 60 supplementary protocols on "normalization" have remained largely on paper. The protocols provided for exchanges and cooperation in all fields—youth programs and joint sporting events, linked power grids, trade, cultural exchanges, mixed economic commissions, cooperation between universities, scientific institutes, artists, students, teachers, and other professionals. To date, the only cooperation that resulted was between agronomists.

Little peace has been achieved since the 1978 signing, in particular, on a people-to-people basis. For example, tourism between the two countries remains one-sided. Only 12,000 Egyptians requested Israeli visas in 1996, compared with 320,000 Israelis who visited Egypt. The combined volume of trade (excluding oil) is only $85 million. Israel's per capital gross national product of $17,000 when compared to Egypt's at only $610, where the former continues to grow and the latter declines.)

See Begin, Menachim; Carter, Jimmy; chronology (September 17, 1978), Egyptian-Israeli Peace Treaty, Sadat, Anwar.

Canada-Israel Free Trade Agreement: Signed in the fall 1996 and implemented in January 1997, it is expected to provide the necessary nourishment to expand trade beyond the 1995 volume of $477 million. Israel traditionally buys Canadian machinery, paper, aluminum, and cereals and sells Canadian diamonds and electrical machinery.

capital flows: See Euro-Mediterranean agreements.

capital markets: See Amman Middle East/North Africa Economic Summit, Banking and Financial Services (Center of the Middle East/North Africa Community), Middle East Peace Process.

capital movements: See Banking and Financial Services (Center of the Middle East/North Africa Community).

Caradon (Lord): See Resolution 242.

Carter, Jimmy: James Carter, born in 1924, was the 39th President of the United States. The Camp David talks and eventual Accord negotiated during Carter's administration brought him international fame as a negotiator and forceful supporter of global peace.

See Camp David Accords, Carter Doctrine, Egyptian-Israeli Peace Treaty.

Carter Doctrine: U.S. policy reacting to soviet military intervention in Afghanistan in December 1979. President Jimmy Carter addressing Congress in January 1980 stated: "An attempt by any outside force to gain control of the Persian Gulf region will be regarded as an assault on the vital interests of the United States. It will be repelled by use of any means necessary, including military force." This doctrine has been followed by future U.S. presidents.

Casablanca Declaration: Given on November 1, 1994 at the Middle East/North Africa Economic Summit in Casablanca, Morocco. It reads:

1. At the invitation of His Majesty King Hassan II of Morocco and with the support and endorsement of Presidents William Clinton of the United States and Boris Yeltsin of the Russian Federation, the representatives of 61 countries and 1,114 business leaders from all the regions of the world, gathered for a Middle East/North Africa Economic Summit in Casablanca from October 30 to November 1, 1994. The participants paid tribute to His Majesty, King Hassan II, in his capacity as President Host of the conference and praised his role in promoting dialogue and understanding between the parties in the Middle East conflict. They also expressed their appreciation to the government and people of Morocco for their hospitality and efforts to ensure the success of the summit.

2. The summit leaders feel united behind the vision that brought them to Casablanca, that of a comprehensive peace and a new partnership of business and government dedicated to furthering peace between Arabs and Israelis.

3. Government and business leaders entered into this new partnership with a deeper understanding of their mutual dependence and common goals. Business leaders recognized that governments should continue to forge peace agreements and create foundations and incentives for trade and investment. They further recognize the responsibility of the private sector to apply its new international influence to advance the diplomacy of peace in the Middle East and beyond. Governments affirmed the indispensability of the private sector in marshaling, quickly, adequate resources to demonstrate the tangible benefits of peace. Together, they pledged to show that business can do business and contribute to peace as well; indeed, to prove that profitability contributes mightily to the economic scaffolding for a durable peace.

4. The summit commended the historic political transformation of the region as a consequence of significant steps toward a just, lasting, and comprehensive peace, based on U.N. Security Council Resolutions 242 and 338, a process begun with the 1979 Treaty of Peace between Egypt and Israel and enlarged dramatically by the Madrid Peace Conference

three years ago. That process has born fruit in Israel-Palestine Liberation Organization Declaration of Principles. The recent signing of the Treaty of Peace between Israel and Jordan gave a new dimension to the process. The decisions of Morocco and Tunisia to establish, respectively, liaison offices and liaison channels with Israel constituted another new positive development. These accomplishments and the next stages of rapid movement toward a comprehensive peace in the region, including Syria and Lebanon, need to be powerfully reinforced by solid economic growth and palpable improvement of the life and security of the peoples of this region. The summit stressed that Syria and Lebanon have an important role to play in the development of the region. The summit expressed a strong hope that they will soon be able to join the regional economic effort.

5. In this connection, the participants note that the urgent need for economic development in the West Bank and Gaza Strip requires special attention from the international community, both public and private, in order to support the Israel-Palestine Liberation Organization Declaration of Principles and subsequent implementing agreements to enable the Palestinian people to participate on an equal basis in regional development and cooperation. They stressed the equal importance of moving ahead on Jordanian-Israeli projects as well as on cooperative projects between Israel and Jordan in order to advance the Jordanian-Israeli-Treaty of Peace.

6. The participants recognized the economic potential of the Middle East and North Africa and explored how best to accelerate the development of the region and overcome, as soon as possible, obstacles including boycotts and all barriers to trade and investment. All agreed that there is a need to promote increased investment from inside and outside the region. They noted that such investment requires free movement of goods, capital, and labor across borders in accordance with market forces, technical cooperation based on mutual interest, openness to the international economy, and appropriate institutions to promote economic interaction. They also noted that the free flow of ideas and increased dialogue, especially among the business communities in the region, will strengthen economic activity. In this context, the participants noted favorably the decision of the Council for Cooperation of the Gulf States regarding the lifting of the secondary and the tertiary aspects of the boycott of Israel.

7. Based on the agreements between Israel and the PLO, it is important that the borders of the Palestinian territories be kept open for labor, tourism, and trade to allow the Palestinian Authority, in partnership with its neighbors, the opportunity to build a viable economy in peace.

8. The participants paid tribute to the multilateral negotiations initiated in Moscow in 1992 which have significantly advanced the objectives of the peace process. The governments represented at Casablanca will examine ways to enhance the role and activities of the multilateral negotiations, including examining regional institutions which address economic, humanitarian, and security issues. The participants noted that the progresses made in the peace process should go along with a serious consideration of the socioeconomic disparities in the region and require to address the idea of security in the region in all its dimensions; social, economic, and political. In this context they agreed that these issues need to be addressed within the framework of a global approach encompassing socioeconomic dimensions, safety, and welfare of individuals and nations of the region.

9. The participants recognized that there must be an ongoing process to translate the deliberations of Casablanca into concrete steps to advance the twin goals of peace and economic development and to institutionalize the new partnership between governments and the business community. To this end:

a. The governments represented at Casablanca and private-sector representatives stated their intention to take the following steps:
   - Build the foundations for a Middle East and North Africa Economic Community which involves, at a determined stage, the free flow of goods, capital, and labor throughout the region.
   - Taking into account the recommendations of the regional parties during the meeting of the subcommittee on finances of the REDWG monitoring committee, the Casablanca Summit calls for a group of experts to examine the different options for funding mechanisms including the creation of a Middle East and North Africa Development Bank. The group of experts will report on its progress and conclusions within six months in the light of the follow-up summit to the Casablanca Conference.
   - The funding mechanism would include appropriate bodies to promote dialogue on economic reform, regional cooperation, technical assistance, and long-term development planning.
   - Establish a regional tourist board to facilitate tourism and promote the Middle East and North Africa as a unique and attractive destination.
   - Encourage the establishment of a private-sector regional Chamber of Commerce and Business Council to facilitate intra-regional trade relations. Such organizations will be instrumental in solidifying ties between the private and public sectors of the various economies.

b. The participants also intend to create the following mechanisms to implement these understandings and embody the new public-private collaboration: A steering committee, comprised of government representatives, including those represented in the steering committee of the multilateral group of the peace process, will be entrusted with the task of following up all issues arising out of the summit and coordinating with existing multilateral strictures such as the REDWG and other multilateral working groups. The steering committee will meet within one month following the Casablanca Summit to consider follow-up mechanisms. The committee will consult widely and regularly with the private sector.

10. The participants welcomed the establishment of a Middle East/North Africa Economic Strategy Group by the Council on Foreign Relations. This private-sector group will recommend strategies for regional economic cooperation and ways to overcome obstacles to trade and private investment. It will operate in close association with the Secretariat and submit its recommendations to the steering committee.

11. The participants also welcomed the intention of the World Economic Forum to form a business interaction group that will foster increased contacts and exchanges among business communities and submit its recommendations to the steering committee.

12. The participants in the Casablanca Summit pledged to transform this event into lasting institutional and individual ties that will provide a better life for the peoples of the Middle East and North Africa. They resolved that the collaboration of the public and private sectors that constituted the singularity of the Casablanca Summit will serve as a milestone in the historic destiny that is now playing itself out in the Middle East/North Africa region.

13. The participants expressed their appreciation to the Council on Foreign Relations and the World Economic Forum for their substantive contribution to the organization of the Casablanca Summit.

14. The participants expressed their intention to meet again in Amman, Jordan, in the first half of 1995 for a second Middle East/North Africa Economic Summit, to be hosted by His Majesty King Hussein.

See Amman Middle East/North Africa Economic Summit, boycott, Business Council, Casablanca Middle East/North Africa Economic Summit, Chamber of Commerce, Council for Cooperation of the Gulf States, Council on Foreign Relations, Declaration of Principles, Gaza Strip, Middle East and North Africa Business Interaction Group, Middle East and North Africa Development Bank, Middle East/North Africa Economic Strategy Group, Palestinian Authority, REDWG, Secretariat, security council, tourist board, United Nations, West Bank, World Economic Forum.

Casablanca Middle East/North Africa Economic Summit: Hosted by King Hassan II, the Middle East/North Africa Economic Summit held in Casablanca (October 30-November 1, 1994) was a critical watershed in the development of the area. Cosponsored by the Council on Foreign Relations and the World Economic Forum, nearly 2,000 political leaders, business executives, and a handful of economists and other academics from 61 nations were present, including all Arab nations except Syria, Iraq, Yemen, Sudan, and Libya. The Moroccan king graciously and historically declared "The conference will give peace a deeper meaning. It will lay the foundations which will ensure that peace will not be an empty word, but will create an interdependence of interests for the future of all peoples of our region."

The mere holding of the conference resulted in a comparison between the Madrid meeting as a forerunner to peace settlements and the Casablanca Summit as a forerunner of economic cooperation. That one-time adversaries sat together to discuss joint ventures did much to undermine the Arab boycott against Israel. For Foreign Minister Peres of Israel, his vision for a new Middle East was being tested.

Security forces were everywhere; fortunately nothing negative happened. Nevertheless, on a lower plain there were moments of tension. When Arab League Secretary Abdel Maguid spoke it was more of the old grievances repeated, this time for a global audience. PLO Chairman Arafat repeatedly resurrected the issue of Jerusalem as the capital of a Palestinian state. Tossing his prepared statement aside, Israeli Prime Minister Rabin displayed moments of undiplomatic argument.

Off camera the participants moved about no different from any prearranged and often spontaneous business encounter, arranging contacts and networking, setting the path for a joint venture or possible investment. For some Arab participants it was the first Israeli that they had ever met (there were roughly 200 Israelis attending the summit).

Weeks before the summit, the Gulf Corporation Council (Saudi Arabia, Kuwait, the United Arab Emirates, Oman, Qatar, and Bahrain) had renounced the secondary and tertiary boycott against Israel. Israel's leaders devoted time at the summit to overturning the remaining boycott that blacklisted Israeli firms and their products, which with the signing of the concluding *Casablanca Declaration* stated: "the participants ...explored how best to accelerate the development of the region and overcome, as soon as possible, obstacles, including boycotts and all barriers to trade and investments."

U.S. Secretary of State Warren Christopher took charge and stated: "The time has come to dismantle the boycott entirely," to which one delegate added: "We are witnessing here a royal funeral of the boycott." Prince Hassan of Jordan echoed the feeling of most in the amazingly beautiful Great Hall of the Royal Palace that "the New Middle East needs new thinking...our vision is of a Middle East without barriers, with free movement of people and goods."

Problems and challenges await those who must take the results of the conference and massage the 14-point *Casablanca Declaration* to point to a progression of efforts and accomplishments. An *executive secretariat*, with Dr. Mostafa Terrab as its acting head, in part as a way of showing appreciation to King Hassan II of Morocco, was established in her capital of Rabat to serve as coordinator and facilitator and to work with a steering committee comprised of government officials and the Council on Foreign Relations, one of the cosponsors of the summit. A strategy group led by Paul Volcker, chairman of the U.S. Federal Reserve Board was formed.

The summit's proposed $10 billion Middle East Development Bank was later rejected by several Arab Nations (the United States again urged creation of the Bank at the Amman Summit in 1995).

David Kimche, President of the Israel Council on Foreign Relations proclaimed that the Casablanca Summit was a critical turning point: "The seed was sown for a new era in the Middle East. It remains to be seen whether that seed will grow into a strong and healthy plant." Yes indeed, that the October 1994 summit took place was a more than justifiable reason for holding it.

Although the concept for a Middle East Development Bank was initially rejected, it was accepted by most of the decision makers of the Middle East/North Africa region that a "matter of fate" mentality was in operation.

Nevertheless, a giant step was taken toward the creation of a Middle East Free Trade Agreement, to be followed by a Middle East/North Africa Free Trade Agreement, as it showed that people from differing cultures and nations could enter into business agreements and joint ventures, some project as profitable, others as reinforcing community needs, such as water and tourism projects.

The atmosphere created its own momentum. The magic of the summit was less what was accomplished and more of what it announced to the world, that the Middle East and North Africa trade accords are viable, potentially profitable, and will shortly develop. Assurances that the free movements of capital, services, and goods can be set in motion, to eventually be followed by the free movement of labor, the ultimate measure of community building.

See Amman Middle East/North Africa Economic Summit, Casablanca Declaration, Hassan (King), Middle East Free Trade Agreement, Middle East/North Africa Free Trade Agreement; Peres, Shimon; Swiss Trade Initiative for the Middle East and North Africa, Taba Declaration.

CBM (confidence-building measure): See Moscow Multilateral Middle East Conference.

Center for International Relations of the University of California (Los Angeles): See Peace Pulse.

Center for Middle East Competitive Strategy: See Middle East Competitive Advantage project.

Center for Middle East/North Africa Business Studies: See Middle East/North Africa Business Studies (Center for).

Centers (of the Middle East/North Africa Community): A Middle East Community, followed by a Middle East/North Africa Community will require institutional structure and governance. Functional operating units, or centers, covering a full range of activities are eventually to be spread across participating member nations. It can be assumed that a designated responsibility will be more attractive to some nations than others. Therefore, preference of assignment should be based when possible on strengths of request for housing the center in a particular country. (Headquartering a center in a stated nation does not imply any favoritism toward that nation or people.)

As MEC, and then MENAC unfolds these centers will take on their own life, with directors reporting to the president of the Secretariat.

In addition, the evolution of these centers will be determined by recommendations made by the council of ministers and the Parliament to the Secretariat. Ideally no country will be home to more than one center, but this may not be practical during the formative years of the community as there may be far more centers than participating nations. At all times, each center will have an executive committee with representation from every member nation.

Centers could be:

A. Agriculture—providing support and equity programs, research and development, and irrigation.
B. Banking and Financial Services—cross-border banking, insurance, securities, and other activities.

C. Telecommunications—planning and development of enhanced networks and professional training programs.
D. Research and Development—regional technology plan.
E. Energy and Environment—concern for environmental impact in building power stations, pipelines, and electric grids.
F. Education—upgrading quality of regional education.
G. Standards—dealing with barriers toward regional usage.
H. Health and Housing—border controls on animal and plant products and inadequate housing.
I. Industrial Cooperation and Trade—encouraging cross-border activities with reduced government involvement.
J. Water and Desertification—desalination, water-sector training, enhancing water data, waste-water treatment, and desertification projects.
K. Transportation—developing regional upgrades in infrastructure of roads, ports, railways, and air transport.
L. Tourism and Culture—efforts at regional sharing.
M. Social Rights—protection of worker's needs.

With the passage of time, and the ascendence of new nations to the community, these centers will become the primary source of recommendation, action, and accountability for all decisions and projects.

See Amman Middle East/North Africa Economic Summit, Casablanca Middle East/North Africa Economic Summit, Middle East Community, Middle East/North Africa Community, Secretariat, structure. (See fuller descriptions for each center under alphabetical titles.)

Central Intelligence Agency (U.S.): See United States.

certificates of origin: See Protocol on Economic Relations (Article VIII).

certification: See Standards (Center of the Middle East/North Africa Community).

Chamber of Commerce (Regional): See Casablanca Declaration.

charter: See PLO Charter.

children: See Social Rights (Center of the Middle East/North Africa Community).

Choma U'migdal: A name given to a method of settlement development between 1936 and 1939. It came in response to Arab attempts at suppressing the establishment of new Jewish lands, coupled with the fear that the Peel Commission would suggest a division of the Land of Israel according to already existing settlements. The fact that the camp was made of wood made it easy to relocate. Fifty-two settlements used the Choma U'migdal system, which enabled a significant expansion of Jewish settlement.

Christians: While most of the world's attention is on the strains among the Muslim and Jewish communities in the MENA region, Christians in the area are experiencing increased persecution. For example, the Holy Land's 170,000 Christians, most of them Arab—some 130,000 within Israel, including 10,000 in Jerusalem, and another 40,000 under Palestinian Authority rule—have long felt uneasy. They have been caught between the growing fundamentalism of their Muslim neighbors and the growing power of Jewish ultra-Orthodoxy, both of which view Christians with suspicion and hostility.

Few villages in the Holy Land retain Christian majorities. Bethlehem, once majority Christian, is now 70 percent Muslim.

See minorities.

Christopher, Warren: At the opening plenary of the Middle East/North Africa Economic Summit in Morocco, the U.S. Secretary of State (1993-1997) said: "If I may borrow the famous Humphrey Bogart line, this conference could be the beginning of a beautiful friendship."

See Agreement of Understanding between Israel and Lebanon, Casablanca Middle East/North Africa Economic Summit, Gaza Strip, and Jericho Area Agreement, Washington Declaration.

chronology (selective) of the Middle East Peace Process: November 2, 1917: The Balfour Declaration is published, in which the British government declared that it views "with favor the establishment in Palestine of a national home for the Jewish people, and will use their best endeavors to facilitate the achievement of this object, it being clearly understood that nothing shall be done which may prejudice the civil and religious rights of the existing non-Jewish communities in Palestine, or the rights and political status enjoyed by Jews in any other country."

January 3, 1919: The Arab-Jewish Agreement between His Royal Highness the Emir Feisal representing and acting on behalf of the Arab Kingdom of Hedjaz, and Dr. Chaim Weizmann, representing and acting on behalf of the Zionist Organization, signed an agreement at Versailles, France confirming "the good understanding which exists between..."

April 20, 1920: At the San Remo Conference the Principal Allied Powers from World War I decided that the mandate for the government of Palestine would be entrusted to Great Britain.

July 24, 1922: The League of Nations Mandate on the administration of the territory of Palestine was passed in London.

March 22, 1945: Arab League founded in Cairo. Seven heads-of-state were present at the signing of the founding pact: Egypt, Yemen, Trans-Jordan, Iraq, Saudi Arabia, Lebanon, and Syria.

December 19, 1945: The Congress of the United States passes a resolution indicating its support "that Palestine would be opened for free entry of Jews into that country...."

1946: Arab boycott of Palestine was initiated, two years before the establishment of the State of Israel, by the newly formed League of Arab States.

November 29, 1947: The United Nations General Assembly adopts the resolution by over two-thirds majority, according to which Mandated Palestine should be partitioned into an Arab and a Jewish state, with Jerusalem and Bethlehem internationalized.

May 14, 1948: The British Mandate officially ends. A Provisional Council of State, under the chairmanship of Ben Gurion, proclaims the establishment of the State of Israel.

May 15, 1948: Five Arab countries declare war on Israel. The same day President Truman recognizes Israel de facto.

December 11, 1948: The U.N. General Assembly adopted the resolution recognizing the right of the Palestine refugees to return to their homes if they wish to live in peace with their neighbors; or receive compensation for property left behind.

February 1949: Armistice agreements are signed at Rhodes between Israel and the governments of Jordan, Egypt, Lebanon, and Syria.

April 3, 1949: The United Nations brokers an Israeli-Jordanian cease-fire agreement, in which Jordan retains control over the West Bank and the eastern portion of Jerusalem.

May 11, 1949: Israel is admitted, as a member, to the United Nations.

July 20, 1951: Jordan's King Abdullah is assassinated in Jerusalem.

September 1951: The Security Council of the U.N. condemns Egypt for depriving Israel of her right to free passage through the Suez Canal.

July 1952: King Farouk of Egypt is deposed by a group of officers led by Lieutenant Colonel Nasser.

May 2, 1953: Hussein is proclaimed King of Jordan.

October 23, 1956: Cairo announces joint Arab command of Egyptian and Syrian forces under an Egyptian commander in chief.

October 29, 1956: Israeli forces launch Sinai campaign as part of a secret agreement with France and Great Britain.

November 2, 1956: General Assembly of the U.N. orders an immediate cease-fire.

November 6, 1956: Britain, France, and Israel accept cease-fire.

March 8, 1957: Israel withdraws all her troops from Sinai and Gaza Strip.

February 22, 1958: Union of Egypt and Syria proclaimed as the United Arab Republic.

September 29, 1961: Syria breaks away from the union with Egypt and proclaims its independence.

January 13-16, 1964: First summit conference of Arab heads of state is held in Cairo. The communique announced the decisions (a) to create a unified command of the Arab military forces; (b) to create a Palestine-Arab political entity in the framework of the Palestine Liberation Organization; and (c) to divert the headquarters of the River Jordan. All these resolutions are directed against Israel.

January 16, 1965: Premier Eshkol declares that Israel will consider any Arab attempt to divert the headwaters of the Jordan River as "an encroachment on our soil" and that if the Arabs put their plan into operation "a military confrontation would become inevitable."

March 15, 1965: Nasser elected President of Egypt.

May 23, 1967: Nasser announces closing of the Straits of Tiran and the Gulf of Aqaba to Israeli shipping and to Israeli-bound shipping carrying "strategic material."

June 5, 1967: Outbreak of the Six Day War; Israel captures the West Bank and reunites the city of Jerusalem. Jordan continues to maintain social, religious, and economic ties with the West Bank and Muslim religious institutions in Jerusalem. Travel for Palestinians and foreigners is permitted between the West Bank and Jordan through a number of border passages. Israel captures most of the Golan Heights and gains an additional strip in the course of the 1973 war.

November 22, 1967: The United Nations Security Council Resolution 242 is passed.

1968: See 1964.

1973: The Yom Kippur War.

October 22, 1973: The United Nations Security Council passes Resolution 338.

May 1974: Following the Israeli-Syrian disengagement agreement, Israel withdraws from the eastern strip of the Golan Heights captured during the 1973 war, and a few strips conquered in 1967.

June 8, 1974: The Phased Plan, the text of the political plan approved by the PLO Council demands the "restoration of the full national rights of the Palestinian nation, and first and foremost the right of return and self-determination on the homeland's soil entire...."

March 1978: The U.N. Security Council adopted Resolution 425 calling for "strict respect for the territorial integrity, sovereignty, and political independence of Lebanon," and for a withdrawal of Israeli forces from Lebanese territory.

September 17, 1978: The Camp David Accords were signed leading to the Egypt-Israel Peace Treaty and its "Framework for Peace in the Middle East" which later would become the basis of the Declaration of Principles between the PLO and Israel.

March 26, 1979: M. Begin of Israel and A. Sadat of Egypt, witnessed by President Carter of the United States, sign a Treaty of Peace between the Arab Republic of Egypt and the States of Israel in Washington, DC.

December 1981: The Likud government extended Israeli law to the Golan Heights and Israeli citizenship was offered to Arabs who had remained.

April 11, 1987: Shimon Peres and King Hussein meet and submit the agreement (London Agreement) signed on April 11, 1987.

October 1989: See May 1991.

May 1991: The October 1989 Taif Agreement and the May 1991 "Treaty of Brotherhood, Cooperation and Coordination between Syria and Lebanon" severely limit Lebanese independence from Syria. The treaty commits the two countries to permanent and total coordination of political, economic, military, social, international, regional, and domestic policies.

October 18, 1991: U.S. and Soviet send out invitations to the Mideast Peace Conference in Madrid.

October 30, 1991: The Madrid Conference convenes, inaugurating direct talks between Israel and its Arab neighbors. Since then, bilateral negotiations between Israel and Syria, Lebanon, Jordan, and the Palestinians have been held. Multilateral talks on key regional issues are under consideration.

November 3, 1991: Bilateral, face-to-face talks between Israel and Jordan begin in Madrid; others are held in Washington, DC.

September 9, 1993: PLO Chairman Yasser Arafat sends a letter to Prime Minister Rabin stating unequivocally that the PLO recognizes Israel's right to exist in peace and security, renouncing terrorism, and committing itself to a peaceful resolution of the conflict. See May 4, 1994.

September 13, 1993: Israel and the PLO sign the "Declaration of Principles on Interim Self-Government Arrangements" (DOP) in Washington, DC. The arrangements include immediate Palestinian self-rule in Gaza and Jericho, early empowerment for the Palestinians in the West Bank, and the election of a Palestinian council. The interim period is to last five years. The document went into effect on October 13, 1994.

September 14, 1993: Israel and Jordan sign a Common Agenda in Washington. The document constitutes a blueprint for a peace treaty between Israel and Jordan. Prime Minister Rabin and Foreign Minister Shimon Peres meet with Morocco's King Hassan in Rabat.

October 1, 1993: During a meeting at the White House between President Clinton, Israeli Foreign Minister Shimon Peres, and Jordanian Crown Prince Hassan it was agreed to set up a trilateral U.S.-Israeli-Jordanian committee to discuss economic cooperation and development. The trilateral group was first convened in Washington on November 30, 1993.

October 13, 1993: See September 13, 1993.

November 30, 1993: See October 1, 1993.

December 5, 1993: Syrian President Hafez al-Assad meets with U.S. Secretary of State Warren Christopher agreeing to reopen bilateral negotiations with Israel in Washington on January 24, 1994.

December 13, 1993: The scheduled withdrawal of Israeli military forces from the areas of Jericho and the Gaza Strip under the enactment of the Declaration of Principles. Postponed until May 1994.

December 31, 1993: Israel and the Vatican sign the Fundamental Agreement providing the basis for establishing diplomatic ties.

January 16, 1994: President Clinton meets with President Assad of Syria.

January 24, 1994: See December 5, 1993.

February 9, 1994: Syria, Lebanon, Jordan, and the Palestinians suspend their talks with Israel after Baruch Goldstein kills 29 Palestinian worshippers at the Tomb of the Patriarchs in Hebron.

February 9, 1994: Representatives of Israel and the PLO sign the first in a series of agreements (the Cairo Agreement) on the implementation of Palestinian autonomy in Gaza and Jericho.

March 7, 1994: A delegation of 57 Israeli Arabs travel to Syria to extend condolences to President Hafez al-Assad on the death of his son, Basil, marking the first visit of an official delegation from Israel.

March 31, 1994: Israel and the PLO agree to establish a three-month Temporary International Presence in Hebron. The issue was raised by the Palestinians after the Hebron massacre and the TIPH agreement marks the resumption of Israeli-Palestinian talks. The TIPH is deployed on May 5 and leaves Hebron on August 9.

April 5, 1994: The IDF begins evacuating from its headquarters in Gaza City.

April 17-19, 1994: Deputy Foreign Minister Yossi Beilin leads an Israeli delegation on a groundbreaking trip to Oman to participate in the multilateral talks on water.

April 29, 1994: Israel and the PLO sign an economic agreement in Paris (the Paris Economic Agreement) which two weeks later becomes part of the Gaza-Jericho agreement.

May 2, 1994: In Copenhagen the governments of Italy, Denmark, and Norway sign a "memorandum of understanding on the operations of the Temporary International Presence in Hebron.

May 2, 1994: The World Bank agrees on a $1.2 billion, three- year, economic development plan for the West Bank and Gaza. Based on the financial pledges of 40 donor countries, the blueprint offers specific strategies for implementing the economic arrangements agreed upon between Israel and the Palestinians. Plan to be carried out by the PLO's Economic Council for Development and Reconstruction (PECDAR).

May 4, 1994: Israel and the PLO sign the Agreement on the Gaza Strip and the Jericho Area in Cairo. Included in the agreement are provisions for the withdrawal and redeployment of Israeli military forces from Gaza and Jericho, a transfer of civil responsibilities from the Israeli Civil Administration to the newly created Palestinian Authority, creation of a Palestinian police force and relations between Israel and the Palestinian Authority. See September 9, 1993.

May 4, 1994: PLO Chairman Arafat and Prime Minister Rabin exchange letters confirming a number of issues not included in the Agreement on the Gaza Strip and the Jericho Area. Arafat pledges cooperation with Israel to locate and return to Israel soldiers who are missing in action and committed to submit to the Palestinian National Council for formal approval of changes in the covenant, as undertaken in the September 9, 1993 letter.

May 9, 1994: Freih Abu-Medein and Zacharia al-Agha become the first two members appointed to the Palestinian Authority Council.

May 10, 1994: President Ezer Weizman and PLO Chairman Yasser Arafat meet in Johannesburg with South Africa's newly inaugurated president, Nelson Mandela. In addition, the first detachment of 150 members of the Palestinian police force enters the Gaza Strip.

May 13, 1994: The IDF withdraws from Jericho.

May 18, 1994: The IDF completes its withdrawal and redeployment in the Gaza Strip.

June 5, 1994: Israel and the Vatican establish formal diplomatic ties and exchange ambassadors.

July 1, 1994: PLO Chairman Yasser Arafat arrives in the Gaza Strip.

July 13, 1994: The scheduled election to be held nine months after the enactment of the Declaration of Principles, to elect a Palestinian Council by the Palestinian residents of the West Bank, Gaza, and East Jerusalem (postponed).

July 18-19, 1994: The first session of Israeli-Jordanian bilateral talks (the Wadi Araba Bilateral Talks) held in the region is convened at Ein Evrona, located in the boundary area north of Aqaba and Eilat.

July 20, 1994: Israeli Foreign Minister Peres arrives in Jordan for the Dead Sea Trilateral Talks, marking the first ever visit of an Israeli official to Jordan.

July 25, 1994: Israel and Jordan sign The Washington Declaration, ending the state of belligerency that existed between the two countries since 1948.

July 26, 1994: King Hussein and Prime Minister Rabin together address a joint session of the U.S. Congress, the first time in history that two leaders do so.

August 3, 1994 On his way home from Washington, King Hussein overflies Israel and Jerusalem; this first-ever Jordanian civil flight to cross over Israel and the Occupied Territories to Jordan.

August 8, 1994: Israel and Jordan open a new border crossing between Eilat and Aqaba for third-country nationals.

August 29, 1994: Israel and the PLO sign the Early Empowerment Agreement, giving the Palestinian Authority jurisdiction over education, tourism, health, welfare, and taxation throughout the West Bank.

September 1, 1994: Israel and Morocco establish diplomatic relations and agree to open liaison offices in Tel Aviv and Rabat.

October 3, 1994: Tunisia and Israel announce that they were establishing low-level diplomatic relations.

October 17, 1994: A draft Israeli-Jordanian peace treaty is initialed in Amman by Israeli Prime Minister Yitzhak Rabin and Jordanian Prime Minister Abdel Salim Majali.

October 26, 1994: Jordan, following Egypt, becomes the second Arab country to sign a peace treaty with Israel.

October 30-November 1, 1994: Morocco hosts The Middle East/North Africa Economic Summit in Casablanca. Its "Casablanca Declaration" outlined a plan to link Arab-Israeli peace with regional economic growth.

November 10, 1994: His Majesty King Hussein of Jordan ratifies the Peace Treaty with Israel.

August 8, 1995: Lieutenant General Hussein Kamel and Saddam Kamel defected to Jordan from Iraq vowing to topple the government of President Saddam Hussein.

October 29-31,1995: Jordan hosts the Amman Middle East/North Africa Economic Summit.

November 4, 1995: Prime Minister of Israel Yitzhak Rabin is assassinated in Jerusalem. Shimon Peres becomes Prime Minister.

February 12, 1996: Chairman Yasir Arafat was sworn in as the first president of the Palestinian National Council in Gaza.

February 25, 26, March 3, 4, 1996: Hamas terrorists kill more than 60 Israelis. Prime Minister Peres declares war on Hamas.

March 13, 1996: Summit of Peacemakers held in Egypt; first time that Israel and Arab countries openly agreed to become partners on a security issue.

March 18, 1996: Chairman Arafat declares that henceforth the areas under his jurisdiction will be called Palestine.

April 24, 1996: The Palestine National Council voted to revoke the clauses in its 32-year-old charter that called for an armed struggle to destroy the Jewish state.

May 29, 1996: Benjamin Netanyahu narrowly defeats Shimon Peres for Prime Minister of Israel.

June 22-23, 1996: The first Arab summit since the Gulf War held in Cairo, in part as a reaction to the election of Netanyahu.

July 23,1996: Netanyahu sends his Foreign Minister to meet Yasir Arafat in the new government's first high-ranking meeting with Palestinians.

July 24, 1996: President Clinton's Middle East envoy, Dennis Ross, makes a peace trip to Damascus in a bid to revive Israeli-Syrian talks.

August 2, 1996: Government angers Arabs by ending four-year freeze on construction in Jewish settlements on the West Bank and Gaza.

August 12, 1996: Netanyahu proposes a separate peace with Lebanon as a springboard to deal with Syria; Lebanon rejects the offer.

September 4, 1996: Netanyahu meets Arafat; both sides agree to discussions on pulling Israeli soldiers out of Hebron on the West Bank.

September 25-26, 1996: Riots break out on the West Bank and in Gaza City and East Jerusalem; Palestinian police and Israeli soldiers exchange gunfire.

October 6, 1996: Following a White House emergency summit meeting, Israel and the PLO begin weeks of marathon talks on redeploying Israel troops from the West Bank town of Hebron.

January 1, 1997: An off-duty Israeli soldier fires into an Arab market in Hebron, wounding seven Palestinians in what he says is an effort to sabotage the handover of the city to the Palestinians.

January 15, 1997: Prime Minister Netanyahu and Chairman Arafat conclude an accord, initialed by negotiators (signed on January 17), on the pull-back of Israeli troops from the city of Hebron and other areas of the West Bank.

January 17, 1997: Israeli soldiers hand over 80 percent of Hebron, ending 30 years of occupation and formally dividing the volatile town between Arab and Jewish rule.

March 7, 1997: Israeli cabinet approves the handover of more of the West Bank to self-rule. Palestinians voice disappointment that they are to get only a third of what they had expected.

March 18, 1997: Israel defies world opinion by breaking ground for the Jewish settlement of Har Homa on a hilltop—known as Jabal Abut Chneim to Palestinians—at the edge of Arab East Jerusalem. The prospective settlement is named Pisgat Shmu'el.

March 21, 1997: Bomb explosion in a Tel Aviv cafe, kills the Palestinian attacker and three Israeli women and wounds 42.

April 25, 1997: An emergency UN General Assembly session approves a resolution for a halt in construction at Har Homa and an end to all settlement activities in the occupied territories.

May 5, 1997: The Palestinian Authority announced that it would impose the death sentence on Palestinians convicted of selling land to Jews.

July 30, 1997: Two suicide-bomers killed themselves and 14 Jews in Jerusalem, injuring over 170 people.

August 4, 1997: The Israeli government suspended reimbursement of taxes and other fees that it owes to the Palestinian Authority. The money accounts for nearly two-thirds of the Authority's annual revenue.

August 11, 1997: U.S. Representative to the Middle East, Dennis Ross, returned to the region in an attempt to revitalize peace negotiations.

August 25, 1997: Israel plans to build dam on disputed land claimed by Syria.

CIA: See United States.

Ciller, Tansu: Prime minister of Turkey in the early 1990s, and coalition partner with Erbakan. See Turkey.

Citibank: See Bahrain.

cities: The liberalization of world trade and the influence of regional trading groups such as MEFTA and MENAFTA will not only reduce the powers of national governments, but also increase those of cities. This will come about because an open trading system will have the effect of making national economies converge, thus leveling the competitive advantages of nations, while leaving those of cities largely untouched. So in the future the arenas in which firms will compete will be found in cities rather than in nations.

Civil Administration (Gaza Strip and Jericho): See Declaration of Principles.

Civil Affairs Coordination and Cooperation Committee (CAC): See Gaza Strip and the Jericho Area Agreement (Article III).

Civil Aviation Agreement: See Israel-Jordan Peace Treaty.

Civil Powers: See Interim Agreement (in the West Bank and the Gaza Strip).

claims: See Israel-Jordan Peace Treaty.

Claims Commissions: See Camp David Accords, Egyptian-Israeli Peace Treaty.

Clinton, William: President Clinton, throughout his administration has been a promoter of the peace process in the Middle East.

On July 1, 1994 the President of the United States wrote a congratulatory letter to King Hassan II of Morocco prior to the opening of the historic Casablanca Middle East/North Africa Economic Summit. In part he said: "This conference will be an important forum for discussing the region's economic prospects and promoting business and investment opportunities. It will also make a significant contribution to advancing the peace process. President Yeltsin and I have been asked by the Council on Foreign Relations to be Honorary Co-Chairman of the conference. I am pleased to accept this position and hope that this will assist in the success of the meeting."

See Albright, Madeleine K., Amman Middle East/North Africa Economic Summit, Casablanca Declaration, chronology (January 6, 1994), Declaration of Principles, Iranian and Libya Sanction Act, Israel-Jordan Peace Treaty, Middle East/ North Africa Economic Summit, Middle East Peace and Stability Fund, Middle East Summit, terrorism, Trilateral Israel-Jordan-U.S. Economic Committee, Washington Declaration.

closure: A description of Israel's barring of Palestinians from leaving Gaza to work in Israel. Such border closings have been imposed on Gaza's million Palestinians to one degree or another for more than half the time since they were granted self-rule. Such restrictions have prevented thousands of Palestinians in the West Bank from working in Israel, and has been devastating for the Gaza Strip, which is more dependent on Israel as a source of jobs. About 70,000 Gazans made a living doing menial work in Israel. Even when the closings have been lifted, Israel has rarely issued more than about 20,000 work permits.

Cohen, Stephen P.: See Council on Foreign Relations.

collaboration: See Gaza Strip and the Jericho Area Agreement.

collective bargaining: See Social Rights (Center of the Middle East/North Africa Community).

Common Agenda: See Israel-Jordan Common Agenda.

common agricultural policy: See agriculture, Free Trade Agreement (European Union and Israel).

common management: See containment, conflict, and common management of common problems.

common market: See regional economic integration.

common problems: See containment, conflict, and common management of common problems.

comparative advantage: A theory first detailed in 1817, holding that nations should export products that they can product relatively more efficiently than other nations and import products for which they are relatively high-cost producers. See input sharing.

Concurrent Resolution of the Congress of the United States: See United States of America (Concurrent Resolution).

Conference on Security and Cooperation in the Middle East (CSCME): See Israel-Jordan Peace Treaty.

Conference to Support Middle East Peace: See United Nations (Middle East Peace Process).

confidence: Confidence is the prerequisite for a successful MEC/MENAC, for economic growth, for upgraded social welfare programs, for political security, for cultural sovereignty. But confidence is not an inherent right, it must be earned through clear thinking, adjustment, change, and then readjustment, with continuous cycles to follow.

confidence-building measure (CBM): See Gaza Strip and the Jericho Area Agreement (Article XX), Moscow Multilateral Middle East Conference.

conflict: See containment, conflict, and common management of common problems.

conscience: Individual consciences must be the guiding light for citizens of the community. Personal notions of right and wrong, one's dictum on how best to fulfill dreams of freedom and aspiration shall have greater moral precedence over the requirements of the MEC/MENAC. Inevitably, conflicts between the greater society and the individual will evolve. The behavior of citizens cannot be pounced on by any regulation or directive coming from MEC/MENAC officials but will evolve constructively and productively only when openly debated and understood. The final arbiter will be individual conscience, which can counter the needs of society, but which must be protected and be permitted to express itself. Individuality must be supreme to the group. One must not harm the other, nor impose a value system that will be counterproductive and create a lack of goodwill. Today's individual, when snapped, forces guilt and reproachment among members of the group, who tomorrow become the next oppressed individual.

consent: Popular consent is a critical force in sustaining interest and achievement for MEC/MENAC. It is imperative that a permanent communication network be open to the people, not merely officials of the government but also those politicians serving its institutions. Officials who attempt to rush decisions past the eyes of its citizens may initially succeed, but the false security will eventually return to haunt them. Consent to govern is the only winning avenue to change, other than revolution, which requires coercion.

Constantinople Convention: See Camp David Accords, Egyptian-Israeli Peace Treaty.

construction: See Amman Middle East/North Africa Economic Summit.

consumers: See Standards (Center of the Middle East/North Africa Community).

containment: See containment, conflict, and common management of common problems.

containment, conflict, and common management of common problems: The four Cs that were developed following World War II were incorporated into the creation of the European Economic Community.

Jean Monnet, a bureaucrat from Cognac, France, with no elected power base, but with enormous vitality and ideas, was skilled in convincing politicians that his ideas were really theirs. By inducing the instigator of two world wars during the century into a series of countervailing organizations, Germany would be contained from asserting herself again.

His model for inclusion of a former enemy into a community of nations found resistance from those who sought to split Germany into smaller regions (as was done in the breakup of the Ottoman Empire following World War I). Dividing Europe's largest nation, both in population and geography appealed to many leaders, but under the influence of the United States who wanted a strong Germany to serve as a buffer against the intrusive and potentially acquiring Soviet Union, this proposal never materialized. Thus the *containment* principle for dealing with past injustices would cross a line and cope with a perceived nemesis threatening to throw the globe into a new series of *conflicts*

The Marshall Plan of billions of dollars of U.S. aid for dealing with the devastation throughout Europe enlisted support for ways to deal with *common problems* of bordering nations. The 1958 Treaty of Rome created the European Economic Community by establishing a *common management* in Brussels, Belgium that would become its headquarters.

The four Cs have direct application to events within the Middle East and North Africa. The issues of *containment*, threats to security and peace, and overcoming poverty remain central to the creation of a Middle East/North Africa regional community. It establishment would help to cope with rising radical Islam, increased conflicts resulting from surging population growth, and poverty at a time of limited water resources and skyrocketing increases in population.

See regionalism.

Continuing Committee (Gaza Strip and Jericho): See Declaration of Principles.

Cooperation Council for the Arab States of the Gulf: See Gulf Cooperation Council, Middle East/North Africa (MENA).

Copenhagen Action Plan: Created by the Madrid Peace Conference of November 1991, the Working Group on Economic Development established the Copenhagen Action Plan on November 8-9, 1993 to study issues of mutual concern, such as the sharing of agricultural and medical research and technology, and joint projects in communications, electricity, and tourism. The European Union pledged $9.2 million for a preliminary study on the construction of a road from Amman, through Jericho, to Jerusalem, which would eventually be extended to the Mediterranean. The EU will also fund a feasibility study on connecting electricity grids between Egypt and Israel (which would save the countries an estimated $4 billion annually).

See Cairo Declaration, Economic Development, Moscow Multilateral Middle East Conference.

Copenhagen Declaration (1997): A first attempt to form a popular peace movement in the Arab countries was launched in January 1997, under the auspices of the Danish Foreign Ministry and with the encouragement of the European Union. Some 50 politicians and intellectuals from Israel, Egypt, Jordan, and the PA gathered in Denmark with the aim of creating a lobbying group to pressure the Arab governments. The declaration spoke of the need for Palestinian statehood and a resolution of the Jerusalem question. It produced a tentative working plan to foster cooperation between Israeli and Arab cities, make joint TV programs, and organize nonofficial people-to-people gatherings.

The participants created the International Alliance for Arab-Israeli Peace and produced a joint declaration to "spare no effort" to accelerate the peace process. The alliance stated its support for Israeli security along with Palestinian self-determination.

Synonymous with International Alliance for Peace in the Middle East.

Copenhagen Summit (1973): See Euro-Arab dialogue.

Copts: See minorities.

copyright: See intellectual property rights.

core parties: In the Middle East peace process the four "core" parties are Israel, Egypt, Jordan, and the Palestinians.

Council (Gaza Strip and Jericho): See Declaration of Principles.

Council for Cooperation of the Gulf States: See Casablanca Declaration.

Council of Arab Economic Unity (CAEU): Formed in 1957 by the Arab Economic Council of the Arab League to encourage economic cooperation and integration of the Arab national economies. Members are Iraq, Jordan, Kuwait, Libya, Mauritania, Somalia, Sudan, Syria, United Arab Emirates, Yemen, and the Palestine Liberation Organization.

Council on Foreign Relations: In June 1997 the (U.S.) Council on Foreign Relations released an independent task force report entitled "U.S. Middle East Peace Policy and the Peace Process," urging a new American approach to get the Oslo process back on track. Several task force members wrote dissenting opinions. Highlights of the report, prepared by Robert K. Lifton are:

The United States has vested interests in restoring the momentum toward Arab-Israeli peace. While the United States cannot and should not impose a settlement on the parties, the time has come for the Administration to take a new initiative, offering its own road map to peace and using its influence to move the parties in that direction.

American Interests in the Middle East

Reversals in the peace process and the prospect of increased regional instability threaten a variety of American interests, including:

- The stability of friendly, relatively moderate Arab states. These states are essential to the uninterrupted flow of oil without major price fluctuations. They are also needed to bolster the ability of the U.S. military to deter Iraq and Iran.
- The long-term security of Israel, which requires a stable peace with its neighbors.
- The prevention of terrorism, which often targets U.S. citizens and institutions, as well as the nonproliferation of weapons of mass destruction.

Recent U.S. Peace Policy

Two of the major principles that have guided U.S. peace policies since the Oslo accords should change. They are:

- Acceptance of Oslo's incremental, confidence-building approach, which was designed to achieve ever-larger levels of Palestinian self-government to be matched by Palestinian efforts to prevent the impairment of Israel's security.
- Reliance on Israel and the Palestinians to negotiate their own agreements with minimal American intervention, except to help manage crises when they occur, provide moral and political support, and rally international backing.

A New Declaration of Principles

The task force proposes intensive U.S. diplomacy along the Israeli-Palestinian track to get the parties to agree to the broad contours of a final settlement. This effort would culminate in a Washington summit, where the agreement would be concluded with the signing of a new Declaration of Principles (DOP). The DOP would neither replace the Oslo accords, relieve the parties of their Oslo obligations, nor attempt to resolve all the difficult details of the final settlement. Instead, it would address the essential goals of both Israelis and Palestinians and accommodate the basic needs of each side in a tradeoff of tangible benefits. Features include:

- Assurance of the Palestinians that the final status of their territories will be statehood in Gaza and most of the West Bank. This would be coupled with the assurance of enhanced security to Israel. The Palestinian state would be demilitarized and would not have the right to forge military alliances with hostile states. A minimum necessary contingent of Israeli troops would be stationed in the Jordan Valley. And appropriate security arrangements to ensure the personal security of Israelis would be implemented, including benchmarks for Palestinian measures against terrorism.
- The agreement would be based on the principle of maximum territorial contiguity for the Palestinian state on the majority of the West Bank and Gaza, while holding to a minimum the relocation of Israelis now living there.
- The issue of Jerusalem would be postponed until all other issues are resolved. However, such postponement does not imply that the final outcome can be anything less than an undivided city. Any final settlement must recognize Jerusalem as the capital of Israel, and al-Quds, whose location and boundaries are to be negotiated by the parties, as the capital of the new Palestinian state.
- The parties would have to reach some understanding about Palestinian functional rights in Jerusalem (e.g., municipal and religious rights, rights of residency within the continuing status quo); unilateral actions that significantly alter the demographics of the city.
- Both parties must commit to a serious effort to end inflammatory rhetoric. Palestinians must commit to an unending war against terrorism.

Once the parties agree to the DOP and try to work out the details of the arrangements, the United States must refrain from acting as a court of first resort when disputes arise. But Washington must be prepared to speak out forcefully when questions of interpretation and implementation threaten to disrupt the peace process.

Other Steps

In addition, to improve the psychological climate and make substantive progress toward solving another important problem, the United

States could take the lead in organizing funds for the dispossessed of the Middle East conflict. The potential beneficiaries would include both Palestinians and Jewish refugees from Arab countries. The United States could also take steps to bolster the economy of the West Bank and Gaza Strip, such as offering guarantees to private investors and helping and encouraging the Palestinian Authority to create a more suitable climate for such investments.

Dissenting views were given to the council's report. Lester Pollack, former Chairman of the Conference of Presidents of Major American Jewish Organizations and Mortimer Zuckerman, Chairman and Editor-in-Chief of *U.S. News & World Report* did not think the United States should take explicit positions on the issues that divide the parties. Instead, they wrote that the parties should strive to agree on a new Declaration of Principles without specific proposals from the United States, which would play a "constitutive" role as an active intermediary in "back-channel negotiations." Stephen P. Cohen, President of the Institute for Middle East Peace and Economic Development, said: "American activism, American peace leadership, is not to declare the substance of potential agreement. The task force proposal, while overtly rejecting the idea, proposes just such an imposed settlement." He asserted that the United States should take a more active role in leading an "imposing process" that will create more momentum toward peace. Robert Satloff, Executive Director of the Washington Institute for Near East Policy, disagreed that the time has come for a new American initiative. He wrote that "the United States has no particular interest in the 'actual nature' of any agreement except that it be acceptable to the parties and that it terminate the conflict. They are the best judges of how to achieve the latter objective."

See Amman Middle East/North Africa Economic Summit, Cairo Middle East/North Africa Economic Conference, Casablanca Declaration, Casablanca Middle East/North Africa Economic Summit, Declaration of Principles, Oslo accords, United States.

counterfeit goods: MENA will have to confront the impact of unequal economies among the nations resulting in an increase in counterfeit production of items. Cheap copies of valued goods may inundate countries where labor costs result in higher prices, which in part results from payment of royalties. Counterfeited goods can sell at a small fraction of the cost in another country resulting in millions of dollars of lost revenue, especially for small business owners. Rules to protect manufacturers, which indirectly protects inventors, artists, and so on will be needed.

See intellectual property rights.

D

dairy: See Protocol on Economic Relations (Article VIII).

Damascus Declaration: See Gulf Cooperation Council.

Damya Bridge: See Cairo Agreement.

dark channel: See Netanyahu, Benjamin.

dawlatay al-Urdunnwqa-Filastin: See Jordanian-PLO Accord.

Dayan, Moshe: See Beilin-Eitan agreement.

DCO (District Coordination Office): See Agreement on the Temporary International Presence in the City of Hebron (1997), Gaza Strip, and the Jericho Area Agreement, Interim Agreement.

Dead Sea: See Cairo Agreement, water.

Declaration of Principles (DOP): See Declaration of Principles on Inter-Self-Government Arrangements.

Declaration of Principles (DOP) on Inter-Self-Government Arrangements (Israel-Palestinian): On September 13, 1993 in Washington, DC and signed by Prime Minister Rabin from the government of Israel and Chairman Arafat from the PLO, and witnessed by President Clinton of the United States of America and Andrei Kozyrev from the Russian Federation, this critical document states:

The Government of the State of Israel and the PLO team (in the Jordanian-Palestinian delegation to the Middle East Peace Conference) (the "Palestinian Delegation"), representing the Palestinian people, agree that it is time to put an end to decades of confrontation and conflict, recognize their mutual legitimate and political rights, and strive to live in peaceful coexistence and mutual dignity and security and achieve a just, lasting, and comprehensive peace settlement and historic reconciliation through the agreed political process. Accordingly, the two sides agree to the following principles:

Article I—Aim of the Negotiations

The aim of the Israeli-Palestinian negotiations within the current Middle East peace process is, among other things, to establish a Palestinian Interim Self-Government Authority, the elected Council (the "Council"), for the Palestinian people in the West Bank and the Gaza Strip, for a transitional period not exceeding five years, leading to a permanent settlement based on Security Council Resolutions 242 and 338.

It is understood that the interim arrangements are an integral part of the whole peace process and the negotiations on the permanent status will lead to the implementation of Security Council Resolutions 242 and 338.

Article II—Framework for the Interim Period

The agreed framework for the interim period is set forth in this Declaration of Principles.

Article III—Elections

1. In order that the Palestinian people in the West Bank and Gaza Strip may govern themselves according to democratic principles, direct, free, and general political elections will be held for the Council under agreed supervision and international observation, while the Palestinian police will ensure public order.

2. An agreement will be concluded on the exact mode and conditions of the elections in accordance with the protocol attached as Annex I, with the goal of holding the elections not later than nine months after the entry into force of this Declaration of Principles.

3. These elections will constitute a significant interim preparatory step toward the realization of the legitimate rights of the Palestinian people and their just requirements.

Article IV—Jurisdiction

Jurisdiction of the Council will cover West Bank and Gaza Strip territory, except for issues that will be negotiated in the permanent status negotiations. The two sides view the West Bank and the Gaza Strip as a single territorial unit, whose integrity will be preserved during the interim period.

Article V—Transitional Period and Permanent Status Negotiations

1. The five-year transitional period will begin upon the withdrawal from the Gaza Strip and Jericho Area.

2. Permanent status negotiations will commence as soon as possible, but not later than the beginning of the third year of the interim period, between the government of Israel and the Palestinian people representatives.

3. It is understood that these negotiations shall cover remaining issues, including: Jerusalem, refugees, settlements, security arrangements, borders, relations and cooperation with other neighbors, and other issues of common interest.

4. The two parties agree that the outcome of the permanent status negotiations should not be prejudiced or preempted by agreements reached for the interim period.

Article VI—Preparatory Transfer of Powers and Responsibilities

1. Upon the entry into force of this Declaration of Principles and the withdrawal from the Gaza Strip and the Jericho Area, a transfer of authority from the Israeli military government and its Civil Administration to the authorized Palestinians for this task, as detailed herein, will commence. This transfer of authority will be of a preparatory nature until the inauguration of the Council.

2. Immediately after the entry into force of this Declaration of Principles and the withdrawal from the Gaza Strip and Jericho Area, with the view to promoting economic development in the West Bank and Gaza Strip, authority will be transferred to the Palestinians on the following spheres: education and culture, health, social welfare, direct taxation, and tourism. The Palestinian side will commence in building the Palestinian police force, as agreed upon. Pending the inauguration of the Council, the two parties may negotiate the transfer of additional powers and responsibilities, as agreed upon.

Article VII—Interim Agreement

1. The Israeli and Palestinian delegations will negotiate an agreement on the interim period (the "Interim Agreement").

2. The Interim Agreement shall specify, among other things, the structure of the Council, the number of its members, and the transfer of powers and responsibilities from the Israeli military government and its Civil Administration to the Council. The Interim Agreement shall also specify the Council's executive authority, legislative authority in accordance with Article IX below, and the independent Palestinian judicial organs.

3. The Interim Agreement shall include arrangements, to be implemented upon the inauguration of the Council, for the assumption by the Council of all of the powers and responsibilities transferred previously in accordance with Article VI above.

4. In order to enable the Council to promote economic growth, upon its inauguration, the Council will establish, among other things a Palestinian Electricity Authority, a Gaza Sea Port Authority, a Palestinian Development Bank, a Palestinian Export Promotion Board, a Palestinian Environmental Authority, a Palestinian Land Authority and a Palestinian Water Administration Authority, and any other authorities agreed upon, in accordance with the Interim Agreement that will specify their powers and responsibilities.

5. After the inauguration of the Council, the Civil Administration will be dissolved, and the Israeli military government will be withdrawn.

Article VIII—Public Order and Security

In order to guarantee public order and internal security for the Palestinians of the West Bank and the Gaza Strip, the Council will establish a strong police force, while Israel will continue to carry the responsibility for defending against threats, as well as the responsibility for overall security of Israelis for the purpose of safeguarding their internal security and public order.

Article IX—Laws and Military Orders

1. The Council will be empowered to legislate, in accordance with the Interim Agreement, within all authorities transferred to it.

2. Both parties will review jointly laws and military orders presently in force in remaining spheres.

Article X—Joint Israeli-Palestinian Liaison Committee

In order to provide for a smooth implementation of this Declaration of Principles and any subsequent agreements pertaining to the interim period, upon the entry into force of this Declaration of Principles, a Joint Israeli-Palestinian Liaison Committee will be established in order to deal with issues requiring coordination, other issues of common interest, and disputes.

Article XI—Israeli-Palestinian Cooperation in Economic Fields

Recognizing the mutual benefit of cooperation in promoting the development of the West Bank, the Gaza Strip, and Israel, upon the entry into force of this Declaration of Principles, an Israeli-Palestinian Economic Cooperation Committee will be established in order to develop and implement in a cooperative manner the programs identified in the protocols attached as Annex III and Annex IV.

Article XII—Liaison and Cooperation with Jordan and Egypt

The two parties will invite the governments of Jordan and Egypt to participate in establishing further liaison and cooperation arrangements between the government of Israel and the Palestinian representatives, on the one hand, and the governments of Jordan and Egypt, on the other hand, to promote cooperation between them. These arrangements will include the constitution of a Continuing Committee that will decide by agreement on the modalities of admission of persons displaced from the West Bank and Gaza Strip in 1967, together with the necessary measures to prevent disruption and disorder. Other matters of common concern will be dealt with by this Committee.

Article XIII—Redeployment of Israeli Forces

1. After the entry into force of this Declaration of Principles, and not later than the eve of elections for the Council, a redeployment of Israeli military forces in the West Bank and Gaza Strip will take place, in addition to withdrawal of Israeli forces carried out in accordance with Article XIV.

2. In redeploying its military forces, Israel will be guided by the principle that its military forces should be redeployed outside populated areas.

3. Further redeployments to specified locations will be gradually implemented commensurate with the assumption of responsibility for public order and internal security by the Palestinian police force pursuant to Article VIII above.

Article XIV—Israeli withdrawal from the Gaza Strip and Jericho Area

Israel will withdraw from the Gaza Strip and Jericho Area, as detailed in the protocol attached as Annex II.

Article XV—Resolutions of Disputes

1. Disputes arising out of the application or interpretation of this Declaration of Principles, or any subsequent agreements pertaining to the interim period, shall be resolved by negotiations through the Joint Liaison Committee to be established pursuant to Article X above.

2. Disputes which cannot be settled by negotiations may be resolved by a mechanism of conciliation to be agreed upon by the parties.

3. The parties may agree to submit to arbitration disputes relating to the interim period, which cannot be settled through conciliation. To this end, upon the agreement of both parties, the parties will establish an Arbitration Committee.

Article XVI—Israeli-Palestinian Cooperation Concerning Regional Programs

Both parties view the multilateral working groups as an appropriate instrument for promoting a "Marshall Plan," the regional programs and other programs, including special programs for the West Bank and Gaza Strip, as indicated in the protocol attached as Annex IV.

Article XVII—Miscellaneous Provisions

1. This Declaration of Principles will enter into force one month after its signing.

2. All protocols annexed to this Declaration of Principles and Agreed Minutes pertaining thereto shall be regarded as an integral part hereof.

Done at Washington, DC, this thirteenth day of September 1993.

Annex I—Protocol on the Mode and Conditions of Elections

1. Palestinians of Jerusalem who live there will have the right to participate in the election process, according to an agreement between the two sides.

2. In addition, the election agreement should cover, among other things, the following issues:

a. the system of elections;
b. the mode of the agreed supervision and international observation and their personal composition; and
c. rules and regulations regarding election campaigns, including agreed arrangements for the organizing of mass media, and the possibility of licensing a broadcasting and TV station.

3. The future status of displaced Palestinians who were registered on June 4, 1967 will not be prejudiced because they are unable to participate in the election process due to practical reasons.

Annex II—Protocol on Withdrawal of Israeli Forces from the Gaza Strip and Jericho Area

1. The two sides will conclude and sign within two months from the date of entry into force of this Declaration of Principles, an agreement on the withdrawal of Israeli military forces from the Gaza Strip and Jericho Area. This agreement will include comprehensive arrangements to apply in the Gaza Strip and the Jericho Area subsequent to the Israeli withdrawal.

2. Israel will implement an accelerated and scheduled withdrawal of Israeli military forces from the Gaza Strip and Jericho Area, beginning immediately with the signing of the agreement on the Gaza Strip and Jericho Area and to be completed within a period not exceeding four months after the signing of this agreement.

3. The above agreement will include, among other things:

a. Arrangements for a smooth and peaceful transfer of authority from the Israeli military government and its Civil Administration to the Palestinian representatives.
b. Structure, powers, and responsibilities of the Palestinian Authority in these areas, except: external security, settlements, Israelis, foreign relations, and other mutually agreed matters.
c. Arrangements for the assumption of internal security and public order by the Palestinian police force consisting of police officers recruited locally and from abroad (holding Jordanian passports and Palestinian documents issued by Egypt). Those who will participate in the Palestinian police force coming from abroad should be trained as police and police officers.
d. A temporary international or foreign presence, as agreed upon.
e. Establishment of a joint Palestinian-Israeli Coordination and Cooperation Committee for mutual security purposes.
f. An economic development and stabilization program, including the establishment of an Emergency Fund, to encourage foreign investment, and financial and economic support. Both sides will coordinate and cooperate jointly and unilaterally with regional and international parties to support these aims.
g. Arrangements for a safe passage for persons and transportation between the Gaza Strip and Jericho Area.

4. The above agreement will include arrangements for coordination between both parties regarding passages:

a. Gaza-Egypt; and
b. Jericho-Jordan.

5. The offices responsible for carrying out the powers and responsibilities of the Palestinian Authority under this Annex II and Article VI of the Declaration of Principles will be located in the Gaza Strip and in the Jericho Area pending the inauguration of the Council.

6. Other than these agreed arrangements, the status of the Gaza Strip and Jericho Area will continue to be an integral part of the West Bank and Gaza Strip, and will not be changed in the interim period.

Annex III—Protocol on Israeli-Palestinian Cooperation in Economic and Development Programs

The two sides agree to establish an Israeli-Palestinian Continuing Committee for Economic Cooperation, focusing, among other things, on the following:

1. Cooperation in the field of water, including a Water Development Program prepared by experts from both sides, which will also specify the mode of cooperation in the management of water resources in the West Bank and Gaza Strip, and will include proposals for studies and plans on water rights of each party, as well as on the equitable utilization of joint water resources for implementation in and beyond the interim period.

2. Cooperation in the field of electricity, including an Electricity Development Program, which will also specify the mode of cooperation for the production, maintenance, purchase, and sales of electricity resources.

3. Cooperation in the field of energy, including an Energy Development Program, which will provide for the exploitation of oil and gas for industrial purposes, particularly in the Gaza Strip and in the Negev, and will encourage further joint exploration of other energy resources. This program may also provide for the construction of a Petrochemical industrial complex in the Gaza Strip and the construction of oil and gas pipelines.

4. Cooperating in the field of finance, including a Financial Development and Action Program for the encouragement of international investment in the West Bank and Gaza Strip, and in Israel, as well as the establishment of a Palestinian Development Bank.

5. Cooperation in the field of transport and communications, including a program which will define guidelines for the establishment of a Gaza Sea Port area, and will provide for the establishing of transport and communication lines to and from the West Bank and the Gaza Strip to Israel and to other countries. In addition, this program will provide for carrying out the necessary construction of roads, railways, communications lines, and so forth.

6. Cooperation in the field of trade, including studies, and Trade Promotion Programs, which will encourage local, regional, and interregional trade, as well as a feasibility study of creating free trade zones in the Gaza Strip and in Israel, mutual access to these zones, and cooperation in other areas related to trade and commerce.

7. Cooperation in the field of industry, including Industrial Development Programs, which will provide for the establishment of a joint Israeli-Palestinian Industrial Research and Development Centers, will promote Palestinian-Israeli joint ventures, and provide guidelines for cooperation in the textile, food, pharmaceutical, electronics, diamonds, computer, and science-based industries.

8. A program for cooperation in, and regulation of, labor relations and cooperation in social welfare issues.

9. A Human Resources Development and Cooperation Plan, providing for joint Israeli-Palestinian workshops and seminars, and for the

establishment of joint vocational training centers, research institutes, and data banks.

10. An Environmental Protection Plan, providing for joint and/or coordinated measures in this sphere.

11. A program for developing coordination and cooperation in the field of communication and media.

12. Any other programs of mutual interest.

Annex IV—Protocol of Israeli-Palestinian Cooperation Concerning Regional Development Programs

1. The two sides will cooperate in the context of the multilateral peace efforts in promoting a Development Program for the region, including the West Bank and the Gaza Strip, to be initiated by the G-7. The parties will request the G-7 to seek the participation in this program of other interested states, such as members of the Organization for Economic Cooperation and Development, regional Arab states and institutions, as well as members of the private sector.

2. The Development Program will consist of two elements:

a. An Economic Development Program for the West Bank and the Gaza Strip.
b. A Regional Economic Development Program.

A. The Economic Development Program for the West Bank and the Gaza Strip will consist of the following elements:

(1) A Social Rehabilitation Program, including a Housing and Construction Program.
(2) A Small and Medium Business Development Plan.
(3) An Infrastructure Development Program (water, electricity, transportation and communications, etc.).
(4) A Human Resources Plan.
(5) Other programs.

B. The Regional Economic Development Program may consist of the following elements:

(1) The establishment of a Middle East Development Fund, as a first step, and a Middle East Development Bank, as a second step.
(2) The development of a joint Israeli-Palestinian-Jordanian Plan for coordinated exploitation of the Dead Sea area.
(3) The Mediterranean Sea (Gaza)-Dead Sea Canal.
(4) Regional Desalinization and other water development projects.
(5) A regional plan for agricultural development, including a coordinated regional effort for the prevention of desertification.

(6) Interconnection of electricity grids.
(7) Regional cooperation for the transfer, distribution, and industrial exploitation of gas, oil, and other energy resources.
(8) A regional Tourism, Transportation, and Telecommunications Development Plan.
(9) Regional cooperation in other spheres.

3. The two sides will encourage the multilateral working groups, and will coordinate toward their success. The two parties will encourage intersessional activities, as well as pre-feasibility and feasibility studies, within the various multilateral working groups.

Agreed Minutes to the Declaration of Principles on Interim Self-Government Arrangements

Any powers and responsibilities transferred to the Palestinians pursuant to the Declaration of Principles prior to the inauguration of the Council will be subject to the same principles pertaining to Article IV, as set out in these Agreed Minutes below.

Specific Understanding and Agreements Article IV. It is understood that:

(1) Jurisdiction of the Council will cover West Bank and Gaza Strip territory, except for issues that will be negotiated in the permanent status negotiations: Jerusalem, settlements, military locations, and Israelis.
(2) The Council's jurisdiction will apply with regard to the agreed powers, responsibilities, spheres, and authorities transferred to it.

Article VI (2). It is agreed that the transfer of authority will be as follows:

(1) The Palestinian side will inform the Israeli side of the names of the authorized Palestinians who will assume the powers, authorities, and responsibilities that will be transferred to the Palestinians according to the Declaration of Principles in the following fields: education and culture, social welfare, direct taxation, tourism, and other authorities agreed upon.
(2) It is understood that the rights and obligations of these offices will not be affected.
(3) Each of the spheres described above will continue to enjoy existing budgetary allocations in accordance with arrangements to be mutually agreed upon. These arrangements also will provide for the necessary adjustments required in order to take into account the taxes collected by the direct taxation office.

(4) Upon the execution of the Declaration of Principles, the Israeli and Palestinian delegations will immediately commence negotiations on a detailed plan for the transfer of authority on the above offices in accordance with the above understandings.

Article VII (2). The Interim Agreement will also include arrangements for coordination and cooperation.

Article VII (5). The withdrawal of the military government will not prevent Israel from exercising the powers and responsibilities not transferred to the Council.

Article VIII. It is understood that the Interim Agreement will include arrangements for cooperation and coordination between the two parties in this regard. It is also agreed that the transfer of powers and responsibilities to the Palestinian police will be accomplished in a phased manner, as agreed in the Interim Agreement.

Article X. It is agreed that, upon entry into force of the Declaration of Principles, the Israel and Palestinian delegations will exchange the names of the individuals designated by them as members of the Joint Israeli-Palestinian Liaison Committee. It is further agreed that each side will have an equal number of members in the Joint Committee. The Joint Committee will reach decisions by agreement. The Joint Committee may add other technicians and experts, as necessary. The Joint Committee will decide on the frequency and place or places of its meetings.

Annex II. It is understood that, subsequent to the Israeli withdrawal, Israel will continue to be responsible for external security, and for internal security and public order of settlements and Israelis. Israeli military forces and civilians may continue to use roads freely within the Gaza Strip and the Jericho Area.

See Agreement on Security Arrangements for Hebron, Cairo Agreement, Casablanca Declaration, chronology (September 13, 1953), Council on Foreign Relations, Gaza Strip and the Jericho Area Agreement, Interim Agreement, interim arrangement, Israel-Jordan Common Agenda, Letters of Mutual Recognition, Moscow Multilateral Middle East Conference, Oslo Agreement, PLO Charter, Preparatory Transfer of Powers and Responsibilities, Protocol on Economic Relations, settlements (and the Oslo accords), United Nations (Middle East Peace Process).

Declaration of the Establishment of the State of Israel: See Israel (Declaration of Establishment).

Declaration on Trade in Services: See Free Trade Agreement (United States and Israel).

Democratic Front for the Liberation of Palestine (DFLP): A breakaway Palestinian political organization; founded in 1969 over a split with PFLP. First to advocate two-state solution in 1974. Expelled from Amman and then Beirut in 1982, moving to Damascus. Worked to frustrate the plans of Arafat and King Hussein to enter into peace talks with Israel. Opposed the PLO-Israeli accord signed in September 1993.

desalination: See Water & Desertification (Center of the Middle East/North Africa Community), Declaration of Principles, Middle East Desalination Research Center, Moscow Multilateral Middle East Conference, water.

desertification: Most of the region can be characterized as Saharo-Arabian geobotanical system, in which average annual rainfall tends to fall short of 200 mm.

See environment, Moscow Multilateral Middle East Conference, Water & Desertification (Center of the Middle East/North Africa Community).

desert rehabilitation: See desertification.

DFLP: See Democratic Front for the Liberation of Palestine.

dinar: The monetary unit of Abu Dhabi, Aden, Algeria, Bahrain, Iraq, Jordan, Kuwait, Libya, South Yemen, Tunisia, and former Yugoslavia.

direct taxation: See Protocol on Economic Relations (Article V).

dirham: The monetary unit of Morocco.

disabled persons: See Israel-Jordan Common Agenda, Social Rights (Center of the Middle East/North Africa Community).

disarmament: See United Nations entries.

discrimination: See Social Rights (Center of the Middle East/ North Africa Community).

diseases: See Health and Housing (Center of the Middle East/ North Africa Community), Protocol on Economic Relations (Article VIII).

displaced persons: See Israel-Jordan Peace Treaty, refugees.

disputes: See Israel-Jordan Peace Treaty.

distance learning: See Secretariat (Middle East/North Africa Economic Summits).

District Coordination Offices (DCO): See Gaza Strip and the Jericho Area Agreement, Interim Agreement.

Doha Economic Conference: See Doha Middle East/North Africa Economic Summit.

Doha Middle East/North Africa Economic Conference: See Doha Middle East/North Africa Economic Summit.

Doha Middle East/North Africa Economic Summit: Scheduled to be held in Doha, Qatar, November 16-18, 1997. His Highness Sheikh Hamad bin Khalifa al-Thani, Emir of the State of Qatar will host the session whose theme is "Creating a new private/public partnership for trade and economic growth beyond the year 2000." Cosponsored by the presidents of the United States and the Russian Federation, with the support and endorsement of Canada, the European Union, and Japan. The initial announcement from Qatar stated:

"The 1997 summit in Doha, Qatar will build on the success and momentum of the previous meetings, illustrating that the region is now on the threshold of rapid development and expansion, offering a tremendous untapped commercial potential.

It is appropriate that the first MENA summit in the Gulf region should take place in Qatar. Rich in natural resources, Qatar has also demonstrated its commitment to developing as a vibrant business and investment center. In addition to geographical proximity to Eastern and Western markets, and a stable economic and political environment, the country has recently introduced new codes and laws to facilitate private sector investment. Qatar is positioning itself to be in the forefront of the regional economic development process and all financial indicators suggest that the country offers sound growth potential."

See Qatar.

donor countries: See Gaza Strip and the Jericho Area Agreement (Article VI).

DOP (Declaration of Principles): See Declaration of Principles on Interim Self-Government Arrangements.

double taxation: See Protocol on Economic Relations (Article V).

drugs: See Hamas Covenant, Israel-Jordan Peace Treaty.

drylands: See Water & Desertification (Center of the Middle East/North Africa Community).

Dunkel, Arthur: See Swiss Trade Initiative for the Middle East and North Africa.

DZD: The international standard code for the currency of Algeria—the Algerian dinar.

E

EAMS: See Euro-Arab Management School.

Early Empowerment Agreement: On August 29, 1994 Israel and the PLO signed the Early Empowerment Agreement, giving the Palestinian Authority jurisdiction over education, tourism, health, welfare, and taxation throughout the West Bank.

earth sciences: See energy.

East Bank (of the Jordan River): The territory on the East Bank belonging to Jordan.

Compare West Bank.

East Jerusalem: Captured by the Arab forces in the Arab-Israeli War in 1948-1949; held by Jordan. It was recaptured by Israel on June 7, 1967 during the Six Day War, making East Jerusalem part of "unified Jerusalem."

Arab residents of East Jerusalem can vote in local elections. More than one dozen Jewish settlements exist in East Jerusalem. Chairman Arafat proclaims this area to be part of the future Palestine.

Compare West Jerusalem.

See Har Homa.

East Mediterranean Region: See investing in peace; New Middle East, South East Mediterranean.

EBA: See Energizing Business Actors.

Eban, Abba: At a meeting of the Council of Europe of Strasbourg in September 1967, Abba Eban, former Deputy Prime Minister and Foreign

Minister of Israel suggested that Israel and its immediate neighbors should explore the European Community idea as a precedent for their own relationships. Little happened.

On December 23, 1973 at the Geneva Peace Conference, speaking in behalf of the government, Eban again called for movement in this area: "The ultimate guarantee of a peace agreement lies in the creation of common regional interests, in such degree of intensity, such entanglement of reciprocal advantage, in such mutual human accessibility as to put future wars beyond rational contingency." Again, little happened.

Shortly after King Hussein of Jordan's call for a Middle East conference Abba Eban called for forming a community with Israel and Jordan based on the European Union model. The European idea began with three small countries agreeing to open their frontiers and not to sign alliances without each other's agreement, he said: "The western powers were willing to give Germany sovereignty on the condition that they join a community with community obligations which would dilute nationalism." Reechoing phrases from the post-Nazi era and attempts to permanently contain German aggression, Mr. Eban spoke "If Israel, Jordan and the new Palestine were members of a community in which none of them could do anything in terms of security without some agreement from the others, boundaries were open and, above all economic investment...proliferates," then Palestinian interests could be realized without being a threat to Israel's security.

See Hussein (King).

EC: See European Community.

Economic Development (Working Group on): The sponsors of the Madrid Peace Conference view economic development as a means to encourage mutual confidence between parties and to foster much needed government and private-sector regional development projects. Economic development was, therefore, included as one of the five issue areas for multilateral discussion. A Working Group on Economic Development was therefore created.

See Amman Middle East/North Africa Economic Summit, Cairo Middle East/North Africa Economic Conference, Casablanca Middle East/North Africa Economic Summit, Copenhagen Action Plan, Moscow Multilateral Middle East Conference, Protocol on Economic Relations.

economic environment: See Amman Middle East/North Africa Economic Summit.

economic freedom: As greater economic freedom evolves in MENAC, there should be greater political freedom. As the private sector flourishes and people accrue money, they also become increasingly independent of government. With time, that is likely to have political consequences.

economic integration: See free trade (theory).

economic reform: See privatization.

economic reform program: See privatization.

Economic Research Forum (ERF) for the Arab Countries, Iran, and Turkey: Based in Cairo, ERF is a nongovernment, nonprofit research networking organization. ERF's mission is to promote independent policy-relevant research, to disseminate the results of research activity to scholars, policymakers, and the business community, and to function as a resource base for researchers through its data bank and documentation library. ERF does not advocate any one policy line, but provides a platform for the exchange of a wide range of views on regional economic issues.

economic summit: See Amman Middle East/North Africa Economic Summit, Cairo Middle East/North Africa Economic Conference, Casablanca Middle East/North Africa Economic Summit.

economic unions: See free trade (theory), regional economic integration.

economies of scale: See Industrial Cooperation and Trade (Center of the Middle East/North Africa Community), Standards (Center of the Middle East/North Africa Community).

education: See Education (Center of the Middle East/North Africa Community), Interim Agreement (on the West Bank and the Gaza Strip).

Education (Center of the Middle East/North Africa Community): Central to the challenge for any nation is to upgrade each citizen's educational opportunity to parallel his or her genetic aptitude and personal drives for success. In particular, the disparity among countries throughout the Middle East/North Africa region is appalling. Nations have their shame to live with, having smothered millions and dismissed them from the ladder of a raised living standard.

An all-out stampede must proceed to overcome the shortcomings and disgrace in restricting educational opportunities so as to reduce cross-fertilization of ideas and skills.

In addition, to understand and study another culture is to fill a void of ignorance for an alternative lifestyle. Much of the ignorance and anger throughout the region over these past decades is triggered by the lack of understanding of traditions generated by history of nations from other parts of the community.

Programs should be instituted in all schools throughout the community where language education will encourage and perhaps demand a required curriculum to include both Arabic and Hebrew. To communicate in a neighbor's language is to offer a commitment to another way of life, to encourage acceptance, to promote a well-being based on a determination to know how to communicate with ones partners.

Furthermore, all members of the region should minimally be taught the history and contributions of each participating community nation. Courses on Islam and Judaism should be available to all throughout the community.

Working together, the community should develop a degree program in which students from the area could complete part of their studies in other nations.

See Centers of the Middle East/North Africa Community, Middle East/ North Africa Community.

EFTA: See European Free Trade Association.

EGP: The international standard code for the currency of Egypt—the Egyptian pound.

Egypt: Official name is Arab Republic of Egypt. The last of British troops left Egypt in June 1956, followed in July 1956 by Colonel Nasser's nationalization of the Suez Canal, a war with Israel, and by Anglo-French military intervention during the Suez Crisis. With the death of Nasser in 1970, Anwar el-Sadat became President and following losses in the October War with Israel commenced closer ties with the West. In 1975 he reopened the Suez Canal to international shipping. In October 1981, two years following the signing of a peace treaty with Israel, Muslim fanatics assassinated Sadat. President Hosni Mubarak became his successor.

Egypt has been opening its doors wider to foreign trade and more aggressively seeking foreign investment since launching its economic reform program in 1991. In 1995 the business climate improved, with a number of laws and new incentive programs already in place or under consideration. Egypt's reforms have created exchange rate stability and allowed the central bank to accumulate $18 billion in reserves. The Egyptian Pound (LE) is fully convertible and investors report no difficulty in repatriating profits. Most multinational investors claim to be making money and Egypt is one of a handful of markets worldwide where the United States has a multibillion dollar trade surplus.

By far the largest Arab country in population, Egypt is in the heart of the Middle East and boasts a reasonably well-educated and English-speaking labor force. The market is large: anywhere from five to 10 million of Egypt's population of 60 million enjoy Western consumption patterns. The rest of the population is poor, but habits and consumption are changing rapidly: fast-food licensing has been a particular area of recent growth. Moreover, given its strategic position in the region, Egypt continues to benefit from strong donor support, including about $2.3 billion annually in U.S. economic and military assistance. Progress in the Arab-Israeli peace process is beginning to have an effect on opening regional markets. Increased tourism is evident and new regional energy projects are under active discussion.

The main impediment to doing business in Egypt continues to be red tape: U.S. exporters and investors still face requirements for a multiplicity of permits, long delays in clearing goods through customs, arbitrary decision making, and an unresponsive commercial court system. Other weaknesses of the market include the difficulty of obtaining financing beyond five years and non-transparent and difficult government tending rules, which often put competing bidding parties in a "Dutch auction" after final tenders have been submitted.

On balance, Egypt's annual imports of $10 billion, modest annual growth rates of 2 percent, openness to foreign goods, market reforms, growing private sector, and diverse manufacturing base contribute to a commercial challenge which holds significant promise for alert investors. The U.S.-Egyptian Partnership for Economic Growth and Development launched in September 1994, sets as its goal the fostering of higher levels of sustainable economic growth, improvement of Egyptian standards of living, job creation, increased productivity, and a mature international trading system.

Arguably, Egypt is rapidly establishing itself as the region's economic success story in the mid-1990s with a stable climate for undertaking a radical economic reform program. While the government is prepared to shoulder the social costs arising from privatization, this was not the case with trade liberalization and tariff reforms in 1997.

Cairo's stock market has increased the free float of equities and attracted 190 foreign mutual funds into their market in 1996. Daily turnover has lifted to above $5 million on some trading days and total market capitalization has been boosted by 25 percent to over $10 billion.

By 1997 Egypt's growth was running at an annual rate of 5 percent, for the first time since 1990. Inflation has fallen from over 20 percent to 5.4 percent. The government's budget deficit is down from 15 percent of GDP to under 1 percent, thanks to reductions in energy subsidies and the introduction of a 10 percent sales tax. The current-account deficit is negligible and foreign reserves are close to $20 billion, enough to cover 15 months of imports. Foreign investment has tripled and is projected to reach $2.1 billion in 1997.

Business between Israel and Egypt has improved after a slow five years since the 1979 peace treaty. Bilateral non-oil trade more than doubled to $43.9 million in 1994 and soared to $35.4 million for the first six months of 1995, a rise of 160 percent over the year before. In fact, 1994 only saw $43 million in direct trade altogether. A huge Egyptian-Israeli investment project, a $1.2 billion oil refinery to be built at Alexandria was recently launched. Israel, on the other hand, imports roughly $400 million worth of Egyptian oil each year.

Egyptian tourism to Israel was also up slightly, to 30,000—8,000 more than the previous year. Some 250,000 Israelis visited Egypt in 1994, with most going to vacation resorts in the Sinai rather than Egypt proper.

In February 1996 the Israel Chambers of Commerce and the Egypt Business Association signed an agreement aimed at expanding business activity between the two countries. After a long period of stagnation the two economies are beginning to supplement each other. The two groups will work toward encouraging business activity between the two countries, including transfer of information, joint ventures and hosting of delegations.

Total imports from Egypt to Israel were $47.8 million in 1995 from $23.8 million in 1994. Exports were $30.7 million in 1995 from $20.7 million in the previous year.

Egypt hosted the Middle East/North Africa Economic Conference in Cairo, in November 1996.

See Begin, Menachim; Cairo Agreement, Cairo Middle East/North Africa Economic Conference, Camp David Accords; Carter, Jimmy; Egyptian-Israeli Peace Treaty, Middle East Peace Process, Protocol on Economic Relations (Article III); Sadat, Anwar; Suez Canal.

Egyptian-Israeli Peace Treaty: Signed on March 26, 1979 in Washington, DC by President Sadat for the Arab Republic of Egypt and Prime Minister Begin for the State of Israel. President Carter of the United States of America witnessed the document signing.

The document ended the state of war that had existed between the two nations since the founding of Israel in May 1948. The treaty was based on principles outlined in the Camp David Accords.

Within a year of the signing Israel had returned two-thirds of occupied Sinai to Egypt. By April 1982 Israel had withdrawn from the rest of the peninsula. Egypt was suspended from the Arab League; restored in 1989.

The Peace Treaty reads:

TREATY OF PEACE BETWEEN THE ARAB REPUBLIC OF EGYPT AND THE STATE OF ISRAEL

The government of the Arab Republic of Egypt and the government of the State of Israel;

Preamble

Convinced of the urgent necessity of the establishment of a just, comprehensive, and lasting peace in the Middle East in accordance with Security Council Resolutions 242 and 338;

Reaffirming their adherence to the "Framework for Peace in the Middle East Agreed at Camp David," dated September 17, 1978;

Noting that the aforementioned Framework as appropriate is intended to constitute a basis for peace not only between Egypt and Israel but also between Israel and each of its other Arab neighbors which is prepared to negotiate peace with it on this basis;

Desiring to bring to an end the state of war between them and to establish a peace in which every state in the area can live in security;

Convinced that the conclusion of a Treaty of Peace between Egypt and Israel is an important step in the search for a comprehensive peace in the area and for the attainment of the settlement of the Arab-Israeli conflict in all its aspects;

Inviting the other Arab parties to this dispute to join the peace process with Israel guided by and based on the principles of the aforementioned Framework;

Desiring as well to develop friendly relations and cooperation between themselves in accordance with the United Nations Charter and the principles of international law governing international relations in times of peace;

Agree to the following provisions in the free exercise of their sovereignty, in order to implement the "Framework for the Conclusion of a Peace Treaty Between Egypt and Israel."

Article I

1. This state of war between the Parties will be terminated and peace will be established between them upon the exchange of instruments of ratification of this Treaty.

2. Israel will withdraw all its armed forces and civilians from the Sinai behind the international boundary between Egypt and mandated Palestine, as provided in the annexed protocol (Annex I), and Egypt will resume the exercise of its full sovereignty over the Sinai.

3. Upon completion of the interim withdrawal provided for in Annex I, the Parties will establish normal and friendly relations, in accordance with Article III (3).

Article II

The permanent boundary between Egypt and Israel is the recognized international boundary between Egypt and the former mandated territory of Palestine, as shown on the map at Annex II, without prejudice to the issue of the status of the Gaza Strip. The Parties recognize this boundary as inviolable. Each will respect the territory integrity of the other, including their territorial waters and airspace.

Article III

1. The Parties will apply between them the provisions of the Charter of the United Nations and the principles of international law governing relations among states in times of peace. In particular:

a. They recognize and will respect each other's sovereignty, territorial integrity, and political independence;
b. They recognize and will respect each other's right to live in peace within their secure and recognized boundaries;
c. They will refrain from the threat or use of force, directly or indirectly, against each other and will settle all disputes between them by peaceful means.

2. Each Party undertakes to ensure that acts or threats of belligerency, hostility, or violence do not originate from and are not committed from within its territory, or by any forces subject to its control or by any other forces stationed on its territory, against the population, citizens, or property of the other Party. Each Party also undertakes to refrain from organizing, instigating, inciting, assisting, or participating in acts or threats of belligerency, hostility, subversion, or violence against the other Party, anywhere, and undertakes to ensure that perpetrators of such acts are brought to justice.

3. The Parties agree that the normal relationship established between them will include full recognition, diplomatic, economic, and cultural relations, termination of economic boycotts and discriminatory barriers to the free movement of people and goods, and will guarantee the mutual enjoyment by citizens of the due process of law. The process by which they undertake to achieve such relationship parallel to the implementation of other provisions of this Treaty is set out in the annexed protocol (Annex III).

Article IV

1. In order to provide maximum security for both Parties on the basis of reciprocity, agreed security arrangements will be established including limited force zones in Egyptian and Israeli territory, and United Nations forces and observers, described in detail as to nature and timing in Annex I, and other security arrangements the Parties may agree upon.

2. The Parties agree to the stationing of United Nations personnel in areas described in Annex I. The Parties agree not to request withdrawal of the United Nations personnel and that these personnel will not be removed unless such removal is approved by the Security Council of the United Nations, with the affirmative vote of the five permanent members, unless the Parties otherwise agree.

3. A Joint Commission will be established to facilitate the implementation of the Treaty, as provided for in Annex I.

4. The security arrangements provided for in paragraphs 1 and 2 of this Article may at the request of either party be reviewed and amended by mutual agreement of the Parties.

Article V

1. Ships of Israel, and cargoes destined for or coming from Israel, shall enjoy the right of free passage through the Suez Canal and its approaches through the Gulf of Suez and the Mediterranean Sea on the basis of the Constantinople Convention of 1888, applying to all nations. Israeli nationals, vessels, and cargoes, as well as persons, vessels, and cargoes destined for or coming from Israel, shall be accorded nondiscriminatory treatment in all matters connected with usage of the canal.

2. The Parties consider the Strait of Tiran and the Gulf of Aqaba to be international waterways open to all nations for unimpeded and nonsuspendable freedom of navigation and overflight. The Parties will respect each other's rights to navigation and overflight for access to either country through the Strait of Tiran and the Gulf of Aqaba.

Article VI

1. This Treaty does not affect and shall be interpreted as affecting in any way the rights and obligations of the Parties under the Charter of the United Nations.

2. The Parties undertake to fulfill in good faith their obligations under this Treaty, without regard to action or inaction of any other party and independently of any instrument external to this Treaty.

3. They further undertake to take all the necessary measures for the application in their relations of the provisions of the multilateral conventions to which they are parties, including the submission of appropriate notification to the Secretary General of the United Nations and other depositaries of such conventions.

4. The Parties undertake not to enter into any obligations in conflict with this Treaty.

5. Subject to Article 103 of the United Nations Charter, in the event of a conflict between the obligations of the Parties under the present Treaty and any of their other obligations, the obligations under this Treaty will be binding and implemented.

Article VII

1. Disputes arising out of the application or interpretation of this Treaty shall be resolved by negotiations;

2. Any such disputes which cannot be settled by negotiations shall be resolved by conciliation or submitted to arbitration.

Article VIII

The Parties agree to establish a claims commission for the mutual settlement of all financial claims.

Article IX

1. This Treaty shall enter into force upon exchange of instruments of ratification.

2. This Treaty supersedes the Agreement between Egypt and Israel of September 1975.

3. All protocols, annexes, and maps attached to this Treaty shall be regarded as an integral part hereof.

4. The Treaty shall be communicated to the Secretary General of the United Nations for registration in accordance with the provisions of Article 102 of the Charter of the United Nations.

Done at Washington, DC this 26th day of March, 1979, in triplicate in the English, Arabic, and Hebrew languages, each text being equally authentic. In case of any divergence of interpretation, the English text shall prevail.

AGREED MINUTES TO ARTICLES I, IV, V, AND VI AND ANNEXES I AND III OF TREATY OF PEACE

Article I

Egypt's resumption of the exercise of full sovereignty over the Sinai provided for in paragraph 2 of Article I shall occur with regard to each area upon Israel's withdrawal from that area.

Article IV

It is agreed between the parties that the review provided for in Article IV(4) will be undertaken when requested by either party, commencing within three months of such a request, but that any amendment can be made only with the mutual agreement of both parties.

Article V

The second sentence of paragraph 2 of Article V shall not be construed as limiting the first sentence of that paragraph. The foregoing is not to be construed as contravening the second sentence of paragraph 2 of Article V, which reads as follows:

> The Parties will respect each other's rights to navigation and overflight for access to either country through the Strait of Tiran and the Gulf of Aqaba.

Article VI (2)

The provisions of Article VI shall not be construed in contradiction to the provisions of the framework for peace in the Middle East agreed at Camp David. The foregoing is not to be construed as contravening the provisions of Article VI (2) of the Treaty, which reads as follows:

> The Parties undertake to fulfill in good faith their obligations under this Treaty, without regard to action or inaction of any other Party and independently of any instrument external to this Treaty.

Article VI (5)

It is agreed by the Parties that there is no assertion that this Treaty prevails over other treaties or agreements or that other treaties or agreements prevail over this Treaty. The foregoing is not to be construed as contravening the provisions of Article VI (5) of the Treaty, which reads as follows:

> Subject to Article 103 of the United Nations Charter, in the event of a conflict between the obligations of the Parties under the present Treaty and any of their other obligations, the obligations under this Treaty will be binding and implemented.

Annex I

Article VI, paragraph 8, of Annex I provides as follows:

> The Parties shall agree on the nations from which the United Nations force and observers will be drawn. They will be drawn from nations other than those which are permanent members of the United Nations Security Council.

The Parties have agreed as follows:

> With respect to the provisions of paragraph 8, Article VI, of Annex I, if no agreement is reached between the Parties, they will accept or support a U.S. proposal concerning the composition of the United Nations force and observers.

Annex III

The Treaty of Peace and Annex III thereto provide for establishing normal economic relations between the Parties. In accordance therewith, it is agreed that such relations will include normal commercial sales of oil by Egypt to Israel, and that Israel shall be fully entitled to make bids for Egyptian-origin oil not needed for Egyptian domestic oil consumption, and Egypt and its oil concessionaires will entertain bids made by Israel, on the same basis and terms as apply to other bidders for such oil.

Twelve years after Israel signed the peace treaty with Egypt in 1981, the trade between the two countries was a mere $20 million a year, plus the $200 million in oil that Egypt promised to sell Israel as part of the Treaty. Seventy million dollars a year was traded through third parties to avoid upsetting Arab clients. Since the 1993 agreement between Israel and the PLO, Israel's trade with Egypt has taken off to more than $60 million in overt trade in 1995.

See Camp David Accords.

Egyptian Pound: The currency of Egypt.

EIB: See European Investment Bank.

Eilat (Elat): See Camp David Accords, Israel-Jordan Peace Treaty.

Ein Avrona: Further Washington bilateral negotiations between Israel and Jordan led to the signing on June 7, 1994 of sub-agendas on security, borders, and territorial matters, as well as water and environmental issues. Subcommissions for each of these areas were also established.

The first meeting in the region of the Israeli and Jordanian bilateral delegations took place July 18-19, 1994 at Ein Avrona, located in the boundary area north of Aqaba and Eilat. At this meeting the delegations confirmed the sub-agendas signed in Washington on June 7, 1994.

At the conclusion of the talks at Ein Avrona the subcommission on boundaries, security, and water and environment issued a joint statement. The main points are:

a. joint team of experts will prepare maps delineating borders
b. confidence-building measures will be implemented in the area of security
c. the parties agree to negotiate rightful water allocations and cooperate on increasing available water, environmental protection, and energy generation.

The parties agreed that talks would be held on a continuous basis, alternately on the Israeli and the Jordanian sides of the border.

These developments culminated in the first public meeting between King Hussein and Prime Minister Rabin in Washington on July 25, 1994. Out of this session emerged the Washington Declaration.

See Israel-Jordan Common Agenda, Washington Declaration.

Eitan, Michael: See Beilin-Eitan agreement.

El Al: See aviation.

El Arish: See Camp David Accords.

Elat: See Eilat.

elections (Gaza Strip and Jericho): See Declaration of Principles, Interim Agreement (on the West Bank and the Gaza Strip).

electric grids: Linking transmission grids throughout MENA can result in significant economic benefits for all involved parties:

a. At times of peak demand in a given country, the country would be able to buy electricity from the existing generating capacity of the other country, thus obviating the need to actuate the more expensive production reserves of its own.
b. The overall amount of installed electrical power reserves which either country would have to maintain over time would be lower, enabling all countries to reduce their investments in the production infrastructure in the long run.

A European Union project will identify and select the most feasible interconnection solution for the national power systems of Egypt, Israel, Jordan, and the Palestinian Authority. The potential economic benefit to each party from the interexchange of electricity will be established, together with the sharing of the spinning reserve in the interconnected system and the potential to avoid or postpone investments in additional capacity for generation and transmission.

See energy, Energy and Environment (Center of the Middle East/ North Africa Community).

electricity grids: See electric grids.

el-Sadat, Anwar: See Sadat, Anwar.

EMA: See Euro-Mediterranean Agreement.

Emirates: See United Arab Emirates.

Energizing Business Actors (EBA): One of the Middle East/North Africa Economic Summit Secretariat's Programs; EBA's goal is twofold:

a. to provide a user-friendly mechanism for promoting greater participation by small and intermediate-sized firms from the Middle East peace process member countries in regional cooperative projects, and
b. to facilitate for all-sized firms continuous access to information concerning the large infrastructure projects proposed at the MENA economic summits.

Large, multinational firms need little assistance in establishing contacts and financing initial investment analysis, but they do benefit from timely access to information about upcoming major capital investment projects (dams, downstream irrigation systems, telecommunications, ports, etc.).

At the same time, because of the high overhead of information gathering and administrative red tape, small and intermediate-sized firms do not always have access to the information about partners, investment opportunities, or sources of financing. In contrast to large firms, however, small and intermediate firms, once they are convinced of the feasibility of an idea and have access to financing, are frequently able to move more rapidly than larger companies to initiate the investment. EBA intends to buy down some of the costs of information gathering and contacts for small and intermediate-sized firms which are the prerequisite of sustainable participation by those firms in cooperative regional economic development. EBA will also assure that firms interested in large, capital intensive, regional investment projects have ready access to the information which is critical to maintaining commitment to those efforts.

EBA will carry out three separate functions related to its goal:

- establish easy access (via the World Wide Web) to information regarding regional cooperative investment opportunities (the opportunity database);
- establish easy access to firms seeking partners for given investment opportunities (the partner database); and
- establish easy access to sources of expertise and networking expertise, such as the World Economic Forum or the World Trade Center network, and financing, such as the Ex-Im Bank, the MENA Fund of OPIC, European investment bank programs, and specific programs of donors active in the region (the assist database).

The WTCA and the Executive Secretariat have agreed to organize joint programs in the areas of investment and trade promotion and facilitation of business partnerships. With over 20 World Trade Centers already established in the region, the WTCA and its WTC Information Network, is well placed to assist in promoting the goals of the economic summit process.

See Secretariat (Middle East/North Africa Economic Summit).

energy: MENA nations proportionately consume little of the world energy. A sampling of energy usage shows:

Country	*World Population (%)*	*World GDP (%)*	*World Energy (%)*
Iraq	1.02	.30	.12
Egypt	1.02	.17	.36
Israel	.09	.23	.13
Libya	.08	.14	.12
Saudi Arabia	.28	.42	.90
Kuwait	.04	.11	.18

In contrast, the United States with 4.84 percent of the world's population consumes 24.24 percent of world energy and the world GDP at 27.30 percent.

Establishing energy projects across MENA should:

- create and improve the basic infrastructure necessary for economic development (energy, water, employment), which will make it possible to populate peripheral desert areas and provide conditions for their demographic and economic growth;
- boost regional economies and augment their potential for sustained economic growth;
- solidify peaceful relations and political stability in the region;
- preserve the quality of the environment throughout the region; and

- raise the technological standards and quality of life for all inhabitants in the region.

See Amman Middle East/North Africa Economic Summit, electric grids, gas, Israel-Jordan Peace Treaty, oil.

Energy and Environment (Center of the Middle East/North Africa Community):

Energy

A better energy infrastructure raises the standard of living, directly and indirectly, as well as the potential for economic output. It enables the acceleration of domestic growth and encourages foreign investments.

Energy projects within the community will serve the people of the region by:

- creating and improving the basic infrastructure necessary for economic development which will make it possible to populate peripheral desert areas and provide conditions for their demographic and economic growth;
- boosting regional economics and augmenting their potential for sustained economic growth;
- solidifying peaceful relations and political stability in the region;
- preserving the quality of the environment throughout the region; and
- raising the technological standards and quality of life for all inhabitants in the community.

The community could work on a host of energy projects, including the optimum utilization of joint energy facilities, such as:

- erecting joint power stations;
- linking transmission grids;
- laying gas mains and operating plants with natural gas;
- extending electric grids; and
- increasing oil pipelines.

Environment

Further efforts are needed to protect the environment throughout the community. Cooperation should focus on environmental protection, such as preventing environmental pollution, nature conservation in tourist areas, and in preserving beaches.

The community should create a regional information and training unit to evolve programs on protecting the environment and enhancing energy conservation. This facility could act to encourage the development of technological innovation in the community, to distribute information, and to commercially promote mature technologies. Environmental protection in the member states interfaces closely with economic development.

The community would benefit from establishing a regional energy information center common to all regional nations founded on computerized data and data based on consumption production, and energy source reserves in the community. Such a center would make an important contribution to technological research and surveys, and would receive the assistance of business entities and research institutions in the efforts to develop energy sources, to protect and efficiently utilize these natural resources based on cooperation and a concerted effort.

See Centers of the Middle East/North Africa Community, electric grids, energy, environment, Middle East/North Africa Community.

energy usage: See energy.

enforcement provision: See Treaty of the Middle East/North Africa Community (Article 16).

engineering: See Amman Middle East/North Africa Economic Summit.

environment: The environment is an issue that cuts across the political lines of the Middle East. Israel, Egypt, Jordan, and Saudi Arabia must all deal with the floundering ecosystem of the Gulf of Aqaba. Pollution and waste from one country often drift over to its neighbor. Desertification is plaguing formerly fertile land throughout the region. At the Madrid Conference a Working Group on the Environment was set up to deal with these issues. A proposed Energy and Environment Center of the Middle East/North Africa Community is suggested.

See Amman Middle East/North Africa Economic Summit, birds, electric grids, energy, Energy and Environment (Center of the Middle East/North Africa Community), gas, Israel-Jordan Peace Treaty, Japan, Multilateral Working Group on the Environment, oil, Upper Gulf of Aqaba Oil Spill Contingency Project.

Environmental Code of Conduct: See Multilateral Working Group on the Environment, Japan.

Environmental Protection Plan: See Declaration of Principles.

environmental standards: See gas, Multilateral Working Group on the Environment.

Environment Working Group: See Japan, Moscow Multilateral Middle East Conference, Multilateral Working Group on the Environment.

equality: See Social Rights (Center of the Middle East/North Africa Community).

equivalent effect: See Free Trade Agreement (European Union and Israel).

Erbakan, Necmettin: See Turkey.

Eretz Israel: See Israel (Declaration of Establishment).

Erez settlement: See Cairo Agreement.

ES: economic summit.

Eskol, Levi (President): See chronology (January l6, 1965).

Establishment of a nuclear-weapon-free zone in the region of the Middle East: See United Nations (nuclear-weapon-free zone in the region of the Middle East).

Etzel: See Arab-Israeli War (I), Hebrew Insurrection Movement.

EU: See European Union.

Euro-Arab dialogue: Negotiations on economic, financial, trade, and other cooperations between the European Union and the Arab League, all resulting from the Copenhagen Summit of December 1973.

Euro-Arab Management School (EAMS): EAMS is the fruit of an initiative of 1995 by the European Union and the Spanish government; a private nonprofit foundation under Spanish law which, through management education and training activities, aims to contribute to the social and economic development of Arab countries and to promote social and economic collaboration between Europe and the Arab world. EAMS is located in the heart of the historic city of Granada in the Spanish province of Andalusia; a city of historical and cultural meeting points between Arab and European civilizations as well as a center of Euro-Arab scholarship.

The mission that the EAMS has been entrusted with by its founders is:

- Through its different training activities, to create competent managers capable of operating with competence, confidence, and understanding in the diverse cultures of Europe and of the Arab world.

- To create a collaborative network of institutional and corporate partners located both in Europe and in the Arab world and, through this network, to foster the creation of a cross-cultural learning community whose members can work together on a variety of mutually beneficial management development activities.
- Through research and action-oriented pedagogical activities undertaken jointly with EAMS network partners to extend and deepen the understanding of managerial, economic, and social issues that confront the European and Arab worlds in their dealings with each other.
- To make that understanding available to managers and policymakers through conferences, seminars, workshops, and publications.
- Through all of the above, to become a center of excellence and of expertise on Euro-Arab managerial issues and to foster the development of such excellence and expertise among its network partners.
- To strengthen the organizational capacity of Arab management development institutions and, through contributing to an increased supply of well-trained Arab managers, to play a role in the economic development of the Arab world.

Euro-Mediterranean Agreement (EMA): In July 1995 Tunisia signed an EMA with the European Union, and Morocco followed in October of 1996. At present there are active negotiations for an EMA between the EU and Egypt, Jordan, and Lebanon. The intention is that by the year 2010, a wider Euro-Mediterranean free trade area will be created encompassing the 15 member states of the EU and 12 Mediterranean states as well as countries of Central and Eastern Europe.

The growth impact of EMAs is expected to result from gains from increased trade and foreign investment and improved productivity. The agreements are, nonetheless, expected to have far-reaching implications in terms of intersectoral adjustment resulting from trade diversion, some of which will negatively impact the movement of labor and capital among different sectors in the economy.

The International Monetary Fund believes that EMAs seek to attain a number of objectives:

- Increase and enhance the political dialogue between the partners.
- Achieve reciprocal free trade between the MENA countries involved and the European Union in most manufactured goods.
- Grant preferential and reciprocal access for some agricultural products at a later stage.
- Establish conditions for the liberalization of the right of establishment and the supply of services.

- Facilitate free capital flows.
- Encourage southern Mediterranean rim countries to take on board a wide range of trade-related European Union regulations (customs, standards, competition, protection of intellectual property rights, and so on.)
- Increase economic, social, and cultural cooperation.

See European Union.

Euro-Mediterranean Conference: See European Union.

Euro-Mediterranean Partnership: The European Union, as the gavel-holder of REDWG, the largest of the peace process working groups, is a major promoter and builder of regional economic cooperation. The union is, at the same time, seeking to strengthen the economies and societies of the region within the wider context of the Euro-Mediterranean Partnership.

The European Union and 12 Mediterranean partners, which include the four core parties to the peace process, agreed in Barcelona on November 28, 1995 to jointly set up a new policy framework—the Euro-Mediterranean Partnership.

Partners agreed in the progressive establishment of a free trade area by the target date of 2010, through the new Euro-Mediterranean Agreements and free trade agreements between the partners of the European Union.

A number of Euro-Mediterranean projects have been launched. For example, the Euro-Mediterranean Partners are actively encouraging the setting up of networks between operators on both sides of the Mediterranean: banks, chambers of commerce and industry, trade fairs, trade promotion boards, economic research institutes, and statistical institutes. Such networks will allow for an increased mutual understanding, an exchange of information and of best practices, and for mutually beneficial cooperation and training.

See Euro-Mediterranean Stability Charter, European Union, Renovated Mediterranean Policy.

Euro-Mediterranean Space: See Mediterranean Crans Montana Forum.

Euro-Mediterranean Stability Charter: The Euro-Mediterranean Partnership developed at the Barcelona Conference was of a multilateral dimension in establishing global political discussion centered around the Mediterranean, for the first time. The need for such a political dialogue originates from the acknowledgment that the stability and prosperity of the European Union are indissociable from those of its immediate neighbors to the south. A number of principles emerged from the Barcelona Conference:

- to act in accordance with the United Nations Charter and the Universal Declaration of Human Rights;
- to develop a political system in accordance with international, commonly accepted norms in the areas of human rights and democracy;
- to take concrete measures for the effective application of fundamental freedoms;
- to respect diversity and pluralism, and to fight intolerance, particularly racism and xenophobia, and discrimination based on race, religion, and sex.

The European Parliament, following the Barcelona Conference, decided to create the MEDA-Democracy program to support local or regional projects aimed at promoting, in the partners of the Union,

- political rights relating to democracy and the legal state;
- civil rights such as the freedoms of expression, meeting, and association.

See Euro-Mediterranean Partnership, European Union.

Euro-Med Partnership: See Euro-Mediterranean Partnership.

European Community (EC): See European Union.

European Free Trade Association (EFTA): Created in 1959 as a counter-institution to the European Economic Community. In 1995 the remaining members were Norway, Switzerland, and Iceland. Former nations of the Soviet bloc may be considered for membership.

See Free Trade Agreement (European Free Trade Association and Israel).

European Investment Bank (EIB): Established in 1958 to help implement specific economic policy objectives. An independent body with capital contributed by community nations. Its goal is to contribute via investment loans in projects to "the balanced and steady development of the common market in the interest of the Community."

See European Union.

European Recovery Program: See Marshall Plan.

European Union (EU): Formed by the Treaty of Rome (1958) when the European Coal and Steel Community, European Atomic Energy Community, and the European Economic Community merged in 1965 to form the European Community. In 1995 there were 15 members and a changed name to the European Union.

The enlargement of the then EC to include Spain and Portugal was seen by some Mediterranean nonmember countries as a threat to their farm exports; the opening up of the community to the east raised fears of a slackening of cooperation with the south; and the approach of 1993 caused some to fear that the single market would become a fortress Europe. Not the least of the challenges resulting from the onset of economic and political instability in the Mediterranean countries themselves, rendering their position precarious and making cooperation both more necessary than ever.

Each year, the community's imports from its Mediterranean partners amounts to some US$25 billion, and its corresponding exports to over US$30 billion. The region is one of the community's main trading partners, ranking well ahead of Japan. The dependence of these nations is considerable, where the European markets accounts for over 50 percent of their exports (75% for Tunisia).

Financial assistance in the case of the Maghreb and Mashreq countries represents over US$3 billion since 1976, the agreements include especially favorable trade arrangements, without the requirement for benefits to be reciprocated (except in the case of Israel).

In December 1989 the Council adopted the framework for redirecting the Community's Mediterranean policy. Implementation began in 1990 and continued during 1991 when the majority of the new financial protocols with the Mediterranean countries entered into force. This new policy has six main components:

- back-up for the process of economic adjustment
- encouragement of private investment
- increase in bilateral and community financial assistance
- maintenance or improvement of arrangements governing access to the community market.
- close involvement with the progress of the community toward the single market
- strengthening of economic and political dialogue, at the regional level where possible.

The new Mediterranean policy provides an overall aid package of ECU$405 million (approximately US$5,000 million) subdivided as follows:

a. ECU 2,075 million for the financial protocols (with Egypt, Morocco, Algeria, Tunisia, Syria, Jordan, Israel and Lebanon) which runs from November 1, 1991 to October 31, 1996.

b. ECU 2,030 million for more broadly based financial cooperation outside the protocols (including regional cooperation and cooperation of the environment).
c. ECU$300 million to provide back-up for economic reform.

The EU's Redirected Mediterranean Policy to intensify cooperation in the 1990s has witnessed the approval of a substantial package of aid for regional programs and projects.

The commission's MED-Invest Program is an industrial cooperation instrument for nonmember Mediterranean nations, and is designed to provide an even wider range of facilities to encourage the development of an environment conducive to small-medium enterprises and to increase their ability to compete on local and EU markets.

The European Investment Bank provides by far the largest percentage of EU industrial sector financing to MTCs.

At its December 1993 summit conference, the European Union agreed to upgrade economic relations with Israel, noting that the conclusion of a new agreement with Israel "is one of the key elements of the development of cooperation between the European Union and the Middle East region."

In addition, the union agreed to provide ECU$500 million in direct aid to Palestine (the Occupied Territories) for 1994-1998.

The 15-member European Union held the Barcelona Conference (also called the Euro-Mediterranean Conference) with 12 nonmember Mediterranean states on November 27-28, 1995, which gave the countries of the union and their eastern and western Mediterranean partners an unprecedented opportunity to decide together what their future relationship would be. The EU's objective is to ensure stability and prosperity in the Mediterranean and is prepared to support those nations in their efforts to turn the region progressively into an area of peace, stability, prosperity, and cooperation and to establish a Euro-Mediterranean partnership. That called for political dialogue, sustainable and balanced economic and social development, combating poverty, and the need for greater understanding between cultures through a reinforcement of the human dimension in exchanges. The partnership comprised these main aspects:

- political and security
- economic and financial
- social and human

The European Union is offering $6 billion to a dozen countries from Morocco to Turkey. The emergence of the EU policy to the Mediterranean is an implicit recognition that the region to its south is of critical importance to the future of Europe.

In August 1996 the EU made a grant of ECU$25 million to the Palestinian Authority, 15 of which will go on infrastructure projects and the remaining 10 to support the private sector.

In March 1997 the Palestinian Authority Chairman Arafat signed an agreement to remove trade barriers over the next five years and to provide for regular political consultation between the EU and the Palestinians. The EU is the biggest single contributor of economic assistance to the PA, spending $406 million on the West Bank and Gaza Strip since 1993.

In July 1997 Israel and the European Union signed trade agreements on public procurement and telecommunications that will pave the way for Israeli businessmen to compete in Europe. The documents provide for the relaxation for EU suppliers of regulations concerning provisions, local content, and threshold limits of Israeli government procurement tenders. In return Israeli suppliers will be entitled to access what the EU describes as the most liberal and open procurement systems in the world. The signing of the telecommunications agreement will give Israeli high-tech companies a better opportunity to compete for European tenders.

See Agreement of Understanding between Israel and Lebanon, Arab world; Arafat, Yasir (Press Statement of December 14, 1988), containment, conflict, and common management of common problems, Copenhagen Action Plan; Eban, Aba; electric grids, Euro-Arab dialogue, Euro-Arab Management School, Euro-Mediterranean Agreements, Euro-Mediterranean Partnerships, Free Trade Agreement (European Union and Israel), Har Homa, Mashrag states, Mediterranean Agreement, Mediterranean Policy, Middle East Desalination Research Center, Middle East/North Africa (trade), Middle East Regional Transportation Study, MNCs, Northern Mediterranean, Regional Animal Health Cooperation Program, South-East-Mediterranean Development Project, TEAM-Area, tourism, Transport Corridors in the Middle East, Upper Gulf of Aqaba Oil Spill, Uruguay Round, Water Data Banks Project.

exchange controls: See Amman Middle East/North Africa Economic Summit.

Executive Secretariat (of the Middle East/North Africa Summit): See Amman Middle East/North Africa Economic Summit, Casablanca Middle East/North Africa Economic Summit; Terrab, Mostafa.

Existing Fund: See Protocol on Economic Relations (Article XI).

Expired Permit Palestinians: See refugees.

F

Fahd, King Bin Saud: King Fahd of Saudi Arabia is seeking to maintain his personal rule despite reportedly suffering health problems. He serves as prime minister, chairman of the supreme council of petroleum and minerals, and holds the religious title of Custodian of the Two Holy Places. Born in 1923, he succeeded to the throne in June 1982. Fahd has seen oil prices wane from the boom years, resulting in some economic liberalization. Since the Gulf War ended in 1991 the highly conservative ruler has had to be wary of religious radicals, but also has set about a tentative modernization of the political system, notably with the adapted rules of succession in 1992 to promote merit alongside lineage, and the creation of the Majlis as-Shura, a 60-man appointed consultative council, in 1993. He remains at the top of a patronage network of 6,000 or so princes in a demographically youthful country, where 60 percent of the population are under 21.

See Saudi Arabia.

farming: See agriculture, water.

Farouk (King): See chronology (July 1952).

Fatah: (al-Fatah). Reverse acronym in Arabic for Palestinian National Liberation Movement. Formed in the late 1950s as a secret underground cell in Kuwait by Arafat and others.

Fatah remained underground until 1964 when the Arab League formed the Palestine Liberation Organization. Arafat gained leadership of the PLO in 1969. Fatah is the largest organization within the PLO and represents the mainstream of Palestinian thinking.

In August 1989 the Fatah Congress produced the following political program:

Palestine is a part of the Arab homeland.

The Palestinian people are an indivisible part of the Arab nations. The Palestinian people have been living in their homeland Palestine, from time immemorial....Since the dawn of Islam and up to the present, the Palestinian people have been the heart of the Arab and Islamic nation protecting the first Qiblah and the third holiest shrine (Al-Aqsa Mosque)....

The Zionist conquest came at the beginning of this century to uproot the Palestinian people, end their civilization, and impact a Zionist entity designed to carry out the imperialist schemes that seek to control the Arab homeland's resources and make this homeland a part of the colonialist's spheres of influence....

The Palestinian people have held out in the face of this Zionist-colonialist barbarism....The crime was consummated by the partition of Palestine and the establishment of the Zionist entity state in 1948.

Most of the Palestinian people were scattered outside their homeland and forced to live in exile in the diaspora....They were builders of civilization wherever they lived....

Then came the conspiracies hatched by the Syrian regime and carried out by its agent tools, including the dissensions conspiracy, the double siege of Tripoli, and the war of siege and elimination against the sons of our people in the camps. The conspiracy sought to contain the Palestinian revolution and to end its effective role within the equation of the conflict.

The intifada has driven a deep crack in the barbarous Zionist entity, strongly attracted the world's attention to the justice of the cause, and imposed a new political reality that has great impact on all Palestinian, Arab, and international levels....

We convene the Fifth General Congress of our Fatah movement with more resolve and determination to continue the struggle with all possible means—political and military—to liberate our homeland and establish our state, the independent state of Palestine with holy Jerusalem as its capital....

The Fifth General Congress of (the) Fatah Movement specifies our future objectives as follows:

1. The Palestinian question is the core of the Arab-Zionist conflict.
2. Firm adherence to Palestinian Arab national inalienable rights in their homeland Palestine, including their right to repatriation, self-determination without foreign interference, and the establishment of their independent state with holy Jerusalem as its capital....
4. The Fifth Fatah General Congress affirms the historic importance of the resolutions of the 19th PNC session, particularly the documents of independence. The Congress supports the establishment

of the independent state of Palestine and extends thanks to the Arab and friendly that have recognized it. The Congress authorizes the movement's central committee to work on all levels to implement the resolutions of establishing the Palestinian people's national inalienable rights, headed by the right to repatriation, self-determination, and of the establishment of the state of Palestine on Palestinian soil with holy Jerusalem as its capital.

5. Continuing to intensify and escalate armed action and all forms of struggle liquidating the Israel-Zionist occupation of our occupied Palestinian land and guaranteeing our people's rights to freedom and independence.
6. Bolstering national Palestinian unity on the various political and military levels, reiterating the PLO's leading role, and escalating the popular intifada aimed at ending the Zionist Israeli occupation....
9. ...rejects the Shamir plan on elections and affirms that any election in our occupied must take place in a free and democratic atmosphere under international supervision after the withdrawal of the Israeli forces.
13. In light of the significant effect of the demographic factor on our conflict with the Zionist enemy, and in light of the huge efforts exerted by the Zionist movement to encourage Zionist immigration to Palestinian territory, the Congress has decided to set up an ad hoc committee within the revolutionary council to oppose the Zionist immigration to our homeland and to assume all cultural, information, and political tasks to prevent the arrival of Jewish immigrants in our occupied homeland....

The Fatah Congress calls on the Arab countries to abide by and implement all the resolutions and honor their financial commitments.

Respecting the right of the Palestinian revolution to perform its militant tasks through any Arab land....

Calling on the Arab countries, especially those on the confrontation lines, to unify their forces and mobilize their masses in order to confront the Israel aggression....

The Fifth General Congress...emphasized...the Palestinian people's national inalienable rights and the right of all oppressed peoples under occupation to use all forms of struggle for their liberation and national independence....

See Arafat, Asifa, Black September, Force 17, terrorism.

Fatah political program: See Asifa, Fatah, Force 17.

Fatah Provisional Command: Synonymous with Fatah Uprising.

Fatah Revolutionary Council: Libyan-backed anti-Arafat radical faction responsible for numerous terrorist attacks. Formed in 1973 as a split from Fatah. Headquartered in Tripoli. Expelled from Iraq and Syria.

Fatah Uprising: Formed in 1983 with Syrian backing. Military commander attempted to overthrow Arafat after PLO ouster from Lebanon. Based in Damascus.

Synonymous with Fatah Provisional Command.

FDI: See foreign direct investment.

Fedayeen: See PLO Charter (Article 10).

Feisal, Emir: See Arab-Jewish Agreement at Versailles, chronology (January 3, 1919).

fences: During the tense periods of relationships between Israel and her Palestinian neighbors, debates over construction of fences separating the West Bank have been revived.

From north to south the invisible line that marks Israel off from the West Bank runs about 216 miles. If the plans of Israel's Labor government evolve, the free movement of goods and people across the Green Line will end.

As a result of terrorist attacks in the winter 1996, the government plans to construct a system of fences, patrol zones, and electronic monitoring barriers that will block free passage. The goal is to stop terrorists who enter along with 70,000 Palestinian workers each day.

The project will leave only a small portion of the Green Line actually fenced. Down most of its length, Israel will instead create a 1.2 mile wide "security zone" in which it will introduce military patrols and high-tech equipment to detect illegal crossover.

Only in the vicinities of Kalkilya and Tulkarim will Israel construct about 17 miles of fences. Jewish and Palestinian towns in these areas are bunched so close together up against the Green Line, there simply is no room for a wider security zone. Ultimately, the Peres government hopes to have 18 entries through the Green Line.

The 350-km (220-mile) "security seam" announced by Prime Minister Peres in March 1996 would be built along, roughly, the old border—except for the Israeli-expanded limits of Jerusalem—between Israel and the West Bank. It will separate the two by fences, patrols, and electronic surveillance.

See borders, separation, terrorism.

Fez summit: See Jordanian-PLO Accord.

Fifth General Congress of the Fatah Movement: See Fatah.

Final Status Plan: See Allon-Plus Plan.

Financial Services (Center of the Middle East/North Africa Community). See Banking and Financial Services (Center of the Middle East/ North Africa Community).

FIS: See Algeria, Islamic Salvation Front.

fiscal deficits: Fiscal deficits in non-oil MENA countries, according to the International Monetary Fund, have the following common characteristics:

- They have been an important element underlying the low savings rate.
- In many cases, they have been the main factor contributing to inflationary pressures.
- They have resulted in a buildup of domestic and external debt and, in many cases, have led to uncertainties about future macroeconomic policies.
- They have tended to crowd out productive private-sector investment through higher interest rates and limited availability of financing (which has been absorbed by the government).

See investment, Middle East/North Africa (MENA).

Fischer, Franklin M.: See water.

FLN: See Front De Liberation Nationale.

foot and mouth disease: See Regional Animal Health Cooperation Program.

Force 17: Fatah faction within PLO formed in early 1970s. Provides personal security forces for Arafat and other leaders. Involved in terrorist operations against Israelis in Cyprus.
See Fatah.

foreign direct investment (FDI): FDI is transforming the world economy. Yet, MENA nations have not participated to any great extent.

Nevertheless, since the signing of Oslo I, foreign direct investment in the Middle East, and in particular in Israel, has been increasing. Prior to the accord there was no foreign investment to speak of. In 1995 $2.1 billion was pumped into the Israeli economy by overseas investors, more than double the level of 1994. Intel Corporation planned a $1.6 billion semiconductor plant, Bankers Trust announced it was expanding its presence in Israel, 3M explored building a plant in Israel, and Unilever officials reaffirmed their company's commitment to investing in Israel. Foreign investors seek regional political stability and integration.

According to Hafsi and Fauchet the future potential of the MENA nations for FDIs are dependent upon:

a. Societal attractiveness, which includes a society's openness and its general attitude toward foreigners and foreign-owned property.
b. Infrastructural attractiveness, which includes the quality and coverage of infrastructure, but also the general quality of government regulations (e.g., property rights) and their administration as well as the stock of existing FDI (which generally attracts other FDI investors).
c. Factors attractiveness, which include the availability and quality of human resources, capital resources, and raw material.
d. Governability attractiveness, or general ability of the governing elite to achieve consensus within the rule of law. Governability is frequently perceived to be an indicator of long-term stability.
e. Competitive attractiveness, or the importance of the country's situation and capabilities in the competitive game in which transnational corporations are involved. These may include particular arrangements, regulations, and incentives related to a specific investment.

See capital markets development, investment, Regional Investment Promotion Program, Secretariat (Middle East/North Africa Economic Summit).

Foreign Investment Advisory Service: See Secretariat (Middle East/North Africa Economic Summit).

four Cs: See containment, conflict, and common management of common problems.

fourth freedom: See Middle East Community.

FPZ: See free processing zone.

Framework for Peace in the Middle East: See Camp David Accords, Egyptian-Israeli Peace Treaty.

Framework for the Conclusion of a Peace Treaty between Egypt and Israel: See Camp David Accords, Egyptian-Israeli Peace Treaty.

frd: further redeployments.
See redeployment.

freedom of movement (Israeli): See Interim Agreement (on the West Bank and the Gaza Strip).

freedom of passage: See Israel-Jordan Peace Treaty.

Freemasons: See Hamas Covenant.

free movement of goods and people: See Camp David Accords, Egyptian-Israeli Peace Treaty, Treaty of the Middle East/North Africa Community (Article 5).

free processing zone (FPZ): In June 1994 the Israeli Labor party was successful in passing a law creating the country's first free trade zone. Within months of the law's passage more than 100 companies expressed interest in locating in the zone. As of January 1996 nothing is in operation. Arguments against FPZs include nationwide deregulation and tax relief which are preferable to one 500-acre, free trade zone.

FPZs were also designed as part of Shimon Peres's economic development program with the Palestinians.

See input sharing.

free trade (theory): Does free trade mean, for example, that firms from Israel have the right to sell anything in Jordan that conforms to Jordan rules, or does it mean that Israeli firms can sell in Jordan whatever they choose to sell in Israel? It allows governments to block imports on grounds of public policy and, on the other hand, it says such bans must not amount to arbitrary discrimination or disguised restrictions on trade.

The theory of free trade accords necessitates that skilled workers win, while unskilled people lose, since the supply of unskilled laborers expands and that of skilled labor shrinks. The disparities in skill among the peoples of a community are great, and many may be disadvantaged during this period of adjustment.

In addition, the theory suggests that average income rises under free trade since unemployment can be expected not to be present. People terminated in any industry should easily find employment in new or expanding industries, so the theory continues. There is no certainty that average incomes throughout the Middle East and North Africa will increase by shifting to a system of free trade.

The theory also believes that the costs of making structural changes from one industry to another are zero. The recent experience with, for example, Eastern Europe, suggests that these costs are usually great, with old facilities being torn down and new ones being built, infrastructures having to be constructed, and other necessary costly projects.

The theory assumes that people are paid salaries based on the demand for their services, not on the industry in which they are employed. Jumping during this transition period from one industry to another will not guarantee equality of wages. The same task in a different industry will pay a different salary, let alone vary from nation to nation.

Consequently, free trade accords with countries of dissimilar living standards, differing wage scales, varying cultures, as in the Middle East and North Africa, do not readily adapt in a short time period. The Middle East and North Africa Community will be hard-pressed to service the needs of the weaker nations as free trade agreements proliferate.

If free trade areas (where partners agree to scrap tariffs on trade among themselves) are to help world trade they must try to be similarly nondiscriminatory. Entry should be granted willingly, not grudgingly. Members should commit themselves to reducing their external, as well as internal, barriers to trade.

Then there is the largely unanswerable and complex question of whether the world is better off as a series of economic trade regions. With the growth of the global economy and more powerful transnational institutions, nationalism and fundamentalism flourishes. The stress and strain within economic unions exploit nationalist feeling, where so many people seem so remote that they turn to local ethnic groups, often adding friction between neighboring nations and within the boundaries of existing countries. For many in the Middle East and North Africa not sharing in the promise of regional economic integration, separatism appears an attractive alternative.

See Israel-Jordan Peace Treaty, logistic trade centers, regionalism; vision, the.

Free Trade Agreement (European Free Trade Association and Israel): Signed in 1992 and became effective on January 1, 1993.

1. Pursuant to negotiations conducted over a period of approximately one year, the Free Trade Area Agreement between Israel and EFTA (European Free Trade Association) was signed on September 17, 1992 in Geneva, and became effective on January 1, 1993.

2. The Israel-EFTA Free Trade Area Agreement provides for the enhancement of economic cooperation between Israel and EFTA and attempts to remove all forms of unfair competition. This agreement encompasses the basic principles of the Israel-EC (now EU) Free Trade Area Agreement, however, it is broader in application and relates to additional related issues such as government procurement, state aid, and the enforcement of intellectual property rights.
3. In contrast to the Free Trade Area Agreements with the EC and the United States, this agreement provides for the immediate, as opposed to gradual, elimination of most tariffs and duties. Accordingly, as of January 1, 1993, all tariffs and similar trade restrictions on industrial products have been abolished while some, but not all, customs tariffs relating to fresh agricultural products and processed foodstuffs were abolished or reduced. Fish and other marine products are dealt with separately in Annex II to the agreement. Specific details of the products which will be admitted duty-free are provided in the protocols and annexes to the agreement.

4. Summary of the Agreement:

4.1 Preamble and Article 1—set forth the intention of the parties to promote the expansion of reciprocal trade, to expand and diversify cooperation in fields of economic interest, and to remove barriers to trade between EFTA and Israel.

4.2 Article 2—defines the scope of the agreement, ensuring that it applies to most products originating in an EFTA state or Israel.

4.3 Article 3—deals with rules of origin.

4.4 Articles 4-6—eliminate customs duties or charges having equivalent effect on imports and exports and prohibits the introduction of any new customs duties or similar charge.

4.5 Article 7—abolishes the "quantitative restrictions on imports or exports and measures having equivalent effect," defined as prohibitions or restrictions on imports or exports into an EFTA state from Israel, or into Israel from an EFTA state made effective through quotas, import or export licenses, or other administrative measures restricting trade.

4.6 Article 8—permits restrictions on imports and exports for noneconomic reasons such as public morality, public security, and protection of health and life of humans, animals or plants, and of the environment.

4.7 Article 9—stipulates that each party will ensure that no state monopoly of a commercial character will practice any form of discrimination regarding the conditions under which goods are procured and marketed between nationals of Israel and EFTA.

4.8 Article 10—recognizes the importance of developing harmonized international standards and technical regulations, reconfirms the adherence of the parties to the GATT Agreement on Technical Barriers to Trade and to its procedures, and confirms the intention of the parties to increase cooperation in the field of testing and certification.

4.9 Article 11—confirms that each individual EFTA state and Israel have concluded bilateral agreements to facilitate trade in agricultural products and obliges all parties to apply their regulations in veterinary, plant health, and health matters in a nondiscriminating fashion.

4.10 Article 12—prohibits the imposition of any measure or practice of an internal fiscal nature which discriminates, directly or indirectly, between the products originating in an EFTA state or Israel.

4.11 Article 13—ensures that the transfer of payments relating to trade between an EFTA state and Israel shall be free from any restriction.

4.12 Article 14—provides that each party shall grant each other's companies unimpeded access to contract award procedures on their public procurement markets on a reciprocal basis in accordance with the GATT Agreement on government procurement and confirms the intention of the parties to gradually extend the list of the procuring entities to be covered with regard to their procurements in the supplies and utilities sectors above the respective thresholds. Furthermore, the parties undertake to progressively amend the relevant rules governing the participation in contracts awarded by public authorities with a view to eliminating discrimination between the potential suppliers from the parties.

4.13 Article 15—obliges the parties to guarantee efficient and nondiscriminative protection of intellectual property rights including counterfeiting and piracy according to existing multilateral conventions.

4.14 Article 16—authorizes a party which considers that the other party is not fulfilling its obligations under the agreement to take appropriate measures established under Article 23.

4.15 Article 17—provides that all agreements between undertakings and concerted practices between undertakings which prevent or distort competition in so far as they affect trade between an EFTA state and Israel are incompatible with the proper functioning of this agreement.

4.16 Article 18—prohibits that granting of any state aid such as subsidies which distort competition between an EFTA state and Israel.

4.17 Articles 19-22—ensure that in certain cases a party may take special remedial measures. Therefore, in accordance with the procedures established in Article 23, a party may invoke appropriate safeguard measures against the following:

(1) dumping;
(2) an increase in imports of a given product, causing serious injury to domestic producers or resulting in a deterioration of the economic situation of the region;

(3) a serious shortage or threat thereof, of a product essential to the exporting party and caused by compliance to the provisions of Articles 6 and 7;
(4) balance of payments difficulties.

4.18 Article 23—deals with the procedures governing the safeguard measures which may be invoked by any party in the event of a dispute or the existence of a situation covered by Articles 19 to 22. Before initiating such procedures the parties are encouraged to solve any differences between them through direct consultation and thereafter by raising the issue before the Joint Committee. In exceptional circumstances requiring immediate action a party may apply safeguard measures without first consulting the Joint Committee, provided the party refers the matter to the Joint Committee as soon as possible thereafter.

4.19 Article 24—ensures that no party shall be required by this agreement to take any measure which may compromise its essential security interests.

4.20 Article 25—emphasizes that the arrangements applied by each party in respect of the other party shall not give rise to discrimination of any kind.

4.21 Article 26—establishes a Joint Committee comprising representations of each party, to administer the agreement and ensure its proper implementation.

4.22 Article 27—instructs the Joint Committee to meet at least once a year and provides that all decisions and actions of this committee require common agreement.

4.23 Article 28—permits possible extension of the agreement to fields not covered by it.

4.24 Article 29—states the aim of achieving a progressive liberalization and mutual opening of markets for investments and trade in services.

4.25 Article 30—incorporates the protocols and annexes into the agreement.

4.26 Article 31—permits the maintenance or establishment of customs unions, free trade areas, or arrangements for frontier trade provided that such actions do not negatively affect the aims of the agreement.

4.27 Article 32—defines the territorial applicability of the agreement.

4.28 Article 33—states the date of entry into force of the agreement.

4.29 Article 34—permits amendments to the agreement following ratification by all parties.

4.30 Article 35—allows any member of EFTA to accede to the agreement provided that the Joint Committee approves its accession on such terms and conditions as it sees fit.

4.31 Article 36—permits withdrawal from the agreement, effective six months after the date of notification. Any member of EFTA withdrawing from the Convention establishing EFTA shall ipso facto on the same day as withdrawal takes effect cease to be a party to the agreement.

4.32 Article 37—provides that the government of Sweden shall serve as depositary and notify all parties of the entry into force of the agreement, of its expiry or of any withdrawal therefrom.

See Free Trade Agreement (European Union and Israel), Free Trade Agreement (United States and Israel), Rules of Origin.

Free Trade Agreement (European Union and Israel): The FTA signed in 1975 between the European Community and Israel is the first FTA in which Israel became an active partner.

1. This agreement, signed in May 1975, although considerably more comprehensive than the FTA with the United States, gives rise to a less profound regime than the FTA with the United States. For example, processed agricultural products, one of Israel's main export products, are excluded from it, unlike the agreement with the United States. The general layout of this agreement is similar to that of the U.S.-Israel FTA, with the agreement being followed by lengthy protocols dealing with (1) Israel exports to the EC, (2) EC exports to Israel, and (3) Rules of Origin.

2. Summary of the Agreement

2.1 Preamble—sets forth the parties' mutual desire to extend economic relations, to strengthen economic links, and to continue with the progressive elimination of obstacles to "substantially all" their trade in accordance with GATT.

2.2 Article 1—reiterates the aims of the agreement, basically to promote cooperation and expansion of trade.

2.3 Article 2—stipulates that Israeli products imported to the EC shall be governed by Protocol 1, EC products imported into Israel shall be governed by Protocol 2, and products from both countries shall be governed by Protocol 3 laying down Rules of Origin.

2.4 Article 3—forbids imposition of new duties on imports, charges having "equivalent effect," or import quotas on trade between the EC and Israel. Charges having an effect equivalent to duties introduced on or after January 1974 were abolished upon entry into force of the agreement, and higher charges were reduced.

2.5 Article 4—forbids introduction of new duties or charges having "equivalent effect" on exports and abolishes such duties as from July 1, 1977.

2.6 Article 5—states that Articles 3 and 4 shall apply to products other than those listed in Annex II to the treaty establishing the EC.

2.7 Article 6—requires a party which is considering any reduction or suspension of duties respecting third countries with most favored nation treatment to notify the Joint Committee at least 30 days in advance. The Article also states that the Joint Committee may adapt the tariff nomenclature in the event of changes in the nomenclature of the customs tariffs as used in the agreement.

2.8 Article 7—permits amendment of the agreement in the event of implementation of specific rules in a common agricultural policy by one of the parties which shall, however, take the interests of the other party into account.

2.9 Article 8—permits the establishment of other agreements for FTAs, and customs unions so long as they do not alter the provisions of this agreement, especially the provisions regarding Rules of Origin.

2.10 Article 9—contains an undertaking to refrain from internal fiscal measures (i.e., subsidies) which would discriminate between the products of the two parties.

2.11 Article 10—frees payments and transfer of payments relating to trade from restrictions where the trade is covered by this agreement.

2.12 Article 11—permits prohibitions or restrictions on imports and exports on moral, historical, public policy, health, security grounds, and so forth. However, this must not constitute a means of arbitrary discrimination.

2.13 Article 12—defines actions incompatible with the proper working of the agreement, including agreements which distort, prevent, or restrict competition, abuse by undertakings of their dominant position, and public aid which distorts, or threatens to distort, competition. If a party considers that a given practice is incompatible with this Article, it may take the appropriate measures according to Article 16.

2.14 Article 13—permits measures according to Article 16 to be taken in the event that an increase in import of a given product due to partial or total reduction of customs duty or lower duties on raw materials levied by the exporting party is likely to be seriously detrimental to production in one of the parties.

2.15 Article 14—permits anti-dumping measures in accordance with Article VI of GATT.

2.16 Article 15—permits measures in accordance with Article 16 in the event of disturbances in the economy or of difficulties "which could bring about serious deterioration in the economic situation of a region."

2.17 Article 16—is divided into three sections:

Section 1 requires a party putting in force administrative procedures whose aim is to provide rapid information on the trend of trade flows, in the event of the difficulties referred to in Articles 13 and 15, to inform the other party of such fact.

Section 2 requires the party about to take action under Articles 12, 13, 14, 15, and 25 to inform the Joint Committee and to supply it with all relevant information in order to attempt to achieve an acceptable solution. When taking such measures the section provides that those measures which least disturb the functioning of the agreement should be taken and not others.

Section 3 defines the various procedures on measures in accordance with Articles 12, 13, and 14. Section 3 (d) permits immediate action without consultation in exceptional circumstances. In this event the relevant information is to be supplied to the Joint Committee as soon as possible after action has been taken.

2.18 Article 17—permits safeguard measures in the event of balance of payment difficulties of one of the parties.

2.19 Article 18—deals with promotion of economic cooperation.

2.20 Article 19—establishes the Joint Committee mentioned previously which shall be responsible for the implementation of the agreement.

2.21 Articles 20 and 21—deal with membership and procedure of the Joint Committee respectively.

2.22 Article 22—states that the parties shall review the results of the agreement and permits, in certain cases, the slowing down of part of Israel's tariff reduction schedule.

2.23 Article 23—restates the parties' rights to take any measures necessary either to prevent disclosure of security information, essential to security in time of war or serious international tension or relating to arms and munitions trade or defense research and development.

2.24 Article 24—prevents discrimination between member states, nationals, companies, or firms of the EC or of Israeli nationals, companies, or firms.

2.25 Article 25—obligates the parties to take measures required to fulfill the agreement and to refrain from measures that might jeopardize it.

2.26 Article 26—permits possible extension of the agreement to fields not covered by it.

2.27 Article 27—incorporates the protocols into the agreement.

2.28 Article 28—permits recision of the agreement by either party after giving six months notice.

2.29 Article 29—defines the territorial applicability of the agreement.

2.30 Article 30—states the languages and date of entry into force of the agreement.

3. The Protocols to the agreement

The Protocols to this agreement are considerably more complex than the Annexes to the U.S.-Israel FTA since they also provide for variations in the tariff duties and application of internal (national) laws within certain of the EC nations. The protocols dealing with actual tariff reductions rates are, in part:

3.1 Protocol 1 deals with Israeli exports to the EC and provides that except for certain articles (i.e., albumins) and products listed in Annex II to the Treaty of Rome, tariff reductions will be as follows:

60 percent on the date of entry into force of the agreement;
80 percent on January 1, 1976;
100 percent on July 1, 1977.

3.2 Protocol 2 deals with EC exports to Israel and originally provided in two lists, two timetables for phasing out Israeli tariffs on EC products. Although 100 percent reduction of the rates was to be achieved according to the lists, in 1985, in practice, the dates set in one of the lists were extended twice and, in fact, 100 percent of tariff abolition was only reached on January 1, 1989.

In July 1995 the Council of Ministers of the European Union and Israel negotiated a free trade accord. Israel's bid to sit as a nonvoting member in the committees that decide on EU-wide research and development projects was accepted by the Europeans. Israel agreed to cut back its citrus exports from 300,000 tons annually to 200,000 tons. In return, the Europeans agreed to expand quotas on peeled fruits from 16,000 tons a year to 35,000 tons.

The agreement updates the rules of origin that determine the nationality of the products entitled to duty-free treatment, but still ignores Israel's request to exempt from duties the European elements that make up Israeli goods. The agreement stipulates that within six months Israel and the EU will conclude negotiations on opening access to government procurement, with an emphasis on telecommunications.

The agreement was signed in October 1995. It will then be subject to ratification by the European Parliament, followed by individual ratification by each of the 15 member countries' parliaments.

See Free Trade Agreement (European Free Trade Association and Israel), Free Trade Agreement (United States and Israel), Rules of Origin.

Free Trade Agreement (United States and Israel):

1. The agreement on the establishment of a free trade area between the government of Israel and the government of the United States of America was signed on April 22, 1985. This followed approximately 18 months of negotiations after the United States and Israel agreed to commence negotiations on the establishment of such an agreement.

2. The U.S.-Israel Free Trade Area Agreement provides for gradual elimination of all tariffs between the two countries, with full elimination being achieved by 1995. However, most products have been free of tariffs since the beginning of 1989. The details of the products that will be admitted duty-free, immediately and after the various interim periods, are specified in annexes to the agreement.

3. Summary of the Agreement

3.1 Preamble—sets forth the aims of the agreement; to further the historic friendship between the United States and Israel and to strengthen and develop economic relations between the two nations for their mutual benefit.

3.2 Article 1—establishes a "Free Trade Area" consistent with Article XXIV (8) of GATT in which duties and other restrictive regulations of commerce affecting bilateral trade between the two countries will be eliminated.

3.3 Article 2—deals with the four annexes to the agreement, the first two dealing with the phases of duty reduction on the various tariff items, the third setting out the Rules of Origin, and the fourth including Israel's undertakings on export subsidies.

3.4 Article 3—reaffirms rights and obligations under prior treaties, including GATT, and states that in the event of inconsistency between the FTA and prior agreements, the FTA shall prevail.

3.5 Article 4—prohibits introduction of new duties or quantitative restrictions on trade between the parties unless permitted by the terms of the FTA or GATT.

3.6 Article 5—verbatim this article reads:

1. When a product is being imported in such increased quantities as to be a substantial cause of serious injury or the threat thereof to domestic producers of like or directly competitive products, the importing party shall consult with the other party in accordance with Article 18 before taking any actions affecting the trade of the other party.
2. Neither party shall take an action which provides solely for a suspension of the reduction or elimination of any duty provided for by this agreement unless the serious injury or threat thereof which is substantially caused by imports to the domestic producers of like or directly competitive products results from the reduction or elimination of a duty provided for by this agreement.
3. When, in the view of the importing party, the importation of a product from the other party is not a substantial cause of the serious

injury or threat thereof referred to in paragraph 1, the importing party may except the product of the other party from any import relief that may be imposed with respect to imports of that product from third countries, taking into account the objective of achieving bilateral free trade as embodied in this agreement, the domestic laws and international obligations of the parties.

3.7 Article 6—enables import restrictions on agricultural products provided that such restrictions are not duties (i.e., quotas and levies).

3.8 Article 7—incorporates the general and national security exemptions of GATT into the FTA. These exemptions enable trade restrictions to the MFN clause in the event of wars, emergencies, and so on.

3.9 Article 8—permits prohibitions made on religious or ritual grounds under certain circumstances (i.e., prohibition of import of non-kosher meat, etc.).

3.10 Article 9—deals with health and requires the parties to review their veterinary and plant health rules to ensure nondiscrimination, to consult on difficulties involving veterinary and plant health matters, and to establish a joint working group to discuss such matters.

3.11 Article 10—allows Israel to apply duties to protect "infant industries." According to this article, Israel may increase or reintroduce ad-valorem customs duties which shall not exceed 20 percent above the level that would otherwise be in effect. The total value of the products for which these measures can be applied may not exceed 10 percent of the total value of Israel's imports from the United States in 1984. The increase or imposition of the duties requires Joint Committee agreement. The increased duties must be reduced at least 5 percent per annum starting two years after imposition, to be eliminated finally by January 1, 1995.

3.12 Article 11—permits either party to impose temporary restrictions for 150-day periods in the event of serious balance of payments difficulties. In the event such restrictions are import surcharges or deposits, they may be extended for additional 150-day periods, but if they are in the form of quantitative restrictions they may be extended for one additional 150-day period only.

3.13 Article 12—regulates import licensing.

3.14 Article 13—prohibits the imposition of export or domestic-purchase requirements on investments and forbids requirements that investors purchase locally produced goods in order to receive government incentives.

3.15 Article 14—reaffirms the current obligations on intellectual property.

3.16 Article 15—requires both parties to waive their buy- national restrictions on government agency procurements having a contract value of $50,000 or more on an agreed list of governmental entities.

3.17 Article 16—deals with the Trade in Services, in accordance with the Declaration on Trade in Services signed together with the FTA.

3.18 Article 17—establishes the Joint Committee which is to meet at least once a year in order to review the function of the agreement, consult on any matter affecting its operation and interpretation, review its result, and review the Declaration on Trade and Services.

3.19 Article 18—deals with notice and consultation between the parties, requiring written notice of any trade measure to be taken by either party or before either party commits itself to take any action which would reduce trade barriers with third countries except in special, urgent circumstances.

3.20 Article 19—establishes the mechanisms for dispute settlement—first consultation, failing which reference to the Joint Committee and, if this proves unsuccessful, reference to a three member conciliation panel whose findings are not binding.

3.21 Article 20—permits either party to increase its specific duties in order to maintain their value if the party's currency decreases by more than 20 percent. Israel's currency will be measured against the U.S. dollar, and the dollar will be measured against "Special Drawing Rights."

3.22 Article 21—requires the parties to notify one another of changes in their respective tariff schedules. Minor changes will not adversely affect the tariff rates for products listed in Annexes I and II and in the event a change is "major," the balance of tariff concessions must be maintained.

3.23 Article 22—provides for entry into force after the parties have completed the domestic legal procedures necessary therefor, but not before September 1, 1985.

3.24 Article 23—states that the agreement will remain in force until it is terminated. Termination is to be in writing, 12 months after the date of such written notification.

See Free Trade Agreement (European Free Trade Association), Free Trade Agreement (European Union), Rules of Origin.

free trade agreements: See Canada-Israel Free Trade Agreement, free trade (theory), regionalism.

Free Trade Agreement (United States and the West Bank and the Gaza Strip): Agreed to on October 17, 1995 by the Palestinian Authority, the government of Israel, and the office of the U.S. trade representative: The Palestinians agreed to the following:

- that the Palestinian Authority will accord products of the United States, duty-free access to the West Bank and the Gaza Strip, and will provide the same or better treatment to products of the United Staters as that provided to products of the West Bank or the Gaza Strip ("national treatment");
- that the Palestinian Authority will assist the United States in verifying compliance with U.S. trade laws and will prevent unlawful transshipment of products not qualifying for duty-free access; and
- that the PLO and the Palestinian Authority will support all efforts to end the Arab League economic boycott of Israel in all its aspects.

See free trade agreements, regionalism.

Free Trade Area Agreement: See Free Trade Agreement.

free trade area: See regional economic integration.

freight shipment: See Protocol on Economic Relations.

French Revolution: See Hamas Covenant.

Front de Liberation Nationale (FLN): An Algerian radical Muslim independence movement and party (the only political party in Algeria until 1989), emphasizing independence, economic development within a socialist state, nonalignment, and close relations with other Arab nations. See Algeria.

FTA: Free Trade Agreement, free trade area.

Fundamental Agreement: On December 31, 1993, Israel and the Vatican signed the Fundamental Agreement, providing the basis for establishing diplomatic ties. Formal diplomatic ties and an exchange of ambassadors was established on June 5, 1994.

fundamentalism: The term "fundamentalism" is of Protestant theological controversy dating from the beginning of the twentieth century. It was characterized by a doctrine of personal salvation and a literal interpretation of the scriptures. Fundamentalism's most basic characteristic, however, is the belief that the Bible is inerranet and that, in its entirety and in all its parts, it speaks the truth. This central tenant is supported by two others:

- the verbal inspiration of the scriptures of their origin and,
- the literal interpretation of the scriptures in their use.

Broadly speaking, today fundamentalism refers to minority group attempts within religions to lay down fundamentals of belief that should, it is believed, be imposed on all within that faith. It denotes an extremism whose characteristics need to be clearly defined in order to separate it from the more genuine forms of revivalism.

Fundamentalism:

- is uncompromising on religion and its relation to modern society. It is a worldview that perceives religion as the exclusive source of authority and guidance on all areas of personal and social life.
- rejects Western ideology and its concomitant modernity.
- is marked by a withdrawal from society at large and the constitution of a counter-society with strict internal rules of conduct and clear boundaries.
- it offers an alternative totalistic system which, if implemented in letter and spirit, would provide all the relevant answers to all the pressing problems of the day.
- has a profound dependence on the primary holy text, such as the Torah or Qur'an, with market selectivity of secondiary texts. This allows for the by-passing of much of the established tradition.
- is where members of other religions are viewed extremely negatively.
- adopts an apocalyptic worldview that awaits the imminent end of days or the messianic redemption.

See Islam fundamentalists, Jewish fundamentalists, militant Islam.

fundamentalists: See fundamentalism, Jewish fundamentalists, militant Islam.

Fund for the Areas: See Protocol on Economic Relations (Article XI).

Further Transfer of Powers and Responsibilities (Protocol on): See Preparatory Transfer of Powers and Responsibilities.

G

G-7: See Declaration of Principles.

Ganzouri, Prime Minister: See Cairo Middle East/North Africa Conference.

gas: Most countries in MENA relay on natural gas to meet their rapidly growing energy needs. Some countries have domestic gas reserves, others import it, and some rely on a combination.

The potential market for natural gas in MENA could include the following components:

- Future power plants based on the combined cycle technology, together with the conversion of existing gas turbines to combined cycle;
- The conversion to natural gas of power plants presently using fuel oil and, if economically justified, coal;
- The conversion to natural gas of major industries presently using fuel oil such as cement, refining, and chemicals;
- Future petrochemical industries using gas as a feedstock to produce methanol or ammonia;
- Future industries requiring steam and setting up cogeneration units.

In addition:

- As the market for gas will come at the expense of other fossil fuels, its relative price will be an important determinant. Obviously, the lower its relative price, the higher the incentive to convert facilities and build new ones based on natural gas.
- Environmental standards, in particular the permitted sulphur and nitrogen levels, if strict, will encourage the switch to natural gas, particularly if the alternative is to install costly pollution control equipment.

See energy, Energy and Environment (Center of the Middle East/North Africa Community), Qatar, transport.

GATT: See Free Trade Agreement (European Free Trade Association and Israel), Free Trade Agreement (European Union and Israel), Free Trade Agreement (United States and Israel), General Agreement on Tariffs and Trade, Protocol on Economic Relations (Article III).

Gaza: The capital of the Gaza Strip; population approximately 250,000. Became the headquarters of the Palestinian Authority in 1994.

See Gaza Strip.

Gaza Clean-up Project: See Japan.

Gaza-Jericho Agreement: See Cairo Agreement, Gaza Strip and the Jericho Area Agreement, Interim Agreement.

Gaza-Jericho Negotiations: See Agreement on Security Arrangements for Hebron.

Gaza Sea Port Authority: See Declaration of Principles.

Gaza Strip: Palestinian territory with 146 square miles of area. Was allocated to Egypt along with Sinai in 1948. Egypt administered Gaza from 1948 until October 1956 when the Israelis occupied the area during the Sinai Campaign giving the region to a United Nations truce force on March 7, 1957.

In May 1967 the U.N. force departed at Egyptian President Nasser's request. The President then forwarded a military force to the strip. In June 1967 Israel returned following the Six-Day War until its departure with the peace treaty signed jointly by Prime Minister Rabin and Chairman Arafat.

Gaza Strip's population is nearly one million and is expected to double within 15 years. Gazan women have an average of 7.44 children, compared to 2.9 born to Israeli women.

Upon signing of the Israeli-Palestinian Liberation Organization Accord in September 1993, nearly one-third of the area was occupied by 19 Jewish settlements (4,500 settlers) and the out-of-bounds military zones. Upon signing of the May 1994 Israel-PLO agreement, buffer zones were created, resulting in about 60 percent of the Gaza Strip coming under the administration of the Palestinian Authority.

See Gaza, West Bank.

Gaza Strip and the Jericho Area Agreement: Shortly after the signing of the Declaration of Principles, negotiations commenced in Cairo and Paris between Israeli and PLO delegations on the implementation of the first stage of the interim arrangement, namely the Gaza-Jericho aspect. The negotiations were completed after seven months of discussions and signed in Cairo on May 4, 1994 by officials from Israel and the PLO. The agreement is a document containing almost 300 pages (the agreement itself and four annexes) with six maps attached. The agreement addresses four main issues—security arrangements, civil affairs, legal matters, and economic relations. Primary elements of the agreement read:

The government of the State of Israel and the Palestine Liberation Organization (hereinafter "the PLO"), the representative of the Palestinian people;

Preamble

Within the framework of the Middle East peace process initiated at Madrid in October 1991;

Reaffirming their determination to existence, mutual dignity and security, while recognizing their mutual legitimate and political rights;

Reaffirming their desire to achieve a just, lasting, and comprehensive peace settlement through the agreed political process;

Reaffirming their adherence to the mutual recognition and commitments expressed in the letters dated September 9, 1993, signed by and exchanged between the Prime Minister of Israel and the Chairman of the PLO;

Reaffirming their understanding that the interim self-government arrangements, including the arrangements to apply in the Gaza Strip and the Jericho Area contained in this agreement, are an integral part of the whole peace process and that the negotiations on the permanent status will lead to the implementation of Security Council Resolutions 242 and 338;

Desirous of putting into effect the Declaration of Principles on Interim Self-Government Arrangements signed at Washington, DC on September 13, 1993, and the agreed minutes thereto (hereinafter "the Declaration of Principles"), and in particular on withdrawal of Israel forces from the Gaza Strip and the Jericho Area;

Hereby agree to the following arrangements regarding the Gaza Strip and the Jericho Area:

Article I—Definitions for the Purpose of this Agreement

a. the Gaza Strip and the Jericho Area are delineated on map Nos. 1 and 2 attached to this agreement;

b. "the Settlements" means the Gush Katif and Erez settlement areas, as well as the other settlements in the Gaza Strip, as shown on attached map No. 1;
c. "the Military Installation Area" means the Israeli military installation area along the Egyptian border in the Gaza Strip, as shown on map No. 1; and
d. the term "Israelis" shall also include Israel statutory agencies and corporations registered in Israel.

Article II—Scheduled Withdrawal of Israeli Military Forces

1. Israel shall implement an accelerated and scheduled withdrawal of Israeli military forces from the Gaza Strip and from the Jericho Area to begin immediately with the signing of this agreement. Israel shall complete such withdrawal within three weeks from this date.
2. Subject to the arrangements included in the protocol concerning withdrawal of Israeli military forces and security arrangements attached as Annex I, the Israeli withdrawal shall include evacuating all military bases and other fixed installations to be handed over to the Palestinian Police, to be established pursuant to Article IX below (hereinafter "the Palestinian Police").
3. In order to carry out Israel's responsibility for external security and for internal security and public order of settlements and Israelis, Israel shall, concurrently with the withdrawal, redeploy its remaining military forces to the settlements and the military installation area, in accordance with the provisions of this agreement. Subject to the provisions of this agreement, this redeployment shall constitute full implementation of Article XIII of the Declaration of Principles with regard to the Gaza Strip and the Jericho Area only.
4. For the purposes of this agreement, "Israeli military forces" may include Israel police and other Israeli security forces.
5. Israelis, including Israeli military forces, may continue to use roads freely within the Gaza Strip and the Jericho Area. Palestinians may use public roads crossing the settlements freely, as provided for in Annex I.
6. The Palestinian Police shall be deployed and shall assume responsibility for public order and internal security of Palestinians in accordance with this agreement and Annex I.

Article III—Transfer of Authority

1. Israel shall transfer authority as specified in this agreement from the Israeli military government and its civil administration to the Palestinian Authority, hereby established, in accordance with Article

V of this agreement, except for the authority that Israel shall continue to exercise as specified in this agreement.

2. As regards the transfer and assumption of authority in civil spheres, powers and responsibilities shall be transferred and assumed as set out in the protocol concerning civil affairs attached as Annex II.
3. Arrangements for a smooth and peaceful transfer of the agreed powers and responsibilities are set out in Annex II.
4. Upon the completion of the Israeli withdrawal and the transfer of powers and responsibilities as detailed in paragraphs 1 and 2 above and in Annex II, the civil administration in the Gaza Strip and the Jericho Area will be dissolved and the Israeli military government will be withdrawn. The withdrawal of the military government shall not prevent it from continuing to exercise the powers and responsibilities specified in this agreement.
5. A Joint Civil Affairs Coordination and Cooperation Committee (hereinafter "the CAC") and two Joint Regional Civil Affairs Subcommittees for the Gaza Strip and the Jericho Area, respectively, shall be established in order to provide for coordination and cooperation in civil affairs between the Palestinian Authority and Israel, as detailed in Annex II.
6. The offices of the Palestinian Authority shall be located in the Gaza Strip and the Jericho Area pending the inauguration of the council to be elected pursuant to the Declaration of Principles.

Article IV—Structure and Composition of the Palestinian Authority

1. The Palestinian Authority will consist of one body of 24 members which shall carry out and be responsible for all the legislative and executive powers and responsibilities transferred to it under this agreement, in accordance with this article, and shall be responsible for the exercise of judicial functions in accordance with Article VI, subparagraph 1.b. of this agreement.
2. The Palestinian Authority shall administer the departments transferred to it and may establish, within its jurisdiction, other departments and subordinate administrative units as necessary for the fulfillment of its responsibilities. It shall determine its own internal procedures.
3. The PLO shall inform the government of Israel of the names of the members of the Palestinian Authority and any change of members. Changes in the membership of the Palestinian Authority will take effect upon an exchange of letters between the PLO and the government of Israel.
4. Each member of the Palestinian Authority shall enter into office upon undertaking to act in accordance with this agreement.

Article V—Jurisdiction

1. The authority of the Palestinian Authority encompasses all matters that fall within its territorial, functional, and personal jurisdiction, as follows:
   a. The territorial jurisdiction covers the Gaza Strip and the Jericho Area territory, as defined in Article I, except for settlements and the military installation area. Territorial jurisdiction shall include land, subsoil, and territorial waters, in accordance with the provisions of this agreement.
   b. The functional jurisdiction encompasses all powers and responsibilities as specified in this agreement. This jurisdiction does not include foreign relations, internal security, and public order of settlements and the military installation area and Israelis, and external security.
   c. The personal jurisdiction extends to all persons within the territorial jurisdiction referred to above, except for Israeli, unless otherwise provided in this agreement.
2. The Palestinian Authority has, within its authority, legislative, executive, and judicial powers and responsibilities, as provided for in this agreement.
3.
   a. Israel has authority over the settlements, the military installation area, Israelis, external security, internal security and public order of settlements, the military installation area, and Israelis, and those agreed powers and responsibilities specified in this agreement.
   b. Israel shall exercise its authority through its military government, which, for that end, shall continue to have the necessary legislative, judicial, and executive powers and responsibilities, in accordance with international law. This provision shall not derogate from Israel's applicable legislation over Israelis in person.
4. The exercise of authority with regard to the electromagnetic sphere and airspace shall be in accordance with the provisions of this agreement.
5. The provisions of this article are subject to the specific legal arrangements detailed in the protocol concerning legal matters attached in Annex III. Israel and the Palestinian Authority may negotiate further legal arrangements.
6. Israel and the Palestinian Authority shall cooperate on matters of legal assistance in criminal and civil matters through legal subcommittee of the CAC.

Article VI—Powers and Responsibilities of the Palestinian Authority

1. Subject to the provisions of this agreement, the Palestinian Authority, within its jurisdiction:
   a. has legislative powers as set out in Article VII of this agreement, as well as executive powers;
   b. will administer justice through an independent judiciary;
   c. will have, inter alia, power to formulate policies, supervise their implementation, employ staff, establish departments, authorities, and institutions, sue and be sued, and conclude contracts; and
   d. will have, inter alia, the power to keep and administer registers and records of the population, and issue certificates, licenses, and documents.
2.
   a. In accordance with the Declaration of Principles, the Palestinian Authority will not have powers and responsibilities in the sphere of foreign relations, which sphere includes the establishment abroad of embassies, consulates, or other types of foreign missions and posts or permitting their establishment in the Gaza Strip or the Jericho Area, the appointment of or admission of diplomatic and consular staff, and the exercise of diplomatic functions.
   b. Notwithstanding the provisions of this paragraph, the PLO may conduct negotiations and sign agreements with states or international organizations for the benefit of the Palestinian Authority in the following cases only:
      (1) economic agreements, as specifically provided in Annex IV of this agreement;
      (2) agreements with donor countries for the purpose of implementing arrangements for the provision of assistance to the Palestinian Authority;
      (3) agreements for the purpose of implementing the regional development plans detailed in Annex IV of the Declaration of Principles or in agreements entered into in the framework of the multilateral negotiations; and
      (4) cultural, scientific, and educational agreements.
   c. Dealings with the Palestinian Authority and representatives of foreign states and international organizations, as well as the establishment in the Gaza Strip and the Jericho Area of representative offices other than those described in subparagraph 2.a. above, for the purpose of implementing the agreements referred to in subparagraph 2.b. above, shall not be considered foreign relations.

Article VII—Legislative Powers of the Palestinian Authority

1. The Palestinian Authority will have the power, within its jurisdiction, to promulgate legislation, including basic laws, laws, regulations, and other legislative acts.
2. Legislation promulgated by the Palestinian Authority shall be consistent with the provisions of this agreement.
3. Legislation promulgated by the Palestinian Authority shall be communicated to a legislation subcommittee to be established by the CAC (hereinafter "the Legislation Subcommittee"). During a period of 30 days from the communication of the legislation, Israel may request that the Legislation Subcommittee decide whether such legislation exceeds the jurisdiction of the Palestinian Authority or is otherwise inconsistent with the provisions of this agreement.
4. Upon receipt of the Israeli request, the Legislation Subcommittee shall decide, as an initial matter, on the entry into force of the legislation pending its decision on the merits of the matter.
5. If the Legislation Subcommittee is unable to reach a decision with regard to the entry into force of the legislation within 15 days, this issue will be referred to a board of review. This board of review shall be comprised of two judges, retired judges or senior jurists (hereinafter "judges"), one from each side, shall develop written informal rules of procedure.
6. Legislation referred to the board of review shall enter into force only if the board of review decides that it does not deal with a security issue which falls under Israel's responsibility, that it does not seriously threaten other significant Israeli interests protected by this agreement and that the entry into force of the legislation could not cause irreparable damage or harm.
7. The Legislation Subcommittee shall attempt to reach a decision on the merits of the matter within 30 days from the date of the Israeli request. If this subcommittee is unable to reach such a decision within this period of 30 days, the matter shall be referred to the Joint Israeli-Palestinian Liaison Committee referred to in Article XV below (hereinafter "the Liaison Committee"). This Liaison Committee will deal with the matter immediately and will attempt to settle it within 30 days.
8. Where the legislation has not entered into force pursuant to paragraphs 5 or 7 above, this situation shall be maintained pending the decision of the Liaison Committee on the merits of the matter, unless it has decided otherwise.

9. Laws and military orders in effect in the Gaza Strip or the Jericho Area prior to the signing of this agreement shall remain in force, unless amended or abrogated in accordance with this agreement.

Article VIII—Arrangements for Security and Public Order

1. In order to guarantee public order and internal security for the Palestinians of the Gaza Strip and the Jericho Area, the Palestinian Authority shall establish a strong police force, as set out in Article IX below. Israel shall continue to carry the responsibility for defense against external threats, including the responsibility for protecting the Egyptian border and the Jordanian line, and for defense against external threats from the sea and from the air, as well as the responsibility for overall security of Israelis and settlements, for the purpose of safeguarding their internal security and public order, and will have all the powers to take the steps necessary to meet this responsibility.
2. Agreed security arrangements and coordination mechanisms are specified in Annex I.
3. A Joint Coordination and Cooperation Committee for mutual security purposes (hereinafter "the JSC"), as well as three joint District Coordination and Cooperation Offices for the Gaza district, the Khan Yunis district, and the Jericho district, respectively (hereinafter "the DCOs") are hereby established as provided for in Annex I.
4. The security arrangements provided for in this agreement and in Annex I may be reviewed at the request of either party and may be amended by mutual agreement of the parties. Specific review arrangements are included in Annex I.

Article IX—The Palestinian Directorate of Police Force

1. The Palestinian Authority shall establish a strong police force, the Palestinian Directorate of Police Force (hereinafter "the Palestinian Police"). The duties, functions, structure, deployment, and composition of the Palestinian Police, together with provisions regarding its equipment and operation, are set out in Annex I, Article III. Rules of conduct governing the activities of the Palestinian Police are set out in Annex I, Article VIII.
2. Except for the Palestinian Police referred to in this article and the Israeli military forces, no other armed forces shall be established or operate in the Gaza Strip or the Jericho Area.
3. Except for the arms, ammunition, and equipment of the Palestinian Police described in Annex I, Article III, and those of the Israeli military forces, no organization or individual in the Gaza Strip and the Jericho Area shall manufacture, sell, acquire, possess, import, or

otherwise introduce into the Gaza Strip or the Jericho Area any firearms, ammunition, weapons, explosives, gunpowder, or any related equipment, unless otherwise provided for in Annex I.

Article X—Passages

Arrangements for coordination between Israel and the Palestinian Authority regarding the Gaza-Egypt and Jericho-Jordan passages, as well as any other agreed international crossings, are set out in Annex I, Article X.

Article XI—Safe Passage between the Gaza Strip and the Jericho Area

Arrangements for safe passage of persons and transportation between the Gaza Strip and the Jericho Area are set out in Annex I, Article IX.

Article XII—Relations Between Israel and the Palestinian Authority

1. Israel and the Palestinian Authority shall seek to foster mutual understanding and tolerance and shall accordingly abstain from incitement, including hostile propaganda, against each other and, without derogating from the principle of freedom of expression, shall take legal measures to prevent such incitement by any organizations, groups, or individuals within their jurisdiction.
2. Without derogating from other provisions of this agreement, Israel and the Palestinian Authority shall cooperate in combatting criminal activity which may affect both sides, including offenses related to trafficking in illegal drugs and psychotropic substances, smuggling, and offenses against property, including offenses related to vehicles.

Article XIII—Economic Relations

The economic relations between the two sides are set out in the Protocol on Economic Relations signed in Paris on April 29, 1994 and the appendices thereto, certified copies of which are attached as Annex IV, and will be governed by the relevant provisions of this agreement and its annexes.

Article XIV—Human Rights and the Rule of Law

Israel and the Palestinian Authority shall exercise their powers and responsibilities pursuant to this agreement with due regard to internationally accepted norms and principles of human rights and the rule of law.

Article XV—The Joint Israeli-Palestinian Liaison Committee

1. The Liaison Committee established pursuant to Article X of the Declaration of Principles shall ensure the smooth implementation

of this agreement. It shall deal with issues requiring coordination, other issues of common interest, and disputes.

2. The Liaison Committee shall be composed of an equal number of members from each party. It may add other technicians and experts as necessary.
3. The Liaison Committee shall adopt as its rules of procedure, including the frequency and place or places of its meetings.
4. The Liaison Committee shall reach its decisions by agreement.

Article XVI—Liaison and Cooperation with Jordan and Egypt

1. Pursuant to Article XII of the Declaration of Principles, the two parties shall invite the governments of Jordan and Egypt to participate in establishing further liaison and cooperation arrangements between the government of Israel and Palestinian representatives on the one hand, and the governments of Jordan and Egypt on the other hand, to promote cooperation between them. These arrangements shall include the constitution of a Continuing Committee.
2. The Continuing Committee shall decide by agreement on the modalities of admission of persons displaced from the West Bank and the Gaza Strip in 1967, together with necessary measures to prevent disruption and disorder.
3. The Continuing Committee shall deal with other matters of common concern.

Article XVII—Settlement of Differences and Disputes

Any difference relating to the application of this agreement shall be referred to the appropriate coordination and cooperation mechanism established under this agreement. The provisions of Article XV of the Declaration of Principles shall apply to any such difference which is not settled through the appropriate coordination and cooperation mechanism, namely:

1. Disputes arising out of the application or interpretation of this agreement or any subsequent agreements pertaining to the interim period shall be settled by negotiations through the Liaison Committee.
2. Disputes which cannot be settled by negotiations may be settled by a mechanism of conciliation to be agreed between the parties.
3. The parties may agree to submit to arbitration disputes relating to the interim period, which cannot be settled through conciliation. To this end, upon the agreement of both parties, the parties will establish an Arbitration Committee.

Article XVIII—Prevention of Hostile Acts

Both sides shall take all measures necessary in order to prevent acts of terrorism, crime, and hostilities directed against each other, against individuals falling under the other's authority, and against their property, and shall take legal measures against offenders. In addition, the Palestinian side shall take all measures necessary to prevent such hostile acts directed against the settlements, the infrastructure serving them and the military installation area, and the Israeli side shall take all measures necessary to prevent such hostile acts emanating from the settlements and directed against Palestinians.

Article XIX—Missing Persons

The Palestinian Authority shall cooperate with Israel by providing all necessary assistance in the conduct of searches by Israel within the Gaza Strip and the Jericho Area for missing Israelis, as well as by providing information about missing Israelis. Israel shall cooperate with the Palestinian Authority in searching for, and providing necessary information about missing Palestinians.

Article XX—Confidence Building Measures

With a view to creating a positive and supportive public atmosphere to accompany the implementation of this agreement, and to establish a solid basis of mutual trust and good faith, both parties agree to carry out confidence-building measures as detailed herewith:

1. Upon the signing of this agreement, Israel will release, or turn over, to the Palestinian Authority within a period of five weeks, about 5,000 Palestinian detainees and prisoners, residents of the West Bank and the Gaza Strip. Those released will be free to return to their homes anywhere in the West Bank or the Gaza Strip. Prisoners turned over to the Palestinian Authority shall be obliged to remain in the Gaza Strip or the Jericho Area for the remainder of their sentence.
2. After the signing of this agreement, the two parties shall continue to negotiate the release of additional Palestinian prisoners and detainees, building on agreed principles.
3. The implementation of the above measures will be subject to the fulfillment of the procedures determined by Israeli law for the release and transfer of detainees and prisoners.
4. With the assumption of Palestinian authority, the Palestinian side commits itself to solving the problem of those Palestinians who were in contact with the Israeli authorities. Until an agreed solution is found the Palestinian side undertakes not to prosecute these Palestinians or to harm them in any way.

5. Palestinians from abroad whose entry into the Gaza Strip and the Jericho Area is approved pursuant to this agreement, and to whom the provisions of this article are applicable, will not be prosecuted for offenses committed prior to September 13, 1993.

Article XXI—Temporary International Presence

1. The parties agree to a temporary international or foreign presence in the Gaza Strip and the Jericho Areas (hereinafter "the TIP"), in accordance with the provisions of this article.
2. The TIP shall consist of 400 qualified personnel, including observers, instructors, and other experts, from five or six of the donor countries.
3. The two parties shall request the donor countries to establish a special fund to provide finance for the TIP.
4. The TIP will function for a period of six months. The TIP may extend this period, or change the scope of its operation, with the agreement of the two parties.
5. The TIP shall be stationed and operate within the following cities and villages: Gaza, Khan Yunis, Rafah, Deir El Ballah, Jabaliya, Absan, Beit Hanun, and Jericho.
6. Israel and the Palestinian Authority shall agree on a special protocol to implement this article, with the goal of concluding negotiations with the donor countries contributing personnel within two months.

Article XXII—Rights, Liabilities, and Obligations

1.
   a. The transfer of all powers and responsibilities to the Palestinian Authority, as detailed in Annex II, includes all related rights, liabilities, and obligations arising with regard to acts or omissions which occurred prior to the transfer. Israel will cease to bear any financial responsibility regarding such acts or omissions and the Palestinian Authority will bear all financial responsibility for these and for its own.
   b. Any financial claim made in this regard against Israel will be referred to the Palestinian Authority.
   c. Israel shall provide the Palestinian Authority with the information it has regarding pending and anticipated claims brought before any court or tribunal against Israel in this regard.
   d. Where legal proceedings are brought in respect of such a claim, Israel will notify the Palestinian Authority and enable it to participate in defending the claim and raise any arguments on its behalf.

e. In the event that an award is made against Israel by any court or tribunal in respect of such a claim, the Palestinian Authority shall reimburse Israel the full amount of the award.

f. Without prejudice to the above, where a court or tribunal hearing such a claim finds that liability rests solely with an employee or agent who acted beyond the scope of the powers assigned to him or her, unlawfully or with willful malfeasance, the Palestinian Authority shall not bear financial responsibility.

2. The transfer of authority in itself shall not affect rights, liabilities, and obligations of any person or legal entity, in existence at the date of signing of this agreement.

Article XXIII—Final Clauses

1. This agreement shall enter into force on the date of its signing.
2. The arrangements established by this agreement shall remain in force until and to the extent superseded by the interim agreement referred to in the Declaration of Principles or any other agreement between the parties.
3. The five-year interim period referred to in the Declaration of Principles or any other agreement between the parties.
4. The parties agree that, as long as this agreement is in force, the security fence erected by Israel around the Gaza Strip shall remain in place and that the line demarcated by the fence, as shown on attached map No. 1, shall be authoritative only for the purpose of this agreement.
5. Nothing in this agreement shall prejudice or preempt the outcome of the negotiations on the interim agreement or on the permanent status to be conducted pursuant to the Declaration of Principles. Neither party shall be deemed, by virtue of having entered into this agreement, to have renounced or waived any of its existing rights, claims, or positions.
6. The two parties view the West Bank and the Gaza Strip as a single territorial unit, the integrity of which will be preserved during the interim period.
7. The Gaza Strip and the Jericho Area shall continue to be an integral part of the West Bank and the Gaza Strip, and their status shall not be changed for the period of this agreement. Nothing in this agreement shall be considered to change this status.
8. The preamble to this agreement, and all annexes, appendices, and maps attached hereto, shall constitute an integral part hereof.

Annex I—Protocol Concerning Withdrawal of Israeli Military Forces and Security Arrangements contains 12 articles.

Annex II—Protocol Concerning Civil Affairs contains two articles.

On May 4, 1994 there was an exchange of letters between Prime Minister Yitzhak Rabin and PLO Chairman Yasir Arafat. Mr. Arafat's letter reads:

Mr. Prime Minister,

With regard to the Agreement on the Gaza Strip and the Jericho Area, signed in Cairo on May 4, 1994 (hereinafter "the Agreement"), the PLO hereby confirms the following:

1. The PLO undertakes to ensure that the Palestinian Authority, including the Palestinian Police and other Palestinian Authority agencies, will function in accordance with the Agreement, and that the Palestinian Authority will activate the coordination and cooperation mechanism in a timely manner.
2. The PLO undertakes to cooperate with Israel and to assist it, in its efforts to locate and to return to Israel Israeli soldiers who are missing in action and the bodies or killed soldiers which have not been recovered.
3. The PLO undertakes to submit to the next meeting of the Palestinian National Council for formal approval the necessary changes in regard to the Palestinian Covenant, as undertaken in the letter dated September 9, 1993 signed by the Chairman of the PLO and addressed to the Prime Minister of Israel.
4. When Chairman Arafat enters the Gaza Strip and the Jericho Area, he will use the title "Chairman [Ra'ees in Arabic] of the Palestinian Authority" or "Chairman of the PLO," and will not use the title "President of Palestine."
5. Neither side shall initiate or take any step that will change the status of the Gaza Strip and the Jericho Area pending the outcome of the permanent status negotiations.
6. Pursuant to Article IV, paragraph 3, of the Agreement, the PLO shall inform the Government of Israel of the names of the members of the Palestinian Authority in a letter that shall be provided within a week of signing the Agreement. The appointment of these members to the Palestinian Authority shall take effect upon an exchange of letters between the PLO and the Government of Israel. Changes in the membership of the Palestinian Authority will take effect upon an exchange of letters between the PLO and the Government of Israel.
7. Immediately after the conclusion of the Agreement, early empowerment negotiations will commence pursuant to Article VI of the Declaration of Principles, and the two sides will explore possible expansion of the scope of these negotiations beyond the five spheres.
8. The two sides will intensify the negotiations on the interim arrangements consistent with the Declaration of Principles and guided by its target date.
9. The two sides reiterate their commitment to commence permanent status negotiations as soon as possible, but no later than the beginning of the third year of the interim period, as provided for in Article V of the Declaration of Principles.
10. As regards the relations between Israel and the PLO, and without derogating from the commitments contained in the letters dated September 9, 1993 signed by and exchanged between the Prime Minister of Israel and the Chairman of the PLO, the two sides will apply between them the provisions contained in Article XII, paragraph 1, with the necessary changes.
11. The two Parties shall, within one month of signing the Agreement, invite the Governments of Jordan and Egypt to establish the Continuing Committee

referred to in Article XII of the Declaration of Principles and in Article XVI of the Agreement.

12. The Government of Israel and the Palestinian Authority shall pass all necessary legislation to implement the Agreement.
13. The two Parties shall continue discussions on the following issues:
    a. the size of the Jericho Area;
    b. the positioning of a Palestinian official at the bridge;
    c. additional arrangement in the Rafah passage; and
    d. all outstanding issues specified in the Agreement.

Sincerely,
Yasser Arafat
Chairman
The Palestinian Liberation Organization

B. Mr. Rabin's letter reads:

Mr. Chairman,

The Government of the State of Israel has the honor to acknowledge receipt of your letter of today's date which reads as follows:

[Items 1 to 13 are repeated from Chairman Arafat's letter]

The Government of the State of Israel acknowledges and confirms the undertakings and commitments contained in this letter.

Sincerely,
Yitzhak Rabin
Prime Minister of Israel

See Agreement on Security Arrangements for Hebron, Cairo Agreement, Camp David Accords, closure, Declaration of Principles on Interim Self-Government Arrangements, Egyptian-Israeli Peace Treaty, Free Trade Agreement (United States and the West Bank and the Gaza Strip), Interim Agreement (on the West Bank and the Gaza Strip), Intifadah, Palestinian, Palestinian Authority, Pentagon Plan, Protocol on Economic Relations, settlements, unemployment.

GCC: See Gulf Cooperation Council.

GDP: See Gross Domestic Product.

Gelb, Leslie H.: See Council on Foreign Relations.

Gellner, Ernest: See nationalism.

gene banks: See Agriculture (Center of the Middle East/North Africa Community).

General Agreement on Tariffs and Trade (GATT): On January 1, 1948 a multilateral trade treaty was signed embodying reciprocal commercial rights and obligations as a means of expanding and liberalizing world trade. It established common regulations and obligations concerning international trading arrangements and the framework for the negotiation of agreements to liberalize world trade. It was accepted by over 80 fully participating nations, with nearly 30 others signing under special arrangements. These nations accounted for almost 80 percent of world trade.

The short-run impact of the 1994 approved GATT-Uruguay Round suggests that the MENA region will be a net loser, because:

- the region is a net food importer so it is likely to be hurt by adverse terms of trade effect as prices of agricultural goods rise in response to the reduction in agricultural export subsidies.
- the region is likely to lose from the dismantling of the Multi-Fiber Arrangement (MFA) where MENA nations will have to compete against all other large exporters for market share in a free trade environment.

On January 1, 1995 GATT transformed itself into the World Trade Organization (WTO), working with the International Monetary Fund and the World Bank to oversee trade and resolve disputes before they can harm the global economy.

According to the International Monetary Fund, the Uruguay Round incorporates progress in removing nontariff measures facing MENA exporters, especially in agriculture and textiles and clothing. However, elements of the Round could have adverse consequences for MENA, particularly in the short term. For example, the most-favored-nation tariff cuts incorporated in the Round will reduce preference margins that some MENA countries have enjoyed with the European Union, under the Mediterranean agreements, as well as under other preferential trade agreements.

A number of studies have been carried out on the impact of the Round on MENA countries. In general, these studies conclude that there will be some important initial losses for MENA. However, these losses can be mitigated to the extent that MENA countries themselves undertake important domestic policy reforms, leading eventually to the development of more efficient productive capacity as well as a more diversified economic and trade structure.

See Euro-Mediterranean agreements, intellectual property rights, International Monetary Fund, Paris Economic Agreement, trade reform (MENA), Uruguay Round, World Trade Organization.

General Assembly: See United Nation entries.

Geneva Conference: See PLO Phased Plan.

Germany: Germany contributed 15 million DM to the development of the Gaza airport in 1996.

GMR: See Great Man-Made River.

Golan Heights: Occupied by Israel following the 1973 Yom Kippur War, 25 Jewish settlements are now present in the Heights. Syria continues to insist that they would only consider normalization with Israel when she withdraws from the Golan Heights.

See Arab Summit (1996), Japan, perceptions, Resolution 338 (United Nations), United Nations Disengagement Observer Force, Yom Kippur War.

Golan Plateau: See Golan Heights.

golden age: See vision, the.

Goldstein, Baruch: See Gush Emunium, Hebron, Tomb of the Patriarchs.

governance: See forewarning.

Grapes of Wrath: See Agreement of Understanding between Israel and Lebanon, New Middle East.

Great Man-Made River (GMR): GMR is claimed by Libya as the largest civil engineering project ever undertaken in the world, creating multi-million dollar opportunities for foreign contractors.

The Great Man-Made River projects aim to bring an eventual six million cubic meters per day of fossil water from beneath the Sahara for use along the coast. On completion, the scheme could comprise over 4,000 kilometers of pipeline, providing Libya with an integrated national water grid, costing an estimated $25 billion.

See water.

greed: The ever-present state of greed (also confused by varying definitions throughout the participating cultures) tempts the weak. The center of MEC/MENAC will welcome visitors with all types of agenda, hidden and revealed. Plenipotentiaries will visit, along with solicitors and attorneys who will come to plead, to set their own careers on a stronger path and legitimacy. There will be a babble of languages, replete with headphoned translators, an intrusion of governmental representatives, the

shuttle of ministers and diplomats, well-paid lawyers, an aviary of pressure groups and advisors, all pushing ideas forward in praise of community integration progress.
See dedication, lobbying, politicians.

Green Line: Israeli settlements beyond the 1967 armistice lines.
See fences, refugees.

grids: See electric grids.

Gross Domestic Product (GDP): Differences of GDP in nations of MENA are significant. For example, Israel's per capita GDP in 1996 was $16,320, as compared with $6,930 in Saudia Arabia, $2,710 in Lebanon, $1,730 in Algeria, $1,690 in Jordan, $1,450 in Iraq, $910 in Egypt, and $880 in Iran.

gross fixed capital formation: See investment.

Group of Ten: Under a 1962 agreement the principal industrial nations of the world (the United States, the Netherlands, the United Kingdom, Sweden, France, Japan, Germany, Italy, Belgium, and Canada), together with Switzerland (an unofficial member), makes available to the International Monetary Fund their currencies up to specified amounts to lend when supplementary funds are needed by the IMF.
Synonymous with Paris Club.

growth (MENA): The International Monetary Fund believes:

> The current emphasis in MENA on growth and investment has a lot to do with policymakers' desire to reverse the disappointing economic performance of the late 1980s and early 1990s. After sustaining an annual average real growth rate of nearly 5 percent in the 1970s and first half of the 1980s, the region's growth nearly halved and lagged behind population growth rates. As a result, per capita income in 1995 is estimated to have been 4 percent below its level a decade earlier—this during a period when the per capita income of developing countries as a group rose by 40 percent and that of the fast growing countries of East Asia by 80 percent. The combination of improved policy implementation, higher international oil prices, as well as better weather conditions in North Africa is contributing to a pick up in growth in 1996 and the prospects of the first year of positive per capita income since 1992....
>
> Looking forward, most analysts agree that current prospects are that MENA's external environment will not provide a major stimulus to growth. Notwithstanding the recent rise, the outlook for international oil prices is a subdued one, with significant downside risk. The high unemployment in European countries is likely to limit demand for labor flows from the MENA region. Finally, the outlook for official bilateral assistance is uncertain as donors and creditors face their own budget

> consolidation issues. Accordingly, the region's growth stimulus will need to come through appropriate domestic policies. Indeed, the main challenge facing the regions' authorities today is to implement policies aiming at spurring growth and, more generally, raising general living standards. The crucial step in meeting this challenge is to increase both the level and efficiency of the region's capital.

See investment.

Gulf: The extension of the Arabian Sea separating Iran and the Arabian Peninsula; joined to the Gulf of Oman by the Straits of Hormuz. Traditionally, the nations of the Gulf are: Bahrain, Kuwait, Oman, Qatar, Saudi Arabia, and the United Arab Emirates (UAE).

Countries in the Gulf area claim two-thirds of the world's petroleum reserves and produce between a quarter and a third of the world's oil.

See Gulf nations.

Gulf Cooperation Council (GCC): An economic, regional, and trade association formed in 1981 by six states of the Arabian peninsula—Saudi Arabia, Kuwait, Qatar, the United Arab Emirates, Oman, and Bahrain. Following the ceasefire between Iran and Iraq in 1988, a continued exclusion of Iran and Iraq meant that the GCC could not truly be a region-wide forum.

By the early 1990s the expatriate labor force as a percentage of the total workforce ranged from just over half in Bahrain and two-thirds in Oman and Saudi Arabia to over 80 percent in Qatar and Kuwait and 90 percent in the UAE.

By 1996 the population of all the GCC states reached 26 million, with 17.5 million nationals and 8.5 million expatriates. Saudi Arabia, with approximately 14 million citizens and five million foreigners, dominates the statistics.

Following the Gulf War the GCC states plus Egypt and Syria issued the Damascus Declaration, which envisioned a new grouping of six plus two that could provide Persian Gulf security.

However, Iran rejected the Damascus Declaration and the attempt by others to exclude it from a regional security agreement. In addition, Iraq's exclusion from regional affairs, due primarily to U.N. sanctions, has created an unnatural situation.

On October 1, 1994 the Cooperation Council of the Arab States of the Gulf, having actively supported the Middle East peace process ever since the launching of the Madrid Conference, secondary and tertiary boycotts of Israel ceased being a threat to the interests of the participating nations.

The Cooperation Council of the Arab States of the Gulf, having actively supported the Middle East peace process ever since the launching of the Madrid Conference, and being fully aware of the important breakthroughs realized so far, particularly in the Palestinian and Jordanian tracks which comprise agreements covering economic cooperation between the Israelis and both the Jordanians and the Palestinians, seriously recognize the importance of a review of the provisions of the Arab boycott of Israel so as to take into consideration progress achieved and substantive future requirements of the peace process.

The GCC member states have constantly reiterated their determination to enhance cooperation with their trading partners in various spheres. Concerning the application of the Arab boycott of Israel, necessary measures have been taken with a view to protecting the mutual interests of the GCC and its trading partners. As a result of these measures and for all practical purposes, secondary and tertiary boycotts are no longer a threat to the interests of these partners.

Whereas the Arab boycott of Israel was enacted by the League of the Arab States, and its review to take into consideration developments and requirements of the Middle East peace process must take place, the GCC member states will support all or any initiative for such review presented in the League of the Arab States. Further, the GCC believe that a sponsorship of such initiative by Arab parties directly involved in the bilateral negotiations, whether selectively or individually, shall facilitate the required review and ensure a greater chance of success.

By 1996 GCC countries, properly so, began to adjust and reform their economies to increase viability and ensure economic stability. Many nations are encouraging investment by simplifying procedures and reducing protectionist restrictions on direct foreign participation, while a privatization movement is creating opportunities for joint ventures.

In addition, changing demographics are putting pressures on governments to increase employment at home. The population growth rate is outstripping the output growth rate in some nations by 3.5 percent to 2 percent.

It is anticipated that annual Israel-GCC trade will be $1 billion by the end of the millennium. Already, the trade between Israel and Kuwait is $300 a year.

See Amman Middle East/North Africa Economic Summit, Arab (intertrade), Bahrain, boycott, Cairo Middle East/North Africa Economic Conference, Casablanca Declaration, Casablanca Middle East/North Africa (trade), Middle East/North Africa Economic Summit, Gulf nations, Kuwait, League of the Arab States, Oman, Qatar, Saudi Arabia, United Arab Emirates.

Gulf crisis: Kuwait was accused by Iraq in July 1990 of violating its agreement by overproducing its oil. Then on August 2 Iraqi troops invaded

Kuwait. By month's end Saddam Hussein proclaimed Kuwait a province of Iraq.

Along with U.N. economic and political sanctions, the U.S.-led coalition commenced the Gulf War on January 17, 1991. Kuwait was finally liberated.

It now appears certain that the Gulf War led to the Madrid Peace Conference, and spurred the current peace process.

Following the truce, U.N. Security Council Resolution 687 required Iraq to implement a U.N.-monitored program of disarming itself of weapons of mass destruction and intermediate and long-range ground-to-ground missiles before U.N. sanctions against it could be lifted.

The Security Council Resolution, adopted on April 3, 1991 contained the following operative paragraphs:

The Security Council,

1. Affirms all 13 resolutions noted above, except as expressly changed below to achieve the goals of the resolution, including a formal ceasefire;

2-4. Concerning the international boundary between Iraq and Kuwait, calls on the Secretary-General to make arrangements to demarcate the boundary between Iraq and Kuwait.

5-6. Concerning immediate deployment of a United Nations observer unit to monitor the demilitarized zone along the international boundary.

7-14. Requires Iraq to accept the destruction, removal, or rendering harmless under international supervision all chemical and biological weapons, and all research, development, support, and manufacturing facilities; and all ballistic missiles with the range greater than 150 km; Requires the Secretary-General to appoint a special commission to carry out on-site inspection of Iraq's biological, chemical, and missile facilities; requests the Director-General of the International Atomic Energy Agency (IAEA) to carry out on-site inspection of Iraq's nuclear capabilities in cooperation with the special commission and set up a monitoring system.

15. Concerning the return to all Kuwait property seized by Iraq.

16-19. Concerning Iraq's liability for direct loss, damage, including depletion of natural resources, or injury to foreign governments, nationals, and corporations as a result of Iraq's invasion and occupation of Kuwait, with Paragraph 19 directing the Secretary-General to recommend within 30 days to the council the establishment of a fund, financed by Iraq, to meet the requirements of the claimants, and a commission to administer the fund.

20-29. Concerning economic embargo against Iraq and its lifting.

20. The prohibitions against the sale or supply to Iraq of commodities or products, and prohibitions against financial transactions with Iraq, contained in Resolution 661 (August 6, 1990) shall not apply to foodstuffs and medicine;

21. The provisions of paragraph 20 shall be reviewed by the Security Council every 60 days for the purposes of determining whether to reduce or lift the prohibitions referred to therein;

22. Upon the approval of the Council of the program mentioned in paragraph 19, and upon Council agreement that Iraq has completed all actions contemplated in paragraphs 8-13, the prohibitions against the import (by other members of the United Nations) of commodities and products originating in Iraq, and the prohibitions against financial transactions related thereto contained in Resolution 661 (1990) shall have no further force or effect.

30-31. Concerning repatriation or return of all Kuwaiti and third country nationals or their remains present in Iraq on or after August 2, 1990.

32. Requires Iraq to confirm that it will not commit or support any act of international terrorism.

33-34. On Iraq's official acceptance of the above provisions, a formal ceasefire becomes effective between Iraq and Kuwait and the member states cooperating with Kuwait in accordance with Resolution 678 (which authorizes member states to use "all necessary means" to implement Resolution 660 and subsequent resolutions on Kuwait if Iraq failed to withdraw from Kuwait by January 15, 1991).

Another conflict erupted in September 1996.

See Arab League; Hussein (His Majesty King Hussein Bin Talal); Hussein, Saadam; Iraq, Kuwait, Middle East Peace Process, Saudi Arabia.

Synonymous with Gulf War (II).

Gulf nations (states): Defined as a segment of the Middle East. In 1996, despite the rise in crude prices, spending was greater than revenues with these oil-rich nations falling deeper into the red. Though crude oil is now worth 20 percent more than had been forecast for 1996, most governments that rely heavily on oil revenues will end the year with budget deficits, adding to their already sizable debt levels.

For the Gulf nations, today's problems were created in the 1970s, when oil prices, stable for decades, shot up fivefold. Wealthy governments went on a spending spree. Today, these governments have lowered expectations on wages and hiring.

For example, Kuwait has an economy 75 percent dependent on oil. Other Gulf countries are in a similar position. Economies are too dependent on one commodity; governments have failed to diversify.

Because Gulf Cooperation Council countries export most of their oil and import most of their goods and services, a sound way to measure the effect of higher oil prices is through trade figures.

	Current Account Balance (in billions)	
	1995	*1996*
Saudi Arabia	–$4.60	–$3.35
Bahrain	–0.45	–0.35
Kuwait	2.35	2.10
Oman	–1.00	–0.70
Qatar	–1.59	–1.17
UAE	0.45	0.25

See Gulf, Middle East.

Gulf of Aqaba: Created by seismic activity along the Afro-Syrian Rift, the Gulf of Aqaba is a deep narrow body of water, bordered by Israel, Jordan, Egypt, and Saudi Arabia and is one of the hinges connecting the Asian and African continents. The gulf extends 180 km from Eilat and Aqaba and joins the Red Sea at the Straits of Tiran, with its widest point spanning 28 km.

Israel's gulf shore extends only a few kilometers, from the city of Eilat to the border with Egypt at Taba. Jordan's shore reaches some 20 km in length, extending to the Saudi border opposite Marsa al Muqabila in northern Sinai. Egypt enjoys the longest gulf border, which stretches some 170 km between Taba and the Straits of Tiran.

See Camp David Accords, Egyptian-Israeli Peace Treaty, Japan, Upper Gulf of Aqaba Oil Spill Contingency Project.

Gulf of Suez: See Camp David Accords, Egyptian-Israel Peace Treaty.

Gulf region: See Carter Doctrine.

Gulf states: See Gulf nations.

Gulf War: See Gulf crisis.

Gulf War (I): From 1980-1988, the war between Iran and Iraq.

Gulf War (II): See Gulf crisis.

Gush Emunium: Founded in 1974 as an Israeli political-religious movement with fervent historical Jewish claims on Judea and Samaria.

Opposed the Israeli-PLO Accord of September 1993. Extremists in the movement applauded the murder of 29 Palestinians who were praying in Hebron. Baruch Goldstein, a resident of Kiryat Arba, one of the Jewish settlements, encouraged by Gush Emunium, following the killings was himself murdered.

Gush Etzion: See Allon-Plus Plan.

Gush Katif settlement: See Cairo Agreement.

H

Hafez al-Assad (President): See Assad, Hafez (President).

Haganah: See Arab-Israeli War (I), Hebrew Insurrection Movement, Shai.

Halacha: See Jewish fundamentalists.

Hamad, Sheik, Bin Khalifa Al-Thani: See Qatar.

Hamas: Inspired by the Islamic Brotherhoods in several Arab countries, Hamas—an Arabic acronym for Islamic Resistance Movement got its start in 1987. It is dedicated to the triumph of Islam through social work, politics, indoctrination, and force. Hamas is a movement without a cohesive chain of command, estimated to enjoy the direct support of about a quarter of the Palestinian population.

The movement gained strength in Israeli-occupied areas alongside the uprising known as the "intifada."

Hamas opposes the PLO's decision to participate in the Middle East peace process commenced in Madrid in October 1991; rejected the Israel-PLO deal.

Following Hamas terrorist bombings in Israel in February and March 1996 and the idea of the peace process being imperiled, Israel's Prime Minister Shimon Peres declared an open war against Hamas, with special emphasis on the Qassam Bridges—the armed wing of Hamas.

See Hamas Covenant, HizbAllah, Intifadah, Islamic National Salvation Party, militant Islam.

Hamas Covenant: Excerpts from the covenant of the Islamic Resistance Movement (Hamas) are:

Our struggle against the Jews is very great and very serious. It needs all sincere efforts. The Islamic Resistance Movement is but one squadron that should be supported...is realized. It strives to raise the banner of Allah over every inch of Palestine....It is one of the links in the chain of the struggle against the Zionist invaders.

The Prophet, Allah bless him and grant him salvation, has said: The Day of Judgment will not come about until Moslems fight the Jews (killing the Jews), when the Jew will hid behind stones and trees. The stones and trees will..."there is a Jew behind me, come and kill him."...There is no solution for the Palestine questions except through Jihad. Initiatives, proposals, and international conferences are all a waste of time and vain endeavors. Palestine is an Islamic land.

Zionist organizations under various names and shapes, such as Freemasons, Rotary Clubs, espionage groups, and others...are all nothing more than cells of subversion and saboteurs. The Islamic peoples should perform their role in confronting the conspiracies of these saboteurs.

Moslem society confronts a vicious enemy which acts in a way similar to Nazism. He has deprived people of their homeland. In their Nazi treatment, the Jews made no exception for women or children.

Our enemies took control of the world media. They were behind the French Revolution and the Communist Revolution....They were behind World War I, when they were able to destroy the Islamic Caliphate, making financinal gains and controlling resources. They obtained the Balfour Declaration, formed the League of Nations through which they could rule the world. They were behind World War II, through which they made huge financial gains by trading in armaments, and paved the way for the establishment of their state. It was they that instigated the replacement of the League of Nations with the United Nations and Security Council to enable them to rule the world through them. There is no war going on anywhere, without (them) having their finger in it.

We pray to Allah that it will soon be the Islamic Resistance Movement's stand toward the PLO as that of the son toward his father, the brother toward his brother, and relative to relative, suffers his pain and supports him in confronting the enemies, wishing him to be wise and well-guided....

The Zionist invasion is a vicious invasion. It does not refrain from resorting to all methods, using all evil and contemptible ways to achieve its end. It greatly relies on the organizations it gave rise to, such as the Masons, the Rotary and Lions Clubs, and other sabotage groups. All these organizations in the interest of Zionism...behind the drug trade and alcoholism in all its kinds so as to facilitate its control and expansion.

Writers, intellectuals, media people, orators, educators and teachers, and all the various sectors in the Arab and Islamic world—all of them are called upon to perform their role, and to fulfill their duty, because of the ferocity of the Zionist offensive and the Zionist influence in many countries exercised through financial and media control.

The Zionist plan is limitless. After Palestine, the Zionists aspire to expand from the Nile to the Euphrates. When they will have digested the region they overtook they will aspire to further expansion, and so on. Their plan is embodied in the "Protocols of the Elders of Zion," and their present conduct is the best proof of what we are saying. Leaving the circle of struggle with Zionism is high treason, and cursed be he who does that.

See Hamas, PLO Charter (Covenant).

Harawi, Elias: See Hrawi, Elias.

Har Homa (Wall Mountain): Area of Israeli development in East Jerusalem. Protested by Palestinians as one of the last parcels of land that would include a future site for the capital of Palestine.

The 1,850 dunam (4 dunams = 1 acre) site in the southeastern corner of annexed East Jerusalem will eventually house more than 25,000 Israelis living in 6,500 units.

PA Chairman Arafat condemned the decision to build at Har Homa, as did the European Union, which noted that "settlements in the occupied areas contravene international law and are a major obstacle to peace." The U.S. response was considerably more restrained, noting only that it "would have preferred that this decision [to build at Har Homa] not have been taken."

The development was given as the primary cause for outbreaks in Hebron, and a terrorist bombing in Tel-Aviv. In Arabic Jabal Abu Ghneim.

See chronology (March 18, 1997), settlement, and the Oslo accords.

Hariri, Rafiq: See Hrawi, Elias; Lebanon.

harmonization: See regional economic integration, standards.

Hashemite Kingdom of Jordan: Arab claimants to be descendants of the Prophet Muhammad, among whom were nineteenth-century emirs of Mecca, Saudia Arabia. The last of these became King of the Hejaz. The Iraqi branch terminated in 1958 but the dynasty continues in Jordan.

See Amman Middle East/North Africa Economic Summit, Hassan, Bin Talal (HRH Crown Prince); Hussein, (His Majesty King Hussein Bin Talal), Israel-Jordan Peace Treaty, Jordan, Transjordan, Washington Declaration.

Hassan, Bin Talal (HRH Crown Prince): King Hussein's younger brother and constitutional heir to the Hashemite Kingdom. For over 20 years Crown Prince Hassan has been tireless in his capacity as patron of economic development and planning. He has taken a leading role in presenting the Middle East peace process to the West and is an established international statesman in his own right.

He was Chairman of the Amman Middle East/North Africa Economic Summit. At the Opening Plenary Session on October 29, 1995, the Crown Prince said:

> We hope to set in motion a dynamic process to rebuild our shattered region. We must seize the opportunity provided by peace-making, to engage in peace-building.
>
> Let us be clear about our long-term goals. We want to give every man, woman and child in the Middle East, a stake in their common future. We want our region to be known for something other than death and destruction. We want to see the day when a product labelled "made in the Middle East" inspires confidence. We want, in short, to achieve shared prosperity through self-reliance.

See Amman Middle East/North Africa Economic Summit, Casablanca Middle East/North Africa Economic Summit, Hashemite dynasty, Hussein (King), Jordan, Jordanian-PLO Accord, Trilateral Israel-Jordan-U.S. Economic Committee.

Hassan II (King): His Majesty, King Hassan II of Morocco is one of the most progressive and creative rulers of the region. Known as an activist, nationalist, and Muslim he has kept Morocco in the forefront of regional decision making. His Majesty has studied law and holds a Ph.D. from the University of Bordeaux, France.

Since taking the throne in 1961, King Hassan II has moved diligently to modernize the Moroccan economy at the same time that he, as a descendant of Mohammed and the inheritor of the Middle East's oldest monarchy (which dates from the sixteenth century), reinforces the Arab and Islamic character of his nation. He heads a nation that maintains one of the few multiparty states of the region, with a strong commitment to the free enterprise and market system.

During his long career, King Hassan II has played a key role in inter-Arab and Muslim affairs, including as Chairman of the Rabat Arab Summit of 1974, as the founder of the Islamic Conference, as a member of the Supreme Arab Committee for resolving the Lebanon crisis, and as the chair of the Fifth Summit Conference of Islamic Nations at Casablanca in 1984.

In February 1989 he hosted a meeting when the Maghreb Union was established.

At the opening of the Casablanca Middle East/North Africa Economic Summit in the fall 1994, His Majesty King Hassan II of Morocco said:

"In working together within this conference, we assume a responsibility for which we are accountable before many generations to come. Only an unshakable faith and a sound vision of the future can ensure the victory of life over despair."

See Casablanca Declaration, Casablanca Middle East/North Africa Economic Summit, Morocco.

health: Basic Indicators of Health Conditions (1992) are:

Country	*Life Expectancy at birth (years)*	*Under-5 Mortality Rate (per 1000 births)*
Algeria	66	72
Egypt	61	55
Israel	76	11
Jordan	68	30
Lebanon	68	43
Morocco	63	61
Syria	67	40
Tunisia	68	38

See Health and Housing (Center of the Middle East/North Africa Community), Israel-Jordan Peace Treaty, Protocol on Economic Relations (Article VII).

Health and Housing (Center of the Middle East/North Africa Community): The community should secure provisions for continued protection of all citizens of member nations by encouraging higher health and housing standards.

Prophets of the community will realize the injustice of health coverage across member nations and will work ceaselessly to overcome this horrid condition, both in personal health, disease from animals and plants. Controls are needed to upgrade facilities and evolve quality means of animal and plant controls with standardized inspections, and to evolve measures against the introduction of organisms harmful to plant or animal life.

The maintenance of border controls on animal and plant products perpetuates the costs and disadvantages of separate national markets, and increases the risk of diseases and contamination to the general public. A community common market for trade in live animals, and animal and plant products cannot be said to exist if there are hold-ups, administrative burdens, and the substantial costs each time goods cross between community borders.

No symbols of a nation's fear of foreign contamination. The community must develop legislation that is accepted by all involved member nations, providing health controls for agricultural animals, ensuring that food of animal origin was safe for consumers, concerning the breeding of animals, and affected animal welfare.

In addition, medical facilities require upgrading and in many regions are nonexistent. Emphasis and media attention has been placed on economic and trade opportunities. Future acceptance of a prospering Middle East/North Africa Community will in large part be based on how officials of the region set aside funds for upgrading daily life needs such as measured by adequate housing and proper health standards.

See Centers of the Middle East/North Africa Community, health, Middle East/North Africa Community.

Hebrew Insurrection Movement: A phase of the Jewish struggle against British rule in Palestine in which a cooperation agreement was reached between the Haganah—the largest Jewish underground defense organization associated with the Zionist Socialist movement—and the Etzel and the Lehi, two smaller Jewish undergrounds associated with the Zionist Revisionist movement. The November 1945 agreement increased Haganah military activity, while the Etzel and the Lehi promised not to pursue independent plans without authorization from the Insurrection Movement command. During this period the Haganah intensified their efforts to assist illegal immigration, and responded to British activity against illegal immigrants. On November 1, 1945 the Haganah carried out a large-scale mission in which 135 railway locations were blown up, sabotaging railway installations at Lydda, scuttling coast guard boats, and damaging oil refineries in Haifa. On June 17, 1946 Palmach soldiers carried out a mission that became known as the "Night of the Bridges." One month later the Hebrew Insurrection Movement dissolved over the Etzel's bombing of the King David Hotel.

See Shai.

Hebron: Hebron, is located south of Jerusalem in the Judean hills and is home to approximately 120,000 Arabs, 500 Jews, and three Christians. An additional 6,000 Jews reside in the adjacent community of Kiryat Arba.

Hebron is the site of the oldest Jewish community in the world, which dates back to Biblical times. The Book of Genesis relates that Abraham purchased the field where the Tomb of the Patriarchs is located as a burial place for his wife Sarah. According to Jewish tradition, the patriarchs Abraham, Isaac, and Jacob, and the matriarchs Sarah, Rebecca, and Leah are buried in the tomb.

In recent years Hebron has been the site of many violent incidents, two of which stand out. In May 1980 Palestinian terrorists murdered six Jewish students and wounded 20 others, who were returning from prayers at the Tomb of the Patriarchs. In February 1994 Dr. Baruch Goldstein opened fire on Muslim worshippers at the tomb, murdering 29 and wounding 125.

After months of acrimonious negotiations Israeli Prime Minister Netanyahu and Palestinian Authority Chairman Arafat concluded an accord on the pullback of Israeli troops from the city of Hebron and other areas of the West Bank initiated by negotiators on January 15, 1997 and signed on January 17.

The Protocol Concerning the Redeployment in Hebron reads:

In accordance with the provisions of the Interim Agreement and in particular of Article VII of Annex I to the Interim Agreement, both parties have agreed on this protocol for the implementation of the redeployment in Hebron.

SECURITY ARRANGEMENTS REGARDING REDEPLOYMENT IN HEBRON

1. Redeployment in Hebron

The redeployment of the Israeli military forces in Hebron will be carried out in accordance with the Interim Agreement and this protocol. This redeployment will be completed not later than 10 days from the signing of this protocol. During these 10 days both sides will exert every possible effort to prevent friction and any action that would prevent the redeployment. This redeployment shall constitute full implementation of the provisions of the Interim Agreement with regard to the city of Hebron unless otherwise provided for in Article VII of Annex I to the Interim Agreement.

2. Security Powers and Responsibilities

a.
   (1) The Palestinian Police will assume responsibilities in Area H-1 similar to those in other cities in the West Bank; and
   (2) Israel will retain all powers and responsibilities for internal security and public order in Area H-2. In addition, Israel will continue to carry the responsibility for overall security of Israelis.

b. In this context, both sides reaffirm their commitment to honor the relevant security provisions of the Interim Agreement, including the provisions regarding Arrangements for Security and Public Order (Article XII of the Interim Agreement); Security Policy for the Prevention of Terrorism and Violence (Article II of Annex I to the Interim Agreement); Guidelines for Hebron (Article VII of Annex I to the Interim Agreement); and Rules of Conduct in Mutual Security Matters (Article XI of Annex I to the Interim Agreement).

3. Agreed Security Arrangements

a. With a view to ensuring mutual security and stability in the city of Hebron, special security arrangements will apply adjacent to the areas under the security responsibility of Israel, in Area H-1, in the area between the Palestinian Police checkpoints delineated on the map attached to this protocol as Appendix I (hereinafter referred to as "the attached map") and the area under the security responsibility of Israel.
b. The purpose of the abovementioned checkpoints will be to enable the Palestinian Police, exercising their responsibilities under the Interim Agreement, to prevent entry of armed persons and demonstrators or other people threatening security and public order, into the abovementioned area.

4. Joint Security Measures

a. The District Coordinating Office (DCO) will establish a sub-office in the city of Hebron as indicated on the attached map.
b. Joint Mobile Units (JMU) will operate in Area H-2 to handle incidents that involve Palestinians only. The JMU movement will be detailed on the attached map. The DCO will coordinate the JMU movement and activity.
c. As part of the security arrangements in the area adjacent to the areas under the security responsibility of Israel, as defined above, JMU will be operating in this area, with special focus on the following places:
    (1) Abu Sneinah
    (2) Harat a-Sheikh
    (3) Sha'aba
    (4) The high ground overlooking new Route No. 35.
d. Two Joint Patrols will function in Area H-1:
    (1) a Joint Patrol will operate on the road from Ras e-Jura to the north of the Dura junction via e-Salaam Road, as indicated on the attached map; and
    (2) a Joint Patrol will operate on existing Route No. 35, including the eastern part of existing Route No. 35, as indicated on the attached map.
e. The Palestinian and Israeli side of the JMUs in the city of Hebron will be armed with equivalent types of weapons (Mini-Ingraham submachine guns for the Palestinian side and short M-l6s for the Israeli side).

f. With a view to dealing with the special security situation in the city of Hebron, a Joint Coordination Center (hereinafter the "JCC"), headed by senior offices of both sides, will be established in the DCO at har Manoah/Jabel Manoah. The purpose of the JCC will be to coordinate the joint security measures in the city of Hebron. The JCC will be guided by all the relevant provisions of the Interim Agreement, including Annex I and this protocol. In this context, each side will notify the JCC of demonstrations and actions taken in respect of such demonstrations, and of any security activity, close to the areas under the responsibility of the other side, including in the area defined in Article 3(a) above. The JCC shall be informed of activities in accordance with Article 5(d)(3) of this protocol.

5. The Palestinian Police

a. Palestinian Police stations or posts will be established in Area H-1, manned by a total of up to 400 policemen, equipped with 20 vehicles and armed with 200 pistols, and 100 rifles for the protection of the police stations.
b. Four designated Rapid Response Teams (RRTS) will be established and stationed in Area H-1, one in each of the police stations, as delineated on the attached map. The main task of the RRTs will be to handle special security cases. Each RRT shall be comprised of up to 16 members.
c. The abovementioned rifles will be designated for the exclusive use of the RRTs, to handle special cases.
d.
   (1) The Palestinian Police shall operate freely in Area H-1.
   (2) Activities of the RRTs armed with rifles in the Agreed Adjacent Area, as defined in Appendix 2, shall require the agreement of the JCC.
   (3) The RRTs will use the rifles in the rest of the Area H-1 to fulfill their abovementioned tasks.
e. The Palestinian Police will ensure that all Palestinian policemen, prior to their deployment in the city of Hebron, will pass a security check in order to verify their suitability for service, taking into account the sensitivity of the area.

6. Holy Sites

a. Paragraphs 2 and 3(a) of Article 32 of Appendix 1 to Annex III of the Interim Agreement will be applicable to the following Holy Sites in Area H-1:
   (1) The Cave of Otniel Ben-Knaz/El Khalil;
   (2) elonei Mamre/Hambre Er-Rameh;

(3) Eshel Avraham/Balatot Ibrahim; and
(4) Ma'ayan Sarah/Ein Sarah

b. The Palestinian Police will be responsible for the protection of the above Jewish Holy Sites. Without derogating from the above responsibility of the Palestinian Police, visits to the above Holy Sites, by worshipers or other visitors, shall be accompanied by a Joint Mobile Unit, which will ensure free, unimpeded, and secure access to the Holy Sites, as well as their peaceful use.

7. Normalization of Life in the Old City

a. Both sides reiterate their commitment to maintain normal life throughout the city of Hebron and to prevent any provocation or friction they may affect the normal life in the city.

b. In this context, both sides are committed to take all steps and measures necessary for the normalization of life in Hebron, including:
(1) The wholesale market—Habashe—will be opened as a retail market in which goods will be sold directly to consumers from within the existing shops.
(2) The movement of vehicles on the Shuhada Road will be gradualy returned, within four months, to the same situation which existed prior to February 1994.

8. The Imara

The Imara will be turned over to the Palestinian side upon the completion of the redeployment and will become the headquarters of the Palestinian Police in the city of Hebron.

9. City of Hebron

Both sides reiterate their commitment to the unity of the city of Hebron, and their understanding that the division of security responsibility will not divide the city. In this context, and without derogating from the security powers and responsibilities of their side, both sides share the mutual goal and movement of people, goods, and vehicles within and in and out of the city will be smooth and normal, without obstacles or barriers.

CIVIL ARRANGEMENTS REGARDING THE REDEPLOYMENT IN HEBRON

10. Transfer of Civil Powers and Responsibilities

a. The transfer of civil powers and responsibilities that have yet to be transferred to the Palestinian side in the city of Hebron (12 spheres) in accordance with Article VII of Annex I to the Interim Agreement shall be conducted concurrently with the beginning of the redeployment of Israeli military forces in Hebron.

b. In Area H-2, the civil powers and responsibilities will be transferred to the Palestinian side, except for those relating to Israelis and their property, which shall continue to be exercised by the Israeli Military Government.

11. Planning, Zoning, and Building

a. The two parties are equally committed to preserve and protect the historic character of the city in a way which does not harm or change that character in any part of the city.

b. The Palestinian side has informed the Israeli side that in exercising its powers and responsibilities, taking into account the existing municipal regulations, it has undertaken to implement the following provisions:

   (1) Proposed construction of buildings above two floors (six meters) within 50 meters of the external boundaries of the locations specified in the list attached to this protocol as Appendix 3 (hereinafter referred to as "the attached list") will be coordinated through the DCL.

   (2) Proposed construction of buildings above three floors (nine meters) between 50 and 100 meters of the external boundaries of the locations specified in the attached list will be coordinated through the DCL.

   (3) Proposed construction on nonresidential, noncommercial buildings within 100 meters of the external boundaries of the locations specified in the attached list that are designed for uses that may adversely affect the environment (such as industrial factories) or buildings and institutions in which more than 50 persons are expected to gather together will be coordinated through the DCL.

   (4) Proposed construction of buildings above two floors (six meters) within 50 meters from each side of the road specified in the attached list will be coordinated through the DCL.

   (5) The necessary enforcement measures will be taken to ensure compliance on the ground with the preceding provisions.

   (6) This article does not apply to existing buildings or to new construction or renovation for which fully approved permits were issued by the municipality prior to January 15th, 1997.

12. Infrastructure

a. The Palestinian side shall inform the Israeli side, through the DCL, 48 hours in advance of any anticipated activity regarding infrastructure which may disturb the regular flow of traffic on roads in Area H-2 or which may affect infrastructure (such as water, sewage, electricity, and communications) serving Area H-2.

b. The Israeli side may request, through the DCL, that the municipality carry out works regarding the roads or other infrastructure required for the well being of the Israelis in Area H-2. If the Israeli side offers to cover the costs of these works, the Palestinian side will ensure that these works are carried out as a top priority.
c. The above does not prejudice the provisions of the Interim Agreement regarding the access to infrastructure, facilities, and installations located in the city of Hebron, such as the electricity grid.

13. Transportation

The Palestinian side shall have the power to determine bus stops, traffic arrangements, and traffic signalization in the city of Hebron. Traffic signalization, traffic arrangements, and the location of bus stops in Area H-2 will remain as they are on the date of the redeployment in Hebron. Any subsequent change in these arrangements in Area H-2 will be done in cooperation between the two sides in the transportation subcommittee.

14. Municipal Inspectors

a. In accordance with paragraph 4.c of Article VII of Annex I of the Interim Agreement, plainclothes unarmed municipal inspectors will operate in Area H-2. The number of these inspectors shall not exceed 50.
b. The inspectors shall carry official identification cards with a photograph issued by the municipality.
c. The Palestinian side may request the assistance of the Israel Police, through the DCL of Hebron, in order to carry out its enforcement activities in Area H-2.

15. Location of Offices of the Palestinian Council

The Palestinian side, when operating new offices in Area H-2, will take into consideration the need to avoid provocation and friction. Where establishing such offices might affect public order or security the two sides will cooperate to find a suitable solution.

16. Municipal Services

In accordance with paragraph 5 of Article VII of Annex I of the Interim Agreement, municipal services shall be provided regularly and continuously to all parts of the city of Hebron, at the same quality and cost. The cost shall be determined by the Palestinian side with respect to work done and materials consumed, without discrimination.

MISCELLANEOUS

17. Temporary International Presence

There will be a Temporary International Presence in Hebron (TIPH). Both sides will agree on the modalities of the TIPH, including the number of its members and its area of operation.

18. Annex I

Nothing in this protocol will derogate from the security powers and responsibilities of either side in accordance with Annex I to the Interim Agreement.

19. Attached Appendices

The appendices attached to this protocol shall constitute an integral part hereof.

Appendix 1 (Map of Hebron)

Appendix 2 (Article 5)

Agreed Adjacent Area

The Agreed Adjacent Area (AAA) shall including the following:

1. An area defined by a line commencing from AAA Reference Point (RP)100, proceeding along old Route No. 35 until RP 101, continuing by a straight line to RP 102, and from there connected by a straight line to RP 103.
2. An area defined by a line commencing at RP 104, following a straight line to RP 105, from there following a line immediately westward of checkpoints 4, 5, 6, 8, 9, 10, 11, 12, and 13, and from there connected by a straight line to RP 106.
3. An area defined by a line connecting RPs 107 and 108, passing immediately northward of checkpoint 15.

Appendix 3 (Article 12)

List of locations

- The area of Al Haram Al Ibrahimi/the Tomb of the Patriarchs (including the military and police installations in its vicinity)
- Al Hisba/Abraham Avinu
- Osama School/Beit Romano (including the military location in its vicinity
- Al Daboya/Beit Hadassah
- Jabla Al Rahama/Tel Rumeida
- The Jewish Cemeteries

- Dir Al Arbein/the Tomb of Ruth and Yishai
- Tel Al Jaabra/Givaat Avot Neighborhood (including the police station in its vicinity)
- The Road connecting Al Haram Al Ibrahimi/the Tomb of the Patriarchs and Kiryat Arba.

Agreed Minute (Article 7)

January 15, 1997

The leaders agreed that the process of reopening the Shuhada Road will begin immediately, and will be fully completed within four months based on the premise that the preparations agreed between the two sides have been completed in accordance with the American plan.

American Plan on Al-Shuhada Street, January 7, 1997

As part of its ongoing support for the peace process between Israelis and Palestinians, the United States intends to finance and supervise preparations to improve and facilitate the complete reopening of al-Shuhada Street in the city of Hebron. This work will take place in conjunction with the implementation of the "Protocol Concerning the Redemployment in Hebron" concluded between the two sides.

DESCRIPTION OF ACTIVITY

USAID intends to fund and manage the upgrading of utilities, primarily potable water lines and sanitary sewers, in the area of al-Shuhada Street to improve access to service. In conjunction with these improvements to utilities, USAID will direct renovation of the area within al-Shuhada Street to provide a safe and attractive environment for local businesses, pedestrians, and vehicular traffic. The existing street will be completely reconstructed for a distance of approximately 700 meters including Gross Square and also along a 200 meter spur of Rahme Road and a 100 meter spur leading to the Old City Gate. Major elements of this proposed activity as currently envisaged include:

- Reconstruction of water distribution mains, sanitary sewers, and electrical lines.
- Installation of a new water transmission line.
- Installation of a new storm drain system (if appropriate)
- Paving of street with asphalt (typically, width varies from six to nine meters).
- Construction of concrete brick sidewalks (typically, two to three meters in width on each side of the street) with precast concrete curbs.
- Construction of concrete brick median strips with precut concrete curbs in the street.

- Installation of lampposts.
- Erection of safety barriers at selected locations.
- Sandblasting of stone exteriors of streetfront shops, stone fences, and stone masonry retaining walls.
- Painting of shop doors.
- Installation of awnings along streetfront shops and residences.
- Installation of cast-iron fencing along curbs.
- Installation of litter baskets.
- Installation of planters.
- Installation of new traffic signs.
- Painting of curbs.
- Planting of small trees, shrubs, and flowers.

This plan also includes the widening of the al-Shuahada Street in the vicinity of Beit Hadassah and Beit Schnerson [The details of this aspect of the plan included major elements].

USAID expects that the renovation of al-Shuhada Street will commence in mid-January 1997 and will be completed as quickly as technical considerations permit. All activities will be subject to applicable laws and regulations. Continuing consultation will be held to address technical issues that may arise during the implementation of this plan.

In addition, U.S. Ambassador Dennis Ross at the request of Prime Minister Binyamin Netanyahu and Ra'ees Yasser Arafat prepared this "Note for the Record":

Prime Minister Binyamin Netanyahu and Palestinian Authority Chairman Yasser Arafat met on January 15, 1997, in the presence of U.S. Special Middle East Coordinator Dennis Ross. They requested him to prepare this "Note for the Record" to summarize what they agreed upon at their meeting. The documents reads as follows:

Mutual Undertakings

The two leaders agreed that the Oslo peace process must move forward to succeed. Both parties to the Interim Agreement have concerns and obligations. Accordingly, the two leaders reaffirmed their commitment to implement the Interim Agreement on the basis of reciprocity and, in this context, conveyed the following undertakings to each other:

Israeli Responsibilities

The Israeli side reaffirms its commitments to the following measures and principles in accordance with the Interim Agreement:

Issues for implementation:

1. Further redeployment phases. The first phase of further redeployments will be carried out during the first week of March.
2. Prisoner release issues. Prison release issues will be dealt with in accordance with the Interim Agreement's provisions and procedures, including Annex VII.

Issues for negotiation:

3. Negotiations on the following outstanding issues from the Interim Agreement will be immediately resumed. Negotiations on these issues will be conducted in parallel:
   a) Safe Passage
   b) Gaza airport
   c) Gaza port
   d) Passages
   e) Economic, financial, civilian, and security issues
   f) People-to-people
4. Permanent status negotiations

Permanent status negotiations will be resumed within two months after implementation of the Hebron Protocol.

Palestinian Responsibilities

The Palestinian side reaffirms its commitments to the following measures and principles in accordance with the Interim Agreement:

1. Complete the process of revising the Palestinian National Charter.
2. Fighting terror and preventing violence.
   a) Strengthening security cooperation.
   b) Preventing incitement and hostile propaganda, as specified in Article XXII of the Interim Agreement.
   c) Combating systematically and effectively terrorist organizations and infrastructure.
   d) Apprehension, prosecution, and punishment of terrorists.
   e) Requests for transfer of suspects and defendants will be acted upon in accordance with Article II (7)(f) of Annex IV to the Interim Agreement.
   f) Confiscation of illegal firearms.
3. Size of Palestinian Police will be pursuant to the Interim Agreement.
4. Exercise of Palestinian governmental activity, and location of Palestinian governmental offices, will be as specified in the Interim Agreement.

The aforementioned commitments will be dealt with immediately and in parallel.

Other Issues

Either party is free to raise other issues not specified above related to implementation of the Interim Agreement and obligations of both sides arising from the Interim Agreement.

See Agreement on Security Arrangements for Hebron, Hebron Plus, Interim Agreement (on the West Bank and the Gaza Strip), Middle East Summit; Netanyahu, B.; Temporary International Presence, Tomb of the Patriarchs.

Hebron accord: See Hebron, Hebron Plus.

Hebron Plus: Although Israel and the PLO referred to the agreement on Israeli withdrawal from most of Hebron in early 1997 as the "Hebron accord," an American newspaper more accurately described it as "Hebron Plus." Not only did Israel surrender most of Hebron—it went far beyond Hebron, by giving the PLO specific dates in March 1997, September 1997, and March 1998 by which Israel would surrender additional parts of Judea and Samaria.

See Hebron.

Herzl, Theodor (Binyamin Ze'ev): See Israel (Declaration of Establishment), World Zionist Organization, Zionism.

HezbAllah: See HizbAllah, Lebanon.

highways: See Moscow Multilateral Middle East Conference.

historical places: See Israel-Jordan Peace Treaty.

HizbAllah: An Iranian-backed Party of God terrorist movement established and nurtured after the Iranian revolution in the late 1970s. Its primary operations come from Beirut, Lebanon. Its earliest achievements in the 1980s are looked upon by terrorist leaders as proof of the effectiveness of terrorism as a state policy.

HizbAllah is an extremist Shi'ite organization whose aim is to establish an Iranian-inspired Islamic state in Lebanon. The groups most infamous anti-Western activities include the 1983 suicide truck bombing of the U.S. marine barracks in Beirut, killing 241 marines, and the seizure of Western hostages in Lebanon. The release of the Western hostages was reportedly only ensured after $86 million was sent to HizbAllah from Tehran. The organization is also believed to be responsible for a number of terrorist incidents in the Middle East, Western Europe, and Asia against Israeli and Jewish targets, including two bombings in Buenos Aires—the 1993 bombing of the Israeli Embassy and the 1994 bombing of the Jewish community building.

See Agreement of Understanding between Israel and Lebanon, Hamas, Jihad, Lebanon, terrorism.

Hizbollah: See HizbAllah.

holocaust: See Israel (Declaration of Establishment).

Holst, Johan Jorgen: See Letters of Mutual Recognition.

holy shrines: See culture, Washington Declaration.

holy war: See Jihad.

hostile act: See Gaza Strip and the Jericho Area Agreement (Article XVIII).

hotels: See park development, tourism.

House of Saud: The ruling dynasty of Saudi Arabia. By the mid-1990s 6,000 descendants made up the House of Saud.

housing: See Health and Housing (Center of the Middle East/North Africa Community).

Hrawi, Elias (President): He took office in November 1989 in Lebanon when his predecessor Rene Mouwad was killed by a car bomb. He has common interest with his Prime Minister Rafiq Hariri in countering what they see as an overpowerful army and legislature. Harawi is a relatively uncontroversial figure, more a figurehead than a political heavyweight.

See Lebanon.

human resources: See Israel-Jordan Common Agenda.

Human Resources Development and Cooperation Plan: See Declaration of Principles.

human rights: See Gaza Strip and the Jericho Area Agreement (Article XIV), Interim Agreement (on the West Bank and the Gaza Strip).

husbandry: See Agriculture (Center of the Middle East/North Africa Community).

Hussain I: See Hussein.

Hussein, (His Majesty King Hussein Bin Talal) (ibn Talal): The ruler of the Hashemite kingdom of Jordan (born November 14, 1935) ascended to the throne during a period of political turmoil in May 1953. The 18-year-old king did not prove the puppet many expected. Rather, he has skillfully presided over Jordan's transformation from a tribal and nomadic society to a modern state, although divisions within the kingdom still create doubts about the future.

In 1988 he renounced all Jordanian claims on the West Bank.

During the 1990-1991 Gulf crisis, the king attempted to mediate between two close allies, Iraq and the United States. He has been a solid proponent of the Middle East peace, something illustrated by his pivotal role in convening the October 1991 Madrid peace conference and by the peace treaty with Israel in October 1994.

In January 1994 it was reported that Jordan's King Hussein had approved the convening of an economic integration conference between Jordan and Israel to detail the future of a "Middle East Commonwealth" (this was six months before the historic meeting in Washington on July 22, 1994 with President Clinton of the United States and President Rabin of Israel). Initially, it was lost, as a peace treaty had not been signed early enough to have the conference in Jordan, and it was moved to Morocco.

His Majesty hosted the second economic summit held in Amman at the end of October 1995. This meeting once again reinforced the king's determination to play a primary role in the evolution of a community of regional nations. Opening the Plenary Session the king said:

> The international community is able to contribute to the rehabilitation of the economies of the region at minimum cost. To achieve this result, there should be a clear and common outlook for the future which we seek for the region, in partnership and through cooperation with the international community. Each one of us will have a distinctive role to play in making our sub-region an active participant in the regional and world economies....May God help us to achieve that which brings benefits to our peoples, and may God bless you all.

See Amman Middle East/North Africa Economic Summit, Casablanca Middle East/North Africa Economic Summit, chronology (May 2, 1953) (November 10, 1994); Eban, Abba; Ein Avrona, Hashemite dynasty,

Hassan (Crown Prince), Jordan, London Agreement, New Middle East; Peres, Shimon; Washington Declaration.

Hussein, Saddam (President): Born in Takrit 1937, he started as a worker for the Ba'ath Party before rising to become one of the better known twentieth-century dictators, with a track record that includes two major wars (one with Iran and one with much of the world following the invasion of Kuwait in August 1990) and bloody campaigns at home against northern-based Kurds, southern Shia groups, and even his own generals. Became President of Iraq in 1979.

See chronology (August 8,1995), Gulf crisis, Iraq.

I

IAIGC: See Inter-Arab Investment Guarantee Corporation.

ibn ad-da'wa al-Islamiya: Used by many Islamists for their supporters—the sons of the Islamic call.

IDB: See Islamic Development Bank.

ideology: See vision, the.

IDF: Israel Defense Forces.

ijara: See Arab (inter-trade).

illiteracy: See Education (Center of the Middle East/North Africa Community).

ILS: The international standard code for the currency of Israel—the Israeli shekel.

Imara: See Hebron.

imbiriyalun: Used by many Arab nationalists for imperialist nations.

IMF: See International Monetary Fund.

immigration: Immigrants to Israel, between 1948-1995, by country of origin are:

U.S.S.R. (former) and C.I.S.	813,708
Morocco, Algeria, and Tunisia	345,753
Romania	273,957
Poland	171,753
Iraq	130,302

Iran	76,000
United States	71,480
Turkey	61,374
Yemen	51,158
Ethiopia	48,624
Argentina	43,990
Bulgaria	42,703
Egypt and Sudan	37,548
Libya	35,865
France	31,172
Hungary	30,316
India	26,759
United Kingdom	26,236
Czechoslovakia	23,984
Germany	17,912
South Africa	16,277
Yugoslavia (former)	10,141
Syria	10,078

See Israel (economy).

immigration "B": See Mossad L'Aliyah Bet.

implementation agreement: See Cairo Agreement.

import policy: See Protocol on Economic Relations (Article III).

import taxes: See Protocol on Economic Relations (Article III).

income distribution: See income gap.

income gap: Throughout the Middle East and North Africa income distribution varies widely, within each member nation and between member nations. Internally, the income gap is largely due to a weak economy.

Among nations of this region the gap will, and assuredly could, remain for a long time period. This income distribution variance is at the very heart of what attracts nations to sign free trade accords. The transfer of technologies, and so on, are traded for an abundance of a less costly labor market. Once erased, the initial incentives for the community disappear. Only after decades of progress can the income gap be closed to the point of creating a leveling of wages throughout the Middle East and North Africa.

See informal economy.

indirect taxes: See Protocol on Economic Relations (Article VI).

individuality: See conscience.

Industrial Cooperation and Trade (Center of the Middle East/North Africa Community): Increasing industrial cooperation and trade by neighboring nations of the community has enormous potential. For example, an effort by Egypt and Israel in the textile and clothing industries yielded a greater division of labor by stages of production in the textile and clothing industries. Resulting from the expansion of world trade this could be a catalyst of economic relations between the two nations. Egypt produces and exports predominantly cotton and cotton fabric, while Israel's advantage is more in the production of clothing. Egypt has huge resources of inexpensive labor missing in Israel. Israel has skilled labor and access to markets throughout the world, which could be helpful to the Egyptians and others.

Industrial cooperation could be achieved in:

- transportation and tourism
- culture, education, science, and technology
- bilateral trade
- joint production ventures.

Forming a community will demand that firms become more and more involved in all matters of regional operations. There will be an ever-increasing number of links with associated enterprises, creditors, and parties in each nation, with the community having the power to veto and block all mergers based on the global market, with a combined worth exceeding an established amount.

Trade between members of the community would involve:

a. trade expansion
   (1) import/export expansion for goods whose cost of transportation is now lower than to the rest of the world.
   (2) trade center by opening a new market for goods, the export of which could not have been profitable, due to high transportation costs.
   (3) production creation by starting to produce goods which were inexpensive enough to import, and the production of which became profitable due to economies of scale created by the regional market.

b. trade diversion with the redirection of trade with other countries to the newly created market. It can take place in the short term, and unlike trade expansion it does not require changes in production and consumption patterns.

What is needed is an environment that favors the development of cooperation between individual firms crossing borders. Such an ambiance is necessary for several reasons. The elimination of internal borders, the movement of goods and capital, freedom of establishment, and the freedom of supply services are fundamental to the creation of the internal market and will confer enormous benefits on suppliers and consumers of goods and services. It will also create opportunities and incentives for cooperation between businesses.

Regulations within the community will be required in the fields of company law, taxation, and intellectual property to prepare for a more equitable and smooth-flowing path of industrial cooperation.

In addition, in the area of public procurement there is a lack of open and effective competition which will limit the evolution of a unified market within the community. As well as increasing costs for the procuring bodies, the lack of intra-community competition in certain key industries discourages the emergence of community firms which are competitive on world markets.

To a large extent, the success of the community will be determined by the inroads and successes within the sphere of industrial cooperation. Clear evidence of an upward shift in this collaboration is critical for mass support. Any weakening in industrial cooperative procedures will undermine future programs requiring faith, trust, and shared energies.

See Centers of the Middle East/North Africa Community, Middle East/North Africa Community.

industrial estates: See industrial zones.

industrial zones: The World Bank plans to create thousands of jobs for Palestinians by building nine industrial zones along the Green Line, but faces stiff resistance. Some argue that an industrial port is preferred; others say that it is too distant in concept. The industrial zones are a $920 million idea. These "industrial estates" are the bank's solution for the closures that Israel imposes occasionally on Palestinian areas and would attract Israeli-Palestinian joint ventures. The proposal calls for the largest parks to be export-oriented and therefore located as borders, as close as possible to existing ports. Industrial powers would provide guarantees to investors, while Israel would secure trans-shipment routes to the ports. As the World Bank hopes, the zones would be managed by an authority independent of the PA, one which would tax local businesses and provide entrepreneurs with an investment and insurance safety net, and would even be in charge of security.

See input sharing, PADICO.

industry: See Protocol on Economic Relations (Article IX).

Indyk, Martin: In the summer of 1997 he left his post as U.S. Ambassador to Israel (the first Jew to hold the post) to become President Clinton's nominee for Assistant Secretary for Near Eastern Affairs.

inequity: See income gap, informal economy.

informal economy: The Middle East and North Africa community will find an active informal economy throughout the region. The region must take into account its relevance to the way in which governments can function and still keep unemployment within its control. The MEC/MENAC should not cast doubts on its utility, nor on its legality, until the causes are identified and erased. The informal economy in the region creates numerous jobs, feeds families, and generates some of the area's most vibrant economic growth. Most politicians of these nations look upon this economic strategy as a means for buying much-needed time to attract foreign investment, to modernize their industry, and to find new sources of development, free of social pressure that could topple existing governments.

These businesses often pay no taxes, have no licenses, offer no social security or medical benefits, and sometimes deal in smuggled goods, and even illegal drugs. Certainly, the MEC/MENAC cannot overlook these problems. At the same time, economists calculate that if the informal economy did not exist, unemployment in some of these countries would skyrocket.

The informal economy is represented by companies that are involved in legal activities, but are not registered with the government because of excessive red tape, which can take up to one year in some nations. Until governments, under the guidance of the MEC/MENAC learn to reduce this heavy burden, the informal economy will remain an integral way of life for the region.

See income gap.

information highway: See Taba Declaration.

infrastructure: See Amman Middle East/North Africa Economic Summit, Israel-Jordan Common Agenda.

input sharing: Existing trade among neighboring nations of Israel constitutes an insufficient basis for predicting bilateral trade patterns between them. Trade based on input sharing can be an important source of "new trade", that is, trade which is not necessarily related to goods and services currently traded by the countries in question. New trade based on input sharing pertains to the imports by Arab countries of inputs in which Israel has a proven comparative advantage, and to imports by Israel of inputs produced in Arab countries in which the latter have a proven comparative advantage. Branches characterized by comparative advantage in the exporting country can improve the competitive position of the importing country, when incorporated in the latter's final products.

In agricultural produce, food products, and certain sub-branches of the textile and clothing industries, Jordan, Syria, and Egypt, for example, all appear to be potential suppliers to Israel. Inter alia, studies show that the Israel import potential of inputs from Jordan appears to be both larger and more evenly distributed among the different branches than the import potential from Syria and even from Egypt. New trade can be very substantial, and may have a very different composition from current trade of the parties concerned.

Several faculty at Tel-Aviv University in their paper "Arab-Israeli Potential Trade: The Role of Input Sharing" cite the following logic:

> To illustrate this argument consider a three-country world consisting of countries A, B, and R. A and B are neighbors who in the past trade only with R but not with each other. Next consider a specific product containing two inputs—p1 and p2. Country A is well endowed with p1 and poorly endowed with p2, while country B is well endowed with p2 and poorly endowed with p1. As long as trade between A and B is impossible, neither country enjoys a comparative advantage in the product under consideration. If trade between A and B becomes possible, A can import p2 from country B, combine it with p1 and produce an internationally competitive final product. Alternatively, B may import p1 from A and combine it with indigenous p2 to produce a final product which is internationally competitive. Production will eventually be located in the country where total costs are minimized. The possibility of combining inputs from A and B makes it possible to increase the range of internationally competitive products in both countries. These products clearly belong to the category of "new trade" whose composition and volume cannot be derived from the analysis of the two countries' existing trade with the rest of the world.

institutional building and governance: See forewarning.

insurance: See Banking and Financial Services (Center of the Middle East/North Africa Community), logistic trade centers, Protocol on Economic Relations (Article XI).

integration (MENA): See regional economic integration.

intellectual property rights (IPR): throughout the MENAC there is a need to adopt and enforce laws to protect patents, trademarks, and copyrights. At present the presence and strength of such laws varies widely from country to country. While some nations have signed onto the Paris Convention for the protection of industrial property in 1883, others offer little or no protection. For example, by 1996 Qatar had no patent law, Kuwait has only begun to adopt a significant copyright law.

However, even when a country's legal code contains patent, trademark, and copyright laws, there can be glaring omissions. Computer hardware and software are often excluded. In some Arab nations pharmaceuticals cannot be patented; instead, drug firms must patent the process and instruments used in a particular drug's manufacture.

In addition, even if protective laws exist, the protection they afford is usually meaningless without strict enforcement. For those countries that are members of the World Trade Organization or its earlier General Agreement on Tariffs and Trade, any lack of enforcement is a violation of their signing the accord.

Adherence to IPR standards will cost Arab consumers more for a wide range of goods. Nevertheless development of adequate IPR laws and enforcement mechanisms will require the cooperation of Arab governments and companies that want to do business in the MENA/MENAC region.

See counterfeit goods.

Inter-Arab Investment Guarantee Corporation (IAIGC): Established in 1972 with headquarters in Kuwait, the goals are to promote the transfer of capital within the Arab area, in particular, by providing insurance coverage for investments with substantial Arab equity against loss resulting from noncommercial risks.

Inter-Arab trade: See Arab (inter-Arab), input sharing.

interdependence: See Amman Middle East/North Africa Economic Summit; vision, the.

interest charges: See usury.

Interim Agreement (on the West Bank and the Gaza Strip): Officially the Israeli-Palestinian Interim Agreement on the West Bank and the Gaza Strip; also called Oslo II, was signed in Washington, DC on September 28, 1995. It consists of 23 pages of articles followed by 284 pages of annexes and appendices.

The agreement, which marks the conclusion of the first stage in negotiations between Israel and the PLO, incorporates and supersedes the Gaza-Jericho and early empowerment agreements (stages I and II).

The main object of the Interim Agreement is to broaden Palestinian self-government in the West Bank by means of an elected self-governing authority—the Palestinian Council. This will allow the Palestinians to conduct their own internal affairs, reduce points of friction between Israelis and Palestinians, and open a new era of cooperation and coexistence based on common interest, dignity, and mutual respect. At the same time, it protects Israel's vital interests, and in particular its security interests, both with regard to external security as well as the personal security of its citizens in the West Bank.

The Interim Agreement between Israel and the PLO, including its various annexes, comprises over 300 pages. To the main body of the agreement are appended seven annexes dealing with security arrangements, elections, civil affairs (transfer of powers), legal matters, economic relations, Israeli-Palestinian cooperation, and the release of Palestinian prisoners.

The main provisions of the agreement are as follows:

ELECTIONS

Democratic elections will be held to the Palestinian Council for an interim period not to exceed five years from the signing of the Gaza-Jericho Agreement (i.e., no later than May 1999). All Palestinians of the West Bank and the Gaza Strip age 18 or over, who are registered in the population register, will have the right to vote. The elections will take place 22 days after the conclusion of an IDF redeployment from populated areas in the West Bank.

The elections to the council will be personal and by district. A separate election will be held simultaneously for the election of the head of the executive authority of the council.

The candidacy of an individual, a party, or a coalition of parties shall be rejected if the said candidate, party, or coalition professes racist views or acts in an illegal or undemocratic manner.

Palestinian residents of Jerusalem will be able to participate in the elections according to special arrangements detailed in the agreement. The voting will take place in localities outside Jerusalem and by means of special envelopes which will be sent from post offices to the Central Elections Committee. A Palestinian with a Jerusalem address who wishes to stand for election to the Palestinian Council will be able to do so only if he or she has an additional valid address in the West Bank or the Gaza Strip.

All stages of the election process will be open to international observation, to ensure that they are free and fair. At the request of the parties, the European Union has agreed to coordinate the election observation. The observer delegation will be composed of representatives from the following states and international organizations: the EU, the U.N., the United States, the Russian Federation, Canada, Egypt, Japan, Jordan, Norway, South Africa, the Non-Aligned Nations, the OAU, and the Islamic Conference Organization.

THE PALESTINIAN COUNCIL

The council to be established following the elections will assume various powers and responsibilities in security and civil spheres in the West Bank and Gaza, as detailed below. With the establishment of the council, the Israeli military government will be withdrawn and the Civil Administration dissolved. The council will assume responsibility for all rights, liabilities, and obligations in the spheres transferred to it. At the same time Israel will retain those powers and responsibilities not transferred to the council.

The council, which will have 82 members, will have the legislative and executive powers. The agreement provides that the legislative powers will be exercised by the council as a whole, while its executive powers will be exercised by a committee of the council—the Executive Authority. This committee shall comprise council members together with a small number of appointed officials.

The powers of the council will extend to all matters within its jurisdiction. It should be noted that it will not have powers in the sphere of foreign relations. The agreement does, however, provide for a number of areas in which the PLO may, on behalf of the council, conduct negotiations and sign agreements (economic, donor countries, regional development).

REDEPLOYMENT

The IDF will redeploy in the West Bank according to the timetable set out in the agreement. In the first stage, designed to facilitate the holding of elections, the IDF will withdraw from the populated areas of the West Bank: the six cities—Jenin, Nablus, Tulkarem, Kalkilya, Ramallah, and Bethlehem—and 450 town and villages. In the city of Hebron, special arrangement will apply, detailed below.

In addition, the agreement provides that a series of further redeployments are to take place at six-month intervals following the inauguration of the council. In the course of these redeployments, additional parts of Area C will be transferred to the territorial jurisdiction of the council, so that by the completion of the redeployment phases, Palestinian territorial jurisdiction will cover West Bank territory except for the areas where jurisdiction is to be determined under the final status negotiations (settlements, military locations, etc.).

SECURITY

Throughout the West Bank and the Gaza Strip, Israel will have overriding responsibility for external security and for the security of Israelis and settlements.

With regard to internal security and public order, the agreement establishes different arrangements for three types of areas:

- Area A comprises the six cities listed above and the city of Hebron, minus the Old City of Hebron, the Jewish Quarter, and everything that is linked from there to Kiryat Arba and the Tomb of the Patriarchs. In these areas, the Palestinian Council will have full responsibility for internal security and public order, as well as full responsibility for civil affairs.
- Area B comprises the Palestinian towns and villages in the West Bank. In these areas, which contain some 68 percent of the Palestinian population, the council will be granted full civil authority, as in Area A. The council will be charged with maintaining public order, while Israel will have overriding security authority to safeguard its citizens and to combat terrorism. This responsibility shall take precedence over the Palestinian responsibility for public order.

Twenty-five Palestinian police stations will be established in specified town and villages to enable the Palestinian police to exercise its responsibility for public order. The agreement contains provisions requiring that the movement of Palestinian police be coordinated and confirmed with Israel.

- In Area C, which comprises the unpopulated areas of the West Bank, areas of strategic importance to Israel and the Jewish settlements, Israel will retain full responsibility for security and public order. The Council will assume all those civil responsibilities not related to territory, such as economics, health, education, and so on.

THE REVOCATION OF THE PLO CONVENANT

The agreement contains an undertaking on the part of the PLO to revoke those articles of the Palestinian Covenant calling for the destruction of Israel, within two months of the inauguration of the council.

SECURITY POLICY FOR THE PREVENTION OF TERRORISM AND VIOLENCE

The agreement provides for the establishment of a strong police force, 12,000 in number, that will constitute the only Palestinian security force.

The Security Annex specifies the deployment of the police force, the approved equipment, and its modes of action.

The Security Annex specifies the commitment of Israel and the Palestinian Council to cooperate in the fight against terrorism and the prevention of terrorist attacks, according to the following framework:

- The Palestinian Police is the only Palestinian security authority.
- The Palestinian Police will act systematically against all expressions of violence and terror.
- The council will issue permits in order to legalize the possession and carrying of arms by civilians; any illegal arms will be confiscated by the Palestinian Police.
- The Palestinian Police will arrest and prosecute individuals suspected of perpetrating acts of violence and terror.

Both sides, in accordance with this agreement, will act to ensure the immediate, efficient, and effective handling of any incident involving the threat or acts of terrorism, violence, or incitement, whether committed by Palestinians or Israelis. To this end they will cooperate in the exchange of information and coordinate policies and activities.

Joint security committees will be established to coordinate between the IDF and the Palestinian Police. Joint patrols will ensure free and secure movement on designated roads in Area A.

TRANSFER OF CIVIL POWERS AND RESPONSIBILITIES

The agreement sets out the arrangements for the transfer of agreed-upon civil powers and responsibilities from the Civil Administration to the council. In Area C, powers and responsibilities not relating to territory will be transferred to the council; powers and responsibilities relating to territory will be gradually transferred along with the redeployments in these areas. The transfer of further civil powers and responsibilities is subject to detailed provisions ensuring, among other things, the land rights of Israelis and the continued provision of services (electricity, water, telecommunications, etc.) to the settlements.

ISRAELI FREEDOM OF MOVEMENT

The IDF and Israelis will continue to move freely on the roads of the West Bank and Gaza. In Area A Israeli vehicles will be escorted by joint patrols. Israelis may not under any circumstances be arrested or placed in custody by the Palestinian Police, and may only be required to present identity and vehicle documentation. On roads that are jointly patrolled, any request for identification shall only be made by the Israeli side of a joint patrol.

LEGAL ISSUES

The Legal Annex of the agreement sets out arrangements governing the legal relations between Israel and the Palestinian Council. These provisions define the criminal and civil jurisdiction of the council and include arrangements for legal assistance in criminal and legal matters, including cooperation with regard to police investigations.

RELIGIOUS SITES

Responsibility over sites of religious significance in the West Bank and Gaza will be transferred to the Palestinian side. In Area C this will be transferred gradually during the "further redeployment phase," except for the issues which will be negotiated during the permanent status negotiations. Both sides shall respect and protect religious rights of Jews, Christians, Moslems, and Samaritans, to wit:

- protecting the holy sites;
- allowing free access to the holy sites;
- allowing freedom of worship and peace.

The agreement guarantees freedom of access to and freedom of worship at the holy sites, and defines access arrangements for the holy places located in Areas A and B. With regard to Rachel's Tomb in Bethlehem and Joseph's Tomb in Nablus, special arrangements are set out in the agreement which will also guarantee freedom of access and freedom of worship.

HEBRON

In view of the Jewish presence in the heart of Hebron and the sensitive historical and religious aspects involved, special arrangements will apply in this city. These arrangements will enable Palestinian police to exercise responsibilities vis-à-vis Palestinian residents, while at the same time Israel will retain the powers and responsibilities necessary to protect Israeli residents living in Hebron and visiting the holy places. The areas of Hebron to remain under full Israeli control are: the Old City of Hebron, the Jewish Quarter, and everything that is linked from there to Kiryat Arba and the Tomb of the Patriarchs.

There will be a redeployment of Israeli military forces in Hebron, except for places and roads where arrangements are necessary for the security and protection of Israelis and their movement. This redeployment will be completed no later than six months after the signing of this agreement. Israel will continue to carry the responsibility for overall security of Israelis for the purpose of safeguarding their internal security and public order.

The status quo at the Tomb of the Patriarchs will remain unchanged, for the time being.

There will be a temporary international presence in Hebron.

HUMAN RIGHTS

The agreement states that Israel and the council shall carry out their functions and responsibilities while adhering to the international norms of human rights and the rule of law, guided by the obligation to protect the public, respect their fellow men, and prevent harassment.

WATER

The agreement contains an undertaking on the part of Israel to increase the amount of water allocated to the Palestinians by 28 million cu.m. Any further addition to either side will be based on an increase in the available water resources to be developed through international funding and channels, among the tripartite American-Palestinian-Israeli forum which will hold its first meeting after the signing of the Interim Agreement. The agreement provides for the establishment of a joint water committee that will manage water resources and enforce water policies, protecting the interests of both parties by the prevention of uncontrolled drilling by enforcing standards, and so on.

RELEASE OF PRISONERS

In order to foster a positive atmosphere as this agreement is being implemented, and to engender mutual confidence and a basis for cooperation between the two peoples, Israel will release Palestinian prisoners who are in Israeli custody in three stages according to the following format:

Stage 1—upon the signing of the agreement
Stage 2—on the eve of elections for the council
Stage 3—according to other principles which will be established separately.

An Israel-Palestinian Joint Committee will be established to discuss the details of the release of prisoners.

COOPERATION AND ECONOMIC RELATIONS

The Economic Annex of the Gaza-Jericho Agreement has been incorporated into the Interim Agreement. Its articles and appendices, including the establishment of a single economic unit for the purpose of customs and import policy, now apply to the whole of the West Bank and the Gaza Strip.

In addition, an entire annex of the agreement deals with cooperation between Israel and the Palestinian Council. The parties are committed to cooperation programs involving officials, institutions, and the private sector in various fields, such as economics, science, culture, and society. A standing committee will be established to foster such cooperation.

Cooperation will focus on five principal areas:

- Economics (comprising industry, agriculture, energy, transport, and tourism);
- Environmental protection;
- Technology and science;
- Culture and education (comprising culture, media and communication, education, sports and youth); and
- The fostering of dialogue and relations between the two peoples through a people-to-people program.

In this context, the parties will work to develop contacts between the economic, agricultural, scientific, and educational sectors, and to find common solutions for environmental protection, such as waste removal and clean energy sources. They will work to advance tourism through investment in infrastructure and joint initiatives, to advance education by the development of curricula, the training of sports instructors, youth exchange programs, drug prevention, and so on.

EDUCATION FOR PEACE

Both sides have undertaken in the agreement to act to strengthen understanding and tolerance, prevent incitement and hostile propaganda, and will use the legal means at their disposal to prevent incitement on the part of groups or individuals. Both sides have pledged that their educational systems will act to advance peace between Israel and the Palestinians.

See Agreement on Security Arrangements for Hebron (1994), Agreement on the Temporary International Presence in the City of Hebron (1997), Beilin-Eitan Agreement, Declaration of Principles, Hebron, interim arrangement, Oslo Agreement.

interim arrangement (Gaza Strip and Jericho): The Declaration of Principles called for the establishment of a five-year interim arrangement for the West Bank and Gaza Strip, to begin with the withdrawal of the Israeli military from Gaza and Jericho.

See Cairo Agreement, Camp David Accords, Declaration of Principles, Gaza Strip and the Jericho Area Agreement, Hebron, Interim Agreement, Israel-Jordan Peace Treaty, Protocol on Economic Relations.

internal market: See Treaty of the Middle East/North Africa Community (Article 4).

International Alliance for Arab-Israeli Peace: See Copenhagen Declaration (1997).

International Atomic Energy Agency: See United Nations (nuclear-weapon-free zone in the region of the Middle East).

International Bank for Reconstruction: Synonymous with World Bank.

international boundary: See Israel-Jordan Peace Treaty.

International Finance Corporation: See Arab (inter-trade).

International Monetary Fund (IMF): A fund established on December 27, 1945, as an independent international organization of the United Nations. It is authorized to supplement its resources by borrowing; its purposes include promoting international monetary cooperation, expanding international trade and exchange stability, assisting in the removal of exchange restrictions and the establishment of a multilateral system of payments, and alleviating any serious disequilibrium in members' international balance of payments by making the resources of the fund available to them under adequate safeguards.

See Euro-Mediterranean agreements, fiscal deficits (MENA), General Agreement on Tariffs and Trade, growth (MENA), investment, Israel (economy), Middle East/North Africa (MENA), Middle East/North Africa (trade), Middle East/North Africa Community, policy issues (MENA), regional economic integration, trade reform (MENA), Uruguay Round.

Compare Organization for Economic Cooperation and Development.

Intifadah: Translated into English as "protest uprising." In the late 1980s stone-throwing and other incidents occurring in the West Bank and Gaza Strip. More than 300 Palestinians were killed by Israeli forces. Intifadah was supported by both the PLO and Hamas.

Intifada commenced in the Palestinian refugee camp of Jabaliya in the Gaza Strip on December 9, 1987, when thousands marched following the killing of four Palestinians by an Israeli truck.

The Palestinians refused to call off the intifada during the early 1990s, believing that the Israeli government would cease suppressing them and would accept their right to self-rule.

See Arafat, Yasir; Fatah, Hamas, terrorism.

intra-MENA trade: See Middle East/North Africa (trade).

intra-regional trade (MENA): See Middle East/North Africa (trade), regional economic integration, trade reform (MENA).

investing in peace: How to use peace to stimulate regional development was the theme of Dr. Yossi Beilin, who in December 1993 was Minister of Economics and Planning for Israel. He identified numerous possibilities:

Peace will bring about a reduction in the need for arms in countries that were involved in the conflict. It will aid in reducing the armaments which were required for mutual deterrence within the Arab world. Billions of dollars will become available for necessary economic development in industrialization, modernization of infrastructure, and the concurrent rise in the quality of life.

Peace will topple artificial barriers that were put up in an era of hostility. Instead, peace will reinforce the base underneath: opening transportation connections, air travel, and maritime relations, telecommunications; and the establishment of efficient and simplified border crossing for goods and services.

A Middle East without hostility, terror, or political agitation will attract outside investment. World financial markets welcome investment opportunities. The Middle East/East Mediterranean region will be an appealing opportunity for foreign investment, suited to the strategic interests of the United States, Europe, and Japan. The lack of stability that characterized recent decades inhibited potential foreign investors. The stability that will come with peace will remove the hesitation and will bring welcome investment.

A peaceful Middle East/East Mediterranean region will foster joint investment initiatives among the states in the region in many spheres of infrastructure:

a. interconnection of the region's electricity grids will optimize production and save the markets of the region hundreds of millions of dollars.
b. a combined transportation infrasture will reduce costs of transportation and shipping for all users in the region.
c. interconnection of the telecommunications infrastructure will substantively reduce the costs of expanding the system and the introduction of new communication grids into the region.

Peace in the Middle East/East Mediterranean will allow trilateral cooperation on projects related to infrastructure, industry, agriculture, transportation, tourism, energy, water, and science.

investment: In 1996 the International Monetary Fund concluded:

Improving growth performance is the major economic policy challenge facing the countries of the Middle East and North Africa. Since these countries' external environment is expected to remain broadly neutral over the medium term with significant downside risk, sustained economic growth will have to depend primarily on domestic policy efforts. Improving the region's investment performance—in both human and physical assets—is an important determinant of its ability to grow. While there are important differences between countries, investment levels for the region as a whole have been low-both in absolute terms and relative to other developing country regions. Investment in the region has also depended too heavily on the public sector and on external influences.

To meet the growth challenge, MENA must improve its investment performance—by investing more and increasing the return on investment spending. This will also encourage higher inflows of foreign direct investment. The variation in performance within the region reflects a number of factors: MENA economies facing severe conflicts and dislocation have tended to invest little and possessed an unresponsive investment function. The end of hostilities in most such countries should serve as an opportunity for the governments to implement policies necessary for improving investment responsiveness. At the other extreme, growth in MENA economies is coupled with a responsive investment function, thus promising sustainable growth. In between the two extremes lie most of the MENA countries, several of which have failed to devote sufficient portions of their output to investment. For all three groups of countries, maintaining a sound investment performance will depend on stable macroeconomic conditions, accelerating structural reforms, investing more effectively in the social sectors, and strengthening the institutional base.

Like the growth performance, MENA's investment performance weakened in the latter part of the 1970-1995 period. Specifically, after growing sharply in the 1970s, gross fixed capital formation (as a percent of GDP) has hovered just above 20 percent since the early 1980s—a level that is lower than the average for developing countries (over 24 percent in 1995) and sharply lower than the ratios prevailing in the fast growing Asia region (nearly 30 percent).

As a ratio to GDP, public investment in the MENA region is among the highest in the world: after peaking at more than 16 percent in 1982, the ratio has declined to around 10 percent at present; however, even the latter ratio is sharply higher than that prevailing in developing countries (hovering around 6 percent since the late 1980s). Meanwhile, private sector investment expenditures in the MENA countries (at around 10 percent of GDP since the early 1990s) remain lower than the average for developing countries and significantly below the average in the Asia region. It is worth noting, however, that the ratio of private sector investment to GDP in the MENA region had edged upwards in the first half of the 1990s although the increase is accounted for by a small number of countries, most notably Iran and Tunisia.

...To meet the growth challenge, MENA must improve its investment performance—by investing more and increasing the return from a given dollar of investment spending. Unlike the past development strategies adopted in most countries

> in the region, the process must be led by the private sector with the public sector playing a supportive role in providing an enabling environment, addressing market failures, and investing in basic social sectors and key infrastructure.

See Amman Middle East/North Africa Economic Summit, banking and financial services, Cairo Middle East/North Africa Economic Conference, Casablanca Middle East/North Africa Economic Summit, foreign direct investment, growth.

investment and growth (MENA): See growth, investment.

IPF: See Israel Policy Forum.

IPR: See intellectual property rights.

IQD: The international standard code for the currency of Iraq and the Neutral Zone (between Saudi Arabia and Iraq)—the Iraqi dinar.

Iran: Officially, Islamic Republic of Iran; called Persia until 1935. Ayatollah Khomeini forced the Shah to flee Iran on January 16, 1979. A revolution ensued for the next two years. In November 1979 the U.S. Embassy was captured with hostages taken. They were released on January 20, 1981.

A war with Iraq took the lives of several million Iranians. After Khomeini's death in 1989, Hasheni Rafsanjani was elected President.

Muhammad Khatami won the presidential election on May 23, 1997 with 69 percent of the vote. He campaigned for human rights, personal freedom, and the rule of law. Many citizens hope his election will result in significant changes in the way the government is run.

See Hussein, Saddam (President); HizbAllah, Iraq; Pahlavi, Mohammed Reza; Persia, terrorism.

Iranian and Libya Sanctions Act: Signed by President Clinton on August 5, 1996, foreign governments and companies that invest $40 million or more in the oil or gas sectors of those nations will be subject to U.S.-imposed sanctions. Considered controversial by many, including other governments, there is concern that this new law will isolate Iran and Libya, or perhaps even backfire, isolating the United States from the other governments whose help is needed to deal with this international problem. Below are the act's key provisions:

The Act:

Declares that the efforts of Iran and Libya to acquire weapons of mass destruction, their support of international terrorism, and Libya's failure to comply with U.N. Security Council Resolutions 731, 748, and 883 concerning the bombing of Pan Am 103 endanger U.S. and allied national security.

Declares that it is U.S. policy to deny Iran the ability to support international terrorism and to fund its weapons of mass destruction program by limiting the development of its petroleum resources. And, declares that it is U.S. policy to seek full compliance by Libya with U.N. resolutions and an end to its support of international terrorism and its efforts to obtain weapons of mass destruction.

Urges the President to begin immediate negotiations to establish a multilateral sanctions regime against Iran, to include limitations on the development of its petroleum resources.

Requires the President to impose two out of six sanctions on foreign entities that invest more than $40 million in any year in the petroleum sectors of Iran or Libya or that trade with Libya in violation of U.N. Security Council resolutions.

The six sanctions from which the President may select are:

1. No extension of credit to a sanctioned entity from the U.S. Export-Import Bank.
2. No export licenses granted to a sanctioned entity seeking advanced U.S. dual-use technology.
3. No loans or credits in excess of $10 million to a sanctioned entity from U.S. financial institutions.
4. Sanctioned foreign financial institutions cannot serve as primary dealers of U.S. debt instruments and/or cannot serve as agents of the U.S. government or as repositories of U.S. government funds.
5. The U.S. government may not procure goods from the sanctioned entity.
6. The President may restrict or ban imports into the United States from the sanctioned entity.

After one year from the date of enactment, the level of investment in Iran's petroleum sector triggering the imposition of sanctions drops from $40 million to $20 million for foreign entities whose governments are not cooperating with the United States in a multilateral sanctions regime against Iran.

The President may terminate sanctions if he receives assurances that the actions which resulted in the sanctions have ended. He may waive their imposition only if he determines that it is important to the national interest to do so and reports to Congress on the reasons why.

See terrorism.

Iranian Rial: The currency of Iran.

Iraq: Officially, Republic of Iraq; formed by the Mesopotamian plains and the rivers Tigris and Euphrates. Saddam Hussein became President of Iraq in July 1979. Embarked on a full-scale war with Iran on September 22, 1980.

A Gulf war began in January 1991, with Iraq forces being pushed out of Kuwait by United Nations military.

Again, in August-September 1996 Saddam Hussein ordered his troops to northern Iraq to capture towns occupied by the Kurds. The response, especially by the United States, was quick and decisive with the release of missiles on installations within Iraq.

See Arab Summit, Gulf crisis; Hussein, Saddam; Iran; Jaber, Shaikh, al-Ahmad al Jaber (Emir); Saudi Arabia, terrorism, United Nations Special Commission on Iraq.

Iraqi Dinar: The currency of Iraq and the Neutral Zone (between Saudi Arabia and Iraq.)

Iraqi Jews: See Michaelberg Operation.

Iraqi-Saudi Non-Aggression Pact: Signed in Bagdad on March 27, 1979. Identified principles of "non-interference in the internal affairs of the two sisterly countries" and "non-use of force and armies between the two states." After the Iraqi invasion of Kuwait in August 1990, Saudi Arabia participated in the U.N.-U.S. led anti-Iraq coalition, the pact means was reduced.

IRR: The international standard code for the currency of Iran—the Iranian rial.

Islam: Preached by the prophet Muhammad; its tenets found in the Koran (Qur'an). One billion followers believe in its racial tolerance, simplicity of worship, and its strong moral code.

The two major branches of Islam are Sunni (mostly found in Muslim nations) and Shiites (mostly found in Iran and Iraq).

See adu-adu, adu Allah, ard al-Isra, ard al-Masra, Bani sahyun, batil, ibn ad-da'wa al-Islamiya, jahaliya, Kuffar, militant Islam.

Islamic banking: Banking based on the adherence to Islamic principles that forbids usury. This is usually meant to exclude the receipt or payment of interest.

In many Arab banking situations customers leave money on deposit without demanding any interest.

Others invest customers money in commodity trading and differing commercial transactions—then give its clients a share of any profits. Instead of levying interest on borrowers, some firms provide funds to buy goods and rent equipment—then add on its own profit margin as part of the deal. These procedures allow Muslims to do their banking without breaking an Islamic prohibition on usury.

It is estimated that these Islamic principles require the management of between $50 billion and $100 billion. Islamic banks have grown rapidly in the last two decades, control more than $150 billion worldwide, and that sum is growing by 15 percent a year.

The first Western financial institution (Citicorp) to found an Islamic bank in the Persian Gulf region opened in Bahrain in July 1996 with start-up capital of $20 million. Money will be made by financing trade and construction projects in partnership with their depositors. Both the bank and the depositors share in any profits or losses. The bank will also offer equity funds, though refraining from investing in firms involved with gambling or selling alcohol.

There are now more than 100 specialized institutions that invest money according to strict Islamic principles, ranging from mass-market savings banks to pukka private banks.

See Bahrain.

Islamic Caliphate: See Hamas Covenant.

Islamic Conference: See Hassan II (King).

Islamic Corporation for the Insurance of Investment and Export Credit: See Arab (inter-trade).

Islamic Development Bank (IDB): Founded in 1974 and headquartered in Jeddah; assists in financing the development of countries with substantial Islamic populations.

See Arab (inter-trade).

Islamic fundamentalism: See Islam, Islamic fundamentalists, militant Islam.

Islamic fundamentalists: In addition to general fundamentalism characteristics, Islamic fundamentalists support:

- a desire to be free of subjugation to secular authority, which is seen as corrupt and degenerate, and a demand for instant results.

- a commitment to establish a theocratic state based on religious law where God's will is supreme and Muslims are elevated to sovereignty. Western values such as liberty, sovereignty, and supremacy of the law, it is believed, can be more effectively realized under Islamic rule. The West is seen to foster individualism, solidarity, and social responsibility.
- a hatred of the West whose ultimate goal is seen to be the undermining and weakening of Islam.
- a belief that a pristine form of Islam is the only solution to humankind's ills.
- enmity to all within the faith who reject these views. The fundamentalists belief is the only valid one. Other co-religionists are seen as having chosen evil over good and should be corrected.
- ambivalence about technology. While modernity and the technology that accompanies it are rejected, modern arms and electronic devices are accepted to assist their cause.

Compare Jewish fundamentalists.

See fundamentalism, militant Islam, United States.

Islamic International Company for Lease Finance: See Arab (inter-trade).

Islamic militants: See Islamic fundamentalists, militant Islam.

Islamic National Salvation Party: On March 20, 1996 a group of Islamic militants formerly associated with Hamas announced that they had formed a new political party that shares Hamas's broad ideology but rejects the use of violence. Formed with the active support of Yasir Arafat, this new party was proclaimed at a news conference held by 19 members of the "political office." The program they presented generally followed the militant Islamic line, rejecting the "unjust" Israeli-Palestinian treaties, declaring Palestine to be an "Islamic, Arab and Palestinian" land with Jerusalem as its capital, and proclaiming Islam as the "solution for all problems."

The party also declared that its means for achieving its ends was "legal political struggle," and that it acknowledges religious, cultural, and political pluralism.

See Hamas, terrorism.

Islamic Republic of Iran: See Iran.

Islamic Resistance Movement: Synonymous with Hamas.

Islamic Salvation Front (FIS): See Algeria.

Islamic Welfare Party: See Turkey.

Isra: See ard-al-Isra.

Israel (Declaration of Establishment): Officially, the State of Israel. The United Nations Palestine Commission recommended the creation of two successor states in 1947—Jewish and Arab. Formed as a Jewish state on May 14, 1948 when the British mandate terminated. The Declaration of Establishment was signed by 37 supporters for independence on May 14, 1948 (5th Iyar, 5708) in Tel Aviv. It reads:

Eretz Israel was the birthplace of the Jewish people. Here their spiritual, religious, and political identity was shaped. Here they first attained to statehood, created cultural values of national and universal significance, and gave to the world the eternal Book of Books.

After being forcibly exiled from their land, the people kept faith with it throughout their dispersion and never ceased to pray and hope for their return to it and for the restoration in it of their political freedom.

Impelled by this historic and traditional attachment, Jews strove in every successive generation to reestablish themselves in their ancient homeland. In recent decades they returned in their masses. Pioneers, ma'pilim, and defenders, they made deserts bloom, revived the Hebrew language, built villages and towns, and created a thriving community, controlling its own economy and culture, loving peace but knowing how to defend itself, bringing the blessings of progress to all the country's inhabitants, and aspiring toward independent nationhood.

In the year 5657 (1897), at the summons of the spiritual father of the Jewish State, Theodor Herzl, the First Zionist Congress convened and proclaimed the right of the Jewish people to national rebirth in its own country.

This right was recognized in the Balfour Declaration of November 2, 1917, and reaffirmed in the Mandate of the League of Nations which, in particular, gave international sanction to the historic connection between the Jewish people and Eretz-Israel and to the right of the Jewish people to rebuild its national home.

The catastrophe that recently befell the Jewish people—the massacre of millions of Jews in Europe—was another clear demonstration of the urgency of solving the problem of its homelessness by reestablishing in Eretz-Israel the Jewish State, which would open the gates of the homeland wide to every Jew and confer upon the Jewish people the status of a fully privileged member of the comity of nations.

Survivors of the Nazi holocaust in Europe, as well as Jews from other parts of the world, continued to migrate to Eretz-Israel, undaunted by difficulties, restrictions, and dangers, and never ceased to assert their right to a life of dignity, freedom, and honest toil in their national homeland.

In the Second World War the Jewish community of this country contributed its full share to the struggle of freedom—and peace—loving nations against the forces of Nazi wickedness and, by the blood of its soldiers and its war effort, gained the right to be reckoned among the peoples who founded the United Nations.

On November 29, 1947 the United Nations General Assembly passed a resolution calling for the establishment of a Jewish State in Eretz-Israel; the General Assembly required the inhabitants of Eretz-Israel to take such steps as were necessary on their part for the implementation of that resolution. This recognition by the United Nations of the right of the Jewish people to establish their state is irrevocable.

This right is the natural right of the Jewish people to be masters of their own fate, like all other nations, in their own sovereign state.

ACCORDINGLY WE, MEMBERS OF THE PEOPLE'S COUNCIL, REPRESENTATIVES OF THE JEWISH COMMUNITY OF ERETZ-ISRAEL AND OF THE ZIONIST MOVEMENT, ARE HERE ASSEMBLED ON THE DAY OF THE TERMINATION OF THE BRITISH MANDATE OVER ERETZ-ISRAEL AND, BY VIRTUE OF OUR NATIONAL AND HISTORIC RIGHT AND ON THE STRENGTH OF THE RESOLUTION OF THE UNITED NATIONS GENERAL ASSEMBLY, HEREBY DECLARE THE ESTABLISHMENT OF A JEWISH STATE IN ERETZ-ISRAEL, TO BE KNOWN AS THE STATE OF ISRAEL.

WE DECLARE that, with effect from the moment of the termination of the mandate, being tonight, the eve of Sabbath, the 6th Iyar, 5708 (May 15, 1948), until the establishment of the elected, regular authorities of the State in accordance with the Constitution which shall be adopted by the Elected Constituent Assembly not later than the 1st October, 1948, the People's Council shall act as a Provisional Council of State, and its executive organ, the People's Administration, shall be the Provisional Government of the Jewish State, to be called "Israel."

THE STATE OF ISRAEL will be open for Jewish immigration and for the ingathering of the exiles; it will foster the development of the country for the benefit of all inhabitants; it will be based on freedom, justice, and peace as envisaged by the prophets of Israel; it will ensure complete equality of social and political rights to all its inhabitants irrespective of religion, race, or sex; it will guarantee freedom of religion, conscience, language, education, and culture; it will safeguard the Holy Places of all religions; and it will be faithful to the principles of the Charter of the United Nations.

THE STATE OF ISRAEL is prepared to cooperate with the agencies and representatives of the United Nations in implementing the resolution of the General Assembly of the 29th November, 1947, and will take steps to bring about the economic union of the whole of Eretz-Israel.

WE APPEAL to the United Nations to assist the Jewish people in the building up of its State and to receive the State of Israel into the comity of nations.

WE APPEAL—in the very midst of the onslaught launched against us now for months—to the Arab inhabitants of the State of Israel to preserve peace and participate in the upbuilding of the State on the basis of full and equal citizenship and due representation in all its provisional and permanent institutions.

WE EXTEND our hand to all neighboring States and their peoples in an offer of peace and good neighborliness, and appeal to them to establish bonds of cooperation and mutual help with the sovereign Jewish people settled in its own land. The State of Israel is prepared to do its share in common effort for the advancement of the entire Middle East.

WE APPEAL to the Jewish people throughout the Diaspora to rally round the Jews of Eretz-Israel in the tasks of immigration and upbuilding and to stand by them in the great struggle for the realization of the age-old dream—the redemption of Israel.

PLACING OUR TRUST IN THE ALMIGHTY, WE AFFIX OUR SIGNATURES TO THIS PROCLAMATION AT THIS SESSION OF THE PROVISIONAL COUNCIL OF STATE, ON THE SOIL OF THE HOMELAND, IN THE CITY OF TEL AVIV, ON THIS SABBATH EVE, THE 5th DAY OF IYAR, 5708 (May 14, 1948).

See Herzl, Theodor; Zionism.

Israel (economy): Israel was created by the United Nations in 1948. The economic colossus of the region is Israel with its more than five and one-half million people possessing a gross national product (GDP) of nearly $90 billion in 1996. Her per capita income of approximately $16,000 in 1996 makes Israelis as affluent as Spaniards or Irish citizens, and in eighteenth place among the world's wealthiest countries. Her GDP has grown by an average of almost 5 percent a year (7.0% in the 1995); foreign debt has dropped from 80 percent of GDP in 1985 to approximately 25 percent by 1994.

Her service industries presently make up half of her GDP, with construction and manufacturing another 44 percent and agriculture about 5 percent.

Since 1989 Israel has absorbed new immigrants. Israel's population by year's end 1995 was 5.6 million, 20 percent higher than in 1990. Eighty-one percent of the population is Jewish, 14.4 percent Muslim, 2.9 percent Christian, and 1.7 percent Druze and others.

In 1995 Israel's economy grew by 7 percent, while the business product expanded by 9 percent and average consumption rose by 5 percent. Cumulatively, the economy grew 42 percent since the beginning of the decade.

Israel's trade deficit in 1995 hit an all-time high of $11.1 billion, contrasted to $9.4 billion the year before. Exports of goods and services totalled $28.7 billion, a 15 percent increase from 1994. Imports expanded from $33.9 billion to $49.8 billion. Foreign investments also hit a record in 1995, increasing to $2 billion from $600 million. However, the current account deficit—the gap between the combined imports and exports of goods, services, and payments—shot up in 1995 by 61 percent to $4.1 billion from $2.5 billion.

Israeli export of goods and services in 1995 rose by 7.5 percent to $27.2 billion. Exports to Asia also rose by more than 80 percent since 1990. In fact, from 1992 to 1995 exports to Asia soared 187 percent—154 percent to Japan alone.

However, Israelis face an economy in dire need of reform. A growing government deficit (up to 5% of GDP) had raised annual inflation to about 15 percent in 1996 from 8 percent in 1995. The balance-of-payments deficit was nearly $1 billion a month (almost 15% of GDP), and productivity had been stagnant for the past five years. The standard of living had improved dramatically, but mostly for the well-connected and well-to-do, while most Israeli families still have to struggle on an average income of 4,935 shekels ($1,500) per month. An average apartment sells for about 100 months of wages.

Not helped by an increase in interest rates in June 1996 to combat a 15 percent inflation rate, Prime Minister Netanyahu's Cabinet agreed to a $1.5 billion cut in public spending.

Israel's combined 1996 trade deficit totalled $10.5 billion, up 4 percent on 1995. Exports to the Palestinian Authority stood at $1.55 billion, with $235 million imported, giving a $1.315 billion trade surplus. This was the first time that the PA had appeared on Israel's annual balance sheet. The increase in the deficit is smaller than those of the previous two years. The deficit increased $2.8 billion (38%) in 1995, and $1.2 billion (19.5%) in 1994. Imports reached $29.6 billion, 5.7 percent higher than in 1995, after increases of 20 percent in 1995 and 16 percent in 1994.

All is not well. Economic growth was significantly lower in 1996 than in previous years. Israel's Central Bureau of Statistics indicated a 3.8

percent increase in GDP with a 4.1 percent in business sector product compared to 7.1 percent in 1995 and a whopping 8.9 percent in business sector product.

In addition, total goods and service exports pointed to a 3.7 percent increase in 1996, compared to an average annual increase of some 12 percent in the previous four years.

The peace process is producing a new economic landscape in the Middle East. The Israeli economy, already the second largest in the region, is well-positioned to benefit greatly from these changes.

Two major factors influencing Israel's economy since the early 1990s are the influx of nearly 600,000 immigrants from the former Soviet Union and the peace process. Large foreign exchange reserves, the erosion of the Arab boycott, and stabilization in unemployment set the stage for continued expansion through 1995.

The historic ties between the United States and Israel give distinct advantages to U.S. firms wishing to do business in Israel. The U.S.-Israel Free Trade Area Agreement, fully implemented on January 1, 1995 is a key factor in keeping the United States as Israel's largest trading partner. U.S. loan guarantees, a cheaper dollar, and programs that encourage joint research and development have helped maintain an advantage for U.S. firms. Bilateral trade in 1994 surpassed the $10 billion mark. Several trade barriers remain, which the United States and Israel are working together to remove. These include technical trade barriers, intellectual-property rights protection, packaging requirements and standards, and some agricultural policy restrictions facing U.S. exporters of agricultural commodities.

Opportunities for U.S. companies exist in virtually every sector of the Israeli economy. They are particularly strong in telecommunications and power generation as government privatization plans continue to develop; hotel development and supply as tourism grows; and airports/ports and the environmental sector as investment in infrastructure continues. Israel's leading edge high-tech sectors (software, biotech, etc.) beckons new partnerships and new products from the United States.

Israel has an increasingly open, modern, and sophisticated commercial environment with a well-educated population. The market for U.S. goods and services exists now and will continue to expand.

The growth in Israel's gross domestic product slowed sharply in the second half of 1996 to 2 percent from the correspondening period in 1995. For the entire year of 1996 growth slowed to a range of 4-5 percent after hovering at around 6-7 percent since the beginning of the decade (except 1993 when the GDP expanded by 3-4%).

On a per capita basis GDP grew in 1996 by 2 percent—half the pace of the previous year, a figure that was still higher than the 1.5 percent average for the world's most developed economies.

Exports of goods and services rose 5 percent in 1996, a sharp decline from the annual increases of the previous three years, which ranged from 11 to 15 percent. Imports, excluding direct defense-related shipments, expanded by 7 percent after rising by 9 percent the previous year.

The trade deficit, excluding defense-related imports, rose to $11 billion in 1996 from $9.9 billion in 1995 and $7.9 billion in 1994.

The public's available income rose 3 percent in real terms in 1996, after expanding 5 percent in 1995. The average monthly wage rose by 1.2 percent in nominal terms, which means a 9.2 percent erosion in real terms. Unemployment rose 0.4 percent in the fourth quarter of 1996 to 7.3 percent, its highest in the past two years.

While its high-tech industry and GNP are rising rapidly, so is the economic and social disparity among Israel's citizens. Israel used to be one of the Western world's most egalitarian nations in terms of income distribution. Depending on which standard is used, there are between 600,000 and one million people living below the poverty level. Those who suffer the most are new immigrants, women, and Arab Israelis.

In April 1997 the International Monetary Fund added Israel to its list of industrialized countries thus hoping to assist Israel's chances of improving its international credit rating and general economic standing.

See boycott, Egyptian-Israeli Peace Treaty, Free Trade Agreement (United States and Israel), immigration, input sharing, Israel-Canada Free Trade Agreement, Israel-Jordan Peace Treaty, Jordanian-Israeli bilateral accords, Qatar.

See also terms under Israeli.

Israel-Canada Free Trade Agreement: See Canada-Israel Free Trade Agreement.

Israel Defense Forces: See Shai.

Israel-Egypt Peace Treaty: See Camp David Accords, Egyptian-Israeli Peace Treaty.

Israel-EFTA: See Free Trade Agreement (European Free Trade Association and Israel).

Israel-European Free Trade Association: See Free Trade Agreement (European Free Trade Association and Israel).

Israel-European Union Free Trade Agreement: See Free Trade Agreement (European Union and Israel).

Israeli: See also terms under Israel.

Israeli Arabs: Officially, Arabs who chose not to leave Palestine prior to or following the War of Independence in 1948 and secured Israeli nationality. Today, nearly 20 percent of Israel's population are Israeli Arabs (160,000). Most support the establishment of a Palestinian state in the West Bank and Gaza and support the 1993 Israeli-PLO Accord.

Israeli Civil Administration: See chronology of the Middle East Peace Process (May 4, 1994).

Israeli Employment Service: See Protocol on Economic Relations (Article VII).

Israeli freedom of movement: See Interim Agreement (on the West Bank and the Gaza Strip).

Israeli Fund: See Protocol on Economic Relations (Article XI).

Israeli-Jordanian bilateral accords: See Jordanian-Israeli bilateral accords.

Israeli National Insurance Institute: See Protocol on Economic Relations (Article VII).

Israeli-Palestinian Continuing Committee for Economic Cooperation: See Declaration of Principles.

Israeli-Palestinian Economic Cooperation Committee: See Declaration of Principles.

Israeli-Palestinian Industrial Research and Development Centers: See Declaration of Principles.

Israeli-Palestinian Interim Agreement on the West Bank and the Gaza Strip: See Interim Agreement (on the West Bank and the Gaza Strip).

Israeli-Palestinian Trade Accord: See Paris Economic Agreement.

Israeli Shekel: The currency of Israel.

Israel-Jordan bilateral accords: See Jordanian-Israeli bilateral accords.

Israel-Jordan Common Agenda: The bilateral talks, initiated at the Madrid Conference, continued for almost two years in Washington until the signing of the Israel-Jordan Common Agenda on September 14, 1993. The Common Agenda constitutes the blueprint for a peace treaty, including the following components: security, water, refugees and displaced persons, borders, and territorial matters. In addition, the parties agreed to explore the potential for future bilateral cooperation within the regional context, such as economy, tourism, natural and human resources, and infrastructure.

A. Goal

The achievement of just, lasting, and comprehensive peace between the Arab States, the Palestinians, and Israel as per the Madrid invitation.

B. Components of Israel-Jordan Peace Negotiations

1. Searching for steps to arrive at a state of peace based on Security Council Resolutions 242 and 338 in all their aspects.

2. Security:

a. Refraining from actions or activities by either side that may adversely affect the security of the other or may prejudge the final outcome of negotiations.

b. Threats to security resulting from all kinds of terrorism.

c.

 i. Mutual commitment not to threaten each other by any use of force and not to use weapons by one side against the other including conventional and nonconventional mass destruction weapons.

 ii. Mutual commitment, as a matter of priority and as soon as possible, to work toward a Middle East free from weapons of mass destruction, conventional and nonconventional weapons; this goal is to be achieved in the context of a comprehensive, lasting, and stable peace characterized by the renunciation of the use of force, reconciliation, and openness.

 Note: The above (c-ii) may be revised in accordance with relevant agreements to be reached in the Multilateral Working Group on Arms Control and Regional Security.

d. Mutually agreed upon security arrangements and security confidence-building measures.

3. Water

a. Securing the rightful water shares of two sides.
b. Searching for ways to alleviate water shortage.

4. Refugees and Displaced Persons: Achieving an agreed just solution to the bilateral aspects of the problem of refugees and displaced persons in accordance with international law.

5. Borders and Territorial Matters: Settlement of territorial matters and agreed definitive delimitation and demarcation of the international boundary between Israel and Jordan with reference to the boundary definition under the mandate, without prejudice to the status of any territories that came under Israeli Military Government control in 1967. Both parties will respect and comply with the above international boundary.

6. Exploring the potentials of future bilateral cooperation, within a regional context where appropriate, in the following:

a. Natural Resources:
   - Water, energy, and environment
   - Rift Valley development
b. Human Resources:
   - Demography
   - Labor
   - Health
   - Education
   - Drug Control
c. Infrastructure:
   - Transportation: land and air
   - Communication
d. Economic areas including tourism

7. Phasing the discussion, agreement, and implementation of the items above including appropriate mechanisms for negotiations in specific fields.

8. Discussion on matters related to both tracks to be decided upon in common by the two tracks.

It is anticipated that the above endeavor will ultimately, following the attainment of mutually satisfactory solutions to the elements of this agenda, culminate in a peace treaty.

See Cairo Agreement, Declaration of Principles, Israel-Jordan Peace Treaty, Jordanian-Israeli bilateral accords, Madrid Conference, Stockholm PLO-American delegation statement, Washington Declaration.

Israel-Jordan Peace Treaty (Jordan-Israel Peace Treaty): The Treaty of Peace between the State of Israel and the Hashemite Kingdom of Jordan was signed on October 26, 1994 by Prime Minister Yitzhak Rabin and Prime Minister Abdul-Salam Majali, the second Israel has signed since its independence. The establishment of peace between the two countries follows closely the Washington Declaration signed on July 25 which ended the state of belligerency between the two countries. In addition to abolishing all hostile activities, the peace treaty recognizes the importance of cooperation in all spheres, developing good neighborly relations, and respecting the role of human development and dignity in the region.

The peace treaty with Jordan comprises 30 articles, and includes five annexes that address boundary demarcations, water issues, police cooperation, environmental issues and mutual border crossings. In particular the agreement establishes the following points:

PREAMBLE

The government of the Hashemite Kingdom of Jordan and the government of the State of Israel:

- Bearing in mind the Washington Declaration, signed by them on July 25, 1994, and which they are both committed to honor,
- Aiming at the achievement of a just, lasting, and comprehensive peace in the Middle East based on Security Council resolution 242 and 338 in all their aspects;
- Bearing in mind the importance of maintaining and strengthening peace based on freedom, equality, justice, and respect for fundamental human rights, thereby overcoming psychological barriers and promoting human dignity;
- Reaffirming their faith in the purposes and principles of the Charter of the United Nations and recognizing their right and obligation to live in peace with each other as well as with all states, within secure and recognized boundaries;
- Desiring to develop friendly relations and cooperation between them in accordance with the principles of international law governing international relations in time of peace;
- Desiring as well to ensure lasting security for both their states and in particular to avoid threats and the use of force between them;
- Bearing in mind that in their Washington Declaration of July 25, 1994, they declared the termination of the state of belligerency between them;
- Deciding to establish peace between them in accordance with this Treaty of Peace;

Have agreed as follows:

Article 1—Establishment of Peace

Peace is hereby established between the Hashemite Kingdom of Jordan and the State of Israel (the "Parties") effective from the exchange of the instruments of ratification of this treaty.

Article 2—General Principles

The parties will apply between them the provisions of the Charter of the United Nations and the principles of international law governing relations among states in time of peace. In particular:

1. They recognize and will respect each other's sovereignty, territorial integrity, and political independence;
2. They recognize and will respect each other's right to live in peace within secure and recognized boundaries;
3. They will develop good neighborly relations of cooperation between them to ensure lasting security, will refrain from the threat or use of force against each other, and will settle all disputes between them by peaceful means;
4. They respect and recognize the sovereignty, territorial integrity, and political independence of every state in the region;
5. They respect and recognize the pivotal role of human development and dignity in regional and bilateral relationships;
6. They further believe that within their control, involuntary movements of persons in such a way as to adversely prejudice the security of either party should not be permitted.

Article 3—International Boundary

1. The international boundary between Jordan and Israel is delimited with reference to the boundary definition under the mandate as is shown in Annex I (a), on the mapping materials attached thereto and coordinates specified therein.
2. The boundary, as set out in Annex I (a), is the permanent, secure, and recognized international boundary between Jordan and Israel, without prejudice to the status of any territories that came under Israeli military government control in 1967.
3. The parties recognize the international boundary, as well as each other's territory, territorial waters, and airspace, as inviolable, and will respect and comply with them.
4. The demarcation of the boundary will take place as set forth in Appendix (I) to Annex I and will be concluded not later than nine months after the signing of the treaty.

5. It is agreed that where the boundary follows a river, in the event of natural changes in the course of the flow of the river as described in Annex I (a), the boundary shall follow the new course of the flow. In the event of any other changes the boundary shall not be affected unless otherwise agreed.
6. Immediately upon the exchange of the instruments of ratification of this treaty, each party will deploy on its side of the international boundary as defined in Annex I (a).
7. The parties shall, upon the signature of the treaty, enter into negotiations to conclude, within nine months, an agreement on the delimitation of their maritime boundary in the Gulf of Aqaba.
8. Taking into account the special circumstances of the Baqura/Naharayim area, which is under Jordanian sovereignty, with Israeli private ownership rights, the parties agree to apply the provisions set out in Annex I (b).
9. With respect to the Al-Ghamr/Zofar area, the provisions set out in Annex I (c) will apply.

Article 4—Security

1.
   a. Both parties, acknowledging that mutual understanding and cooperation in security-related matters will form a significant part of their relations and will further enhance the security of the region, take upon themselves to base their security relations on mutual trust, advancement of joint interests and cooperation, and to aim toward a regional framework of partnership in peace.
   b. Toward that goal, the parties recognize the achievements of the European Community and European Union in the development of the Conference on Security and Cooperation in Europe (CSCE) and commit themselves to the creation, in the Middle East, of a Conference on Security and Cooperation in the Middle East (CSCME).

   This commitment entails the adoption of regional models of security successfully implemented in the post World War area (along the lines of the Helsinki Process) culminating in a regional zone of security and stability.
2. The obligations referred to in this article are without prejudice to the inherent right of self-defense in accordance with the United Nations Charter.
3. The parties undertake, in accordance with the provisions of this article, the following:

   a. To refrain from the threat of use of force or weapons, conventional, nonconventional, or of any other kind, against each other, or of other actions or activities that adversely affect the security of the other party;
   b. To refrain from organizing, instigating, inciting, assisting, or participating in acts or threats of belligerency, hostility, subversion, or violence against the other party.
   c. To take necessary and effective measures to ensure that acts or threats of belligerency, hostility, subversion, or violence against the other party do not originate from, and are not committed within, through, or over their territory (hereinafter the term "territory" includes the airspace and territorial waters).
4. Consistent with the area of peace and with the efforts to build regional security and to avoid and prevent aggression and violence, the parties further agree to refrain from the following:
   a. Joining or in any way assisting, promoting, or cooperating with any coalition, organization, or alliance with a military or security character with a third party, the objectives or activities of which include launching aggression or other acts of military hostility against the other party, in contravention of the provisions of the present treaty;
   b. Allowing the entry, stationing, and operating on their territory, or through it, of military forces, personnel, or material of a third party, in circumstances which may adversely prejudice the security of the other party.
5. Both parties will take necessary and effective measures, and will cooperate in combating terrorism of all kinds. The parties undertake:
   a. To take necessary and effective measures to prevent acts of terrorism, subversion, or violence from being carried out from their territory or through it and to take necessary and effective measures to combat such activities and all their perpetrators;
   b. Without prejudice to the basic rights of freedom of expression and association, to take necessary and effective measures to prevent the entry, presence, and operation in their territory of any group or organization, and their infrastructure which threatens the security of the other party by the use of, or incitement to the use of, violent means;
   c. To cooperate in preventing and combating cross-boundary infiltrations.
6. Any question as to the implementation of this article will be dealt with through a mechanism of consultations which will include a liaison system, verification, supervision, and where necessary, other

mechanisms, and higher level consultations. The details of the mechanism of consultations will be contained in an agreement to be concluded by the parties within three months of the exchange of the instruments of ratification of this treaty.

7. The parties undertake to work as a matter of priority, and as soon as possible, in the context of the Multilateral Working Group on Arms Control and Regional Security, and jointly, toward the following:
   a. The creation in the Middle East of a region free from hostile alliances and coalitions;
   b. The creation of a Middle East free from weapons of mass destruction, both conventional and nonconventional, in the context of a comprehensive, lasting, and stable peace, characterized by the renunciation of the use of force, and by reconciliation and good will.

Article 5—Diplomatic and other Bilateral Relations

1. The parties agree to establish full diplomatic and consular relations and to exchange resident ambassadors within one month of the exchange of the instruments of ratification of this treaty.
2. The parties agree that the normal relationship between them will further include economic and cultural relations.

Article 6—Water

With the view to achieving a comprehensive and lasting settlement of all the water problems between them:

1. The parties agree mutually to recognize the rightful allocations of both of them in Jordan River and Yarmouk River waters and Araba/Arava ground water in accordance with the agreed acceptable principles, quantities, and quality as set out in Annex II, which shall be fully respected and complied with.
2. The parties, recognizing the necessity to find a practical, just, and agreed solution to their water problems and with the view that the subject of water can form the basis for the advancement of cooperation between them, jointly undertake to ensure that the management and development of their water resources do not, in any way, harm the water resources of the other party.
3. The parties recognize that their water resources are not sufficient to meet their needs. More water should be supplied for their use through various methods, including projects of regional and international cooperation.
4. In light of paragraph 3 of this article, with the understanding that cooperation in water-related subjects would be to the benefit of both

parties, and will help alleviate their water shortages, and that water issues along their entire boundary must be dealt with in their totality, including the possibility of trans-boundary water transfers, the parties agree to search for ways to alleviate water shortages and to cooperate in the following fields:

a. Development of existing and new water resources, increasing the water availability, including cooperation on a regional basis, as appropriate, and minimizing wastage of water resources through the chain of their uses;
b. Prevention of contamination of water resources;
c. Mutual assistance in the alleviation of water shortages;
d. Transfer of information and joint research and development in water-related subjects, and review of the potentials for enhancement of water resources development and use.

5. The implementation of both parties undertakings under this article is detailed in Annex II.

Article 7—Economic Relations

1. Viewing economic development and prosperity as pillars of peace, security, and harmonious relations between states, peoples, and individual human beings, the parties, taking note of understandings reached between them, affirm their mutual desire to promote economic cooperation between them, as well as within the framework if wider regional economic cooperation.
2. In order to accomplish this goal, the parties agree to the following:
   a. To remove all discriminatory barriers to normal economic relations, to terminate economic boycotts directed at the other party, and to cooperate in terminating boycotts against either party by their parties;
   b. Recognizing that the principle of free and unimpeded flow of goods and services should guide their relations, the parties will enter into negotiations with a view to concluding agreements on economic cooperation, including trade and the establishment of a free trade area or areas, investment, banking, industrial cooperation, and labor, for the purpose of promoting beneficial economic relations, based on principles to be agreed upon, as well as on human development considerations on a regional basis. These negotiations will be concluded no later than six months from the exchange of the instruments of ratification of this treaty;
   c. To cooperate bilaterally, as well as in multilateral forums, toward the promotion of their respective economies and of their neighborly economic relations with other regional parties.

Article 8—Refugees and Displaced Persons

Recognizing the massive human problems caused to both parties by the conflict in the Middle East, as well as the contribution made by them toward the alleviation of human suffering, the parties will seek to further alleviate those problems arising on a bilateral level.

Recognizing that the above problems caused by the conflict in the Middle East cannot be fully resolved on the bilateral level, the parties will seek to resolve them in appropriate forums, in accordance with international law, including the following:

a. In the case of displaced persons, in a quadripartite committee together with Egypt and the Palestinians;
b. In the case of refugees,
   (i) In the framework of the Multilateral Working Group on Refugees;
   (ii) In negotiations, in a framework to be agreed, bilateral or otherwise, in conjunction with and at the same time as the permanent status negotiations pertaining to the territories referred to in Article 3 of this treaty;
c. Through the implementation of agreed United Nations programs and other agreed international economic programs concerning refugees and displaced persons, including assistance to their settlement.

Article 9—Places of Historical and Religious Significance and Interfaith Relations

1. Each Party will provide freedom of access to places of religious and historical significance.
2. In this regard, in accordance with the Washington Declaration, Israel respects the present special role of the Hashemite Kingdom of Jordan in Muslim Holy shrines in Jerusalem. When negotiations on the permanent status will take place, Israel will give high priority to the Jordanian historic role in these shrines.
3. The parties will act together to promote interfaith relations among the three monotheistic religions, with the aim of working toward religious understanding, moral commitment, freedom of religious worship, and tolerance and peace.

Article 10—Cultural and Scientific Exchanges

The parties, wishing to remove biases developed through periods of conflict, recognize the desirability of cultural and scientific exchanges in all fields, and agree to establish normal cultural relations between them.

Thus, they shall, as soon as possible and not later than nine months from the exchange of the instruments of ratification of this treaty, conclude the negotiations on cultural and scientific agreements.

Article 11—Mutual Understanding and Good Neighborly Relations

1. The parties will seek to foster mutual understanding and tolerance based on shared historic values, and accordingly undertake:
   a. To abstain from hostile or discriminatory propaganda against each other, and to take all possible legal and administrative measures to prevent the dissemination of such propaganda by any organization or individual present in the territory of either party;
   b. As soon as possible, and not later than thtee months from the exchange of the instruments of ratification of this treaty, to repeal all adverse or discriminatory references and expressions of hostility in their respective legislation;
   c. To refrain in all government publications from any such references or expressions;
   d. To ensure mutual enjoyment by each other's citizens of due process of law within their respective legal systems and before their courts.
2. Paragraph 1 (a) of this article is without prejudice to the right to freedom of expression as contained in the International Covenant on Civil and Political Rights.
3. A joint committee shall be formed to examine incidents where one party claims there has been a violation of this article.

Article 12—Combating Crime and Drugs

The parties will cooperate in combating crime, with an emphasis on smuggling, and will take all necessary measures to combat and prevent such activities as the production of, as well as the trafficking in illicit drugs, and will bring to trial perpetrators of such acts. In this regard, they take note of the understandings reached between them in the above spheres, in accordance with Annex III and undertake to conclude all relevant agreements not later than nine months from the date of the exchange of the instruments of ratification of this treaty.

Article 13—Transportation and Roads

Taking note of the progress already made in the area of transportation, the parties recognize the mutuality of interest in good neighborly relations in the area of transportation and agree to the following means to promote relations between them in this sphere:

1. Each party will permit the free movement of nationals and vehicles of the other into and within its territory according to the general rules applicable to nationals and vehicles of other states. Neither party will impose discriminatory taxes or restrictions on the free movement of persons and vehicles from its territory to the territory of the other.
2. The parties will open and maintain roads and border crossings between their countries and will consider further roads and rail links between them.
3. The parties will continue their negotiations concerning mutual transportation agreements in the above and other areas, such as joint projects, traffic safety, transport standards and norms, licensing of vehicles, land passages, shipment of goods and cargo, and meteorology, to be concluded not later than six months from the exchange of the instruments of ratification of this treaty.
4. The parties agree to continue their negotiations for a highway to be constructed and maintained between Egypt, Jordan, and Israel near Eilat.

Article 14—Freedom of Navigation and Access to Ports

1. Without prejudice to the provisions of paragraph 3, each party recognizes the right of the vessels of the other party to innocent passage through its territorial waters in accordance with the rules of international law.
2. Each party will grant normal access to its ports for vessels and cargoes of the other, as well as vessels and cargoes destined for or coming from the other party. Such access will be granted on the same conditions as generally applicable to vessels and cargoes of other nations.
3. International waterways open to all nations for unimpeded and non-suspendable freedom of navigation and over flight. The parties will respect each other's right to navigation and overflight for access to either party through the Straits of Tiran and the Gulf of Aqaba.

Article 15—Civil Aviation

1. The parties recognize as applicable to each other the rights, privileges, and obligations provided for by the multilateral aviation agreements to which they are both party, particularly by the 1944 Convention on International Civil Aviation (the Chicago Convention) and the 1944 International Air Services Transit Agreement.
2. Any declaration of national emergency by a party under Article 89 of the Chicago Convention will not be applied to the other party on a discriminatory basis.

3. The parties take note of the negotiations on the international air corridor to be opened between them in accordance with the Washington Declaration. In addition, the parties shall, upon the exchange of the instruments of ratification of this treaty enter into negotiations for the purpose of concluding a Civil Aviation Agreement. All the above negotiations are to be concluded not later than six months from the exchange of the instruments of ratification of this treaty.

Article 16—Post and Telecommunications

The parties take note of the opening between them, in accordance with the Washington Declaration, of direct telephone and facsimile lines. Postal links, the negotiations on which having been concluded, will be activated upon the signature of this treaty. The parties further agree that normal wireless and cable communications and television relay services by cable, radio, and satellite, will be established between them, in accordance with all relevant international conventions and regulations. The negotiations on these subjects will be concluded not later than nine months from the exchange of the instruments of ratification of this treaty.

Article 17—Tourism

The parties affirm their mutual desire to promote cooperation between them in the field of tourism. In order to accomplish this goal, the parties—taking note of the understandings reached between them concerning tourism—agree to negotiate, as soon as possible, and to conclude not later than three months from the exchange of the instruments of ratification of this treaty, an agreement to facilitate and encourage mutual tourism and tourism from third countries.

Article 18—Environment

The parties will cooperate in matters relating to the environment, a sphere to which they attach great importance, including conservation of nature and prevention of pollution, as set forth in Annex IV. They will negotiate an agreement on the above, to be concluded not later than six months from the exchange of the instruments of ratification of this treaty.

Article 19—Energy

1. The parties will cooperate in the development of energy resources, including the development of energy related projects such as the utilization of solar energy. The parties having concluded their negotiations the interconnecting of their electric grids in the Eilat-Aqaba

area, will implement the interconnecting upon the signature of this treaty. The parties view this step as a part of a wider binational and regional concept. They agree to continue their negotiations as soon as possible to widen the scope of their interconnected grids.

2. The parties will conclude the relevant agreements in the field of energy within six months from the date of exchange of the instruments of ratification of this treaty.

Article 20—Rift Valley Development

The parties attach great importance to the integrated development of the Jordan Rift Valley area, including joint projects in the economic, environmental, energy-related, and tourism fields. Taking note of the Terms of Reference developed in the framework of the Trilateral Jordan-Israel-U.S. Economic Committee toward the Jordan Rift Valley Development Master Plan, they will vigorously continue their efforts toward the completion of planning and towards implementation.

Article 21—Health

The parties will cooperate in the area of health and shall negotiate with a view to the conclusion of an agreement within nine months of the exchange of the instruments of ratification of this treaty.

Article 22—Agriculture

The parties will cooperate in the areas of agriculture, including veterinary services, plant protection, biotechnology, and marketing, and shall negotiate with a view to the conclusion of an agreement within six months from the date of the exchange of instruments of ratification of this treaty.

Article 23—Aqaba and Eilat

The parties agree to enter into negotiations, as soon as possible, and not later than one month from the exchange of the instruments of ratification of this treaty, on arrangements that would enable the joint development of the towns of Aqaba and Eilat with regard to such matters, inter alia, as joint tourism development joint customs posts, free trade zone, cooperation in aviation, prevention of pollution, maritime matters, police, customs, and health cooperation. The parties will conclude all relevant agreements within nine months from the exchange of instruments of ratification of the treaty.

Article 24—Claims

The Parties agree to establish a claims commission for the mutual settlement of all financial claims.

Article 25—Rights and Obligations

1. This treaty does not affect and shall not be interpreted as affecting, in any way, the rights and obligations of the parties under the Charter of the United Nations.
2. The parties undertake to fulfill in good faith their obligations under this treaty, without regard to action or inaction of any other party and independently of any instrument inconsistent with this treaty. For the purposes of this paragraph, each party represents to the other than in its opinion and interpretation there is no inconsistency between their existing treaty obligations and this treaty.
3. They further undertake to take all the necessary measures for the application in their relations of the provisions of the multilateral conventions which they are parties, including the submission of appropriate notification to the Secretary General of the United Nations and other depositories of such conventions.
4. Both parties will also take all the necessary steps to abolish all pejorative references to the other party, in multilateral conventions to which they are party, to the extent that such references exist.
5. The parties undertake not to enter into any obligation in conflict with this treaty.
6. Subject to Article 103 of the United Nations Charter, in the event of a conflict between the obligations of the parties under the present treaty and any of their other obligations, the obligations under this treaty will be binding and implemented.

Article 26—Legislation

Within three months of the exchange of the instruments of ratification of this treaty, the parties undertake to enact any legislation necessary in order to implement the treaty, and to terminate any international commitments and to repeal any legislation that is inconsistent with the treaty.

Article 27—Ratification and Annexes

1. This treaty shall be ratified by both parties in conformity with their respective national procedures. It shall enter into force on the exchange of the instruments of ratification.
2. The annexes, appendices, and other attachments to this treaty shall be considered integral parts thereof.

Article 28—Interim Measures

The Parties will apply, in certain spheres to be agreed upon, interim measures pending the conclusion of the relevant agreements in accordance with this treaty, as stipulated in Annex V.

Article 29—Settlement of Disputes

Disputes arising out of the application or interpretation of this treaty shall be resolved by negotiations.

Any such disputes that cannot be settled by negotiations shall be resolved by conciliation or submitted to arbitration.

Article 30—Registration

This treaty shall be transmitted to the Secretary General of the United Nations for registration in accordance with the provisions of Article 102 of the Charter of the United Nations.

Done at the Araba/Arava Crossing point this day, Jumada Al-Ula, 21st 1415, Heshvan 21st, 5755 to which corresponds October 26, 1994 in the Arabic, Hebrew, and English languages, all texts being equally authentic. In case of divergence of interpretation, the English text shall prevail.

For the Hashemite Kingdom of Jordan, Abdul Salam Majali, Prime Minister

For the State of Israel, Yitzhak Rabin, Prime Minister

Witnessed by: William J. Clinton, President of the United States of America

ANNEXES, APPENDICES, AND OTHER ATTACHMENTS

(1) Annex I:
- (a) International Boundary
- (b) Baqura/Naharayim Area
- (c) Al-Ghamr/Zofar Area

(2) Appendices (27 sheets)
- Wadi Araba (10 sheets), 1:20,000 orthophoto maps
- Dead Sea (2 sheets), 1:50,000 orthoimages
- Jordan and Yarmouk Rivers (12 sheets), 1:10,000 orthophoto maps
- Baqura Area (1 sheet), 1:20,000 orthophoto maps
- Al-Ghamr Area (1 sheet), 1:20,000 orthophoto maps
- Gulf of Aqaba (1 sheet), 1:50,000 orthoimages

(3) Annex II: Water

(4) Annex III: Crime and Drugs

(5) Annex IV: Environment

(6 Annex V: Interim Measures

(7) Attachments: Agreed Minutes A to D

See Declaration of Principles, Egyptian-Israeli Peace Treaty, Israel-Jordan Common Agenda, Jordanian-Israel bilateral accords, Labor Party (Israel), Stockholm PLO-American delegation statement.

Israel-Jordan Track Common Agenda: See Israel-Jordan Common Agenda.

Israel-Jordan Water agreement: See water.

Israel-Lebanon Agreement of Understanding (of April 26, 1996): See Agreement of Understanding between Israel and Lebanon.

Israel-Morocco Chamber of Commerce: See regional development centers.

Israel-PLO Declaration of Principles: See Declaration of Principles.

Israel-PLO Protocol on Economic Relations: See Protocol on Economic Relations.

Israel-PLO recognition: See Stockholm PLO-American delegation statement.

Israel Policy Forum (IPF): Founded in 1933, just prior to the Israeli-Palestinian agreement signing on the White House lawn, IPF is an independent leadership institution established by members of the American Jewish community. Its mission is to create greater awareness of the Middle East Peace Process, and to encourage and support an active U.S. role in resolving the Arab-Israeli conflict.

IPF programs include:

- Israel Policy Papers—Monthly essays for opinion leaders that analyze issues related to the peace process and provide information not generally available in the American media.
- Peace Pulse—An ongoing assessment of the status of the Mideast peace talks utilizing an innovative index.
- Ongoing surveys of American and American Jewish opinion on the peace process.
- Regular dissemination of timely information on the peace process to policymakers, members of Congress, their staff, and the press, using written materials and private briefings.

See Peace Pulse.

Israel's right to exist in peace: See Arafat, Yasir (letter to Y. Rabin recognizing Israel's right to exist in peace, September 9, 1993).

Israel-United States Free Trade Agreement: See Free Trade Agreement (United States and Israel).

Israel Wing: See Cairo Agreement.

Isra'iliun: Used by many Arab nationalists for Israelis.

ist i'mariun: Used by many Arab nationalists for U.S. or external enemies or colonialists.

J

Jabal Abu Ghneim: See Har Homa.

Jaber, Shaikh, al-Ahmad al Jaber (Emir): The son of the late ruler was born in 1926, became a minister in 1962 after working in the oil sector, prime minister in 1965 and emir in December 1977. Shaikh Jaber became known internationally when Kuwait was invaded by Iraq in August 1990.

See Kuwait..

Japan: Until recently there was little contact between Japan and the nations of the Middle East. With the discovery of oil in the Persian Gulf by the 1930s Japan became increasingly aware of the strategic value of the Middle East.

With the onset of the Arab Oil Crisis in 1973 Japan's position began to shift. Japan, since entering the United Nations, had maintained a neutral, noncommittal policy toward the Arab-Israel conflict. However, today Japan realizes her role as a force in the stability of the region and plays an increasing part in regional disputes.

Jarring, Gunnar (Ambassador): See Resolution 242.

JEC: See Joint Economic Committee.

Jericho: See Cairo Agreement, Protocol on Economic Relations.

Jerusalem: The holy city for three monotheistic religions—Judaism, Christianity, and Islam. In 1967 Israel annexed the entire city and in 1980 it was declared the capital of Israel.

Today Jerusalem is three times bigger than the pre-1967 city—having grown from 38,000 dunams (9,525 acres) in 1964 to 108,000 dunams on June 28, 1967 to 123,000 in 1993.

Jerusalem is Israel's largest city in size and population: 603,000 inhabitants—70 percent Jews, 30 percent Arabs. More people have left Jerusalem than have moved there, each year since 1988—in 1995, 9,000 came, 14,500 left.

Twenty percent of all families and 37 percent of all children live beneath the poverty line.

See Beilin-Eitan agreement, Casablanca Middle East/North Africa Economic Summit, Har Homa, United States.

Jewish fundamentalists: The following are characteristics of Jewish fundamentalism:

- its intellectual and social foundations can be found in the new yeshivot and what can be called "Yeshiva Judaism."
- within this movement, study of Torah has been elevated to supremacy.
- there is a self-conscious and deliberate attempt to establish official dogmas which transform the diversity of tradition into a formalized series of articulated fundamentals of faith and/or practice which is now seen to be binding on all believers as the "old time religion."
- results in an opposition to the inherent diversity of religious traditions within Judaism.
- rejects an immoral and degenerate modernism at war with modernity and seeking to separate its members, the B'nai Torah, from all others.
- the B'nai Torah live in their own communities and intermarry only with other B'nai Torah families. To interact with the outside world is to place onself in spiritual danger. They see a tremendous risk in accepting the modern orthodox model of a Jew who shows it is possible simultaneously to express fidelity to tradition and to non-Jewish society and culture. There is an accompanying temptation to resist the implementation of conflicting state laws.
- the yeshiva's historical relationship to the whole of Judaism yielded to a concern with the well-being of B'nai Torah alone.
- B'nai Torah do not recognize the religious authority of non-yeshiva rabbis. Differing interpretations, including talmudically based and orthodox ones, are seen as apostasy. Halakhah's open-endedness, which had always allowed it to accommodate itself to social and economic circumstances, is rejected by the B'nai Torah.
- the opinions of roshei yeshiva are identified with the divine will. Any deviations from their doctrine are seen as a threat to Judaism itself.
- rejection of both Western values in general and Christianity in particular.

The Six-Day War was to mark the emergence of a new form of Jewish fundamentalism. There were those for whom the return to statehood, followed by this spectacular victory, signified the beginning of messianic redemption, and the end of diaspora existence. Individuals could hasten the messianic coming by working to create a just, merciful, and faithful society in the Land of Israel, based on exclusivist interpretations of Jewish law.

Compare Islamic fundamentalists.

See fundamentalism.

JHC: Joint Hebron Committee.

See Agreement on Security Arrangements for Hebron.

Jihad: Translated into English as Holy War. A global effort against the West and its Judeo-Christian values.

See HizbAllah, terrorism.

JIPER: See Jordanian-Israeli-Palestinian Economic Region.

See Jordanian-Israeli-Palestinian Economic Zone.

JIPEZ: See Jordanian-Isareli-Palestinian Economic Zone.

JIPTA: See Jordanian-Israeli-Palestinian Trade Agreement.

JOD: The international standard code for the currency of Jordan—the Jordanian dinar.

Joint Civil Affairs Coordination and Cooperation Committee: See Gaza Strip and Jericho Area Agreement.

Joint Economic Committee (JEC): See Paris Economic Agreement, Protocol on Economic Relations (Article II).

Joint Hebron Committee (JHC): See Agreement on Security Arrangements for Hebron.

Joint Israeli-Palestinian-American Committee: See Interim Agreement.

Joint Israeli-Palestinian Liaison Committee (Gaza Strip and Jericho): See Agreement on Security Arrangements for Hebron, Declaration of Principles, Gaza Strip and the Jericho Area Agreement (Article XV), Interim Agreement.

Joint Liaison Committee (Gaza Strip and Jericho): See Agreement on Security Arrangements for Hebron, Declaration of Principles.

joint projects: See Amman Middle East/North Africa Economic Summit, Israel-Jordan Peace Treaty.

Joint Security Coordination and Cooperation Committee (JSC): See Gaza Strip and the Jericho Area Agreement.

joint ventures: See Amman Middle East/North Africa Economic Summit.

Jordan: On June 10, 1916 the Arab Legion was formed which launched the Great Arab Revolt against the Ottoman Empire. On May 25, 1946 (Jordan's Independence Day) the end of the British mandate over the Emirate of Transjordan took place to be replaced by the birth of the Hashemite Kingdom of Jordan. On August 11, 1952 King Hussein acceded to the throne of Jordan, succeeding his father, King Talal, who abdicated due to health reasons.

Relations between Israel and Jordan have always moved along two tracks. On one hand, Jordan was an active participant in the military struggle against Israel in 1948 and 1967, and publicly promoted an anti-Israel line in the name of Arab unity. On the other hand, Israeli leaders have conducted an almost continuous dialogue with the Hashemite monarchs since their independence in 1948.

When Israel gained the West Bank (and the Gaza Strip from Egypt) in 1967, it kept much of the Jordanian system in place. King Hussein continued in his role as guardian of Muslim holy places in Jerusalem, including the Al-Aqsa mosque—Islam's third holiest site.

During the Gulf crisis in 1990, Jordan's economy suffered resulting from sanctions against Iraq, its major trading partner.

On August 8, 1994 Crown Prince Hassan and Israeli Prime Minister Rabin inaugurated the first border crossing between Jordan and Israel. The event marked the first visit of an Israeli prime minister to Jordan. On October 26, 1994 the historic Jordan-Israel Treaty of Peace was signed in the Jordanian area of Wadi Araba, officially establishing peace after 46 years of hostilities.

Jordan is one of the more promising emerging markets in the Middle East. An annual population growth rate of 3.8 percent has increased market demand by twofold and pushed up imports by 30 percent since 1990. The small size of the Jordanian market somewhat limits opportunities for American business, although changing political conditions in the region have helped Jordan regain its potential as a supplier of goods and services to the Palestinian Self-Rule Area and the Gulf. Closer cooperation between Jordan and Israel in trade, transportation, energy, and water should lead to more business opportunities in Jordan.

The commercial importance of Jordan is reinforced by its strategic geographic position at the confluence of three continents. The success of the Middle East peace process has stimulated increased interest in the region. Foreign businesses are exploring opportunities for greater investment in and trade with Jordan. In 1994 and 1995 the U.S. government forgave approximately $700 million in Jordanian debt to the United States. The EU, Germany, France, and Japan are also working with the government on addressing Jordan's economic needs.

Jordan is reportedly pursuing the signing of free trade arrangements with potential partners in the industrial world. The government has applied for membership in the World Trade Organization.

With the signing of the Jordan-Israel Peace Treaty, both countries reached agreements on trade, joint development of the Jordan Rift Valley and the Gulf of Aqaba, civil aviation, navigation, telecommunications, transportation, and cooperation in the areas of tourism, energy, agriculture, science, and the environment.

The trade agreement with Israel, initialed October 17, 1995 will activate all other Jordanian-Israeli economic agreements and Jordanian-Palestinian trade and monetary agreements already signed. As required by its peace treaty with Israel, Jordan abolished the Boycott of Israel Law and repealed other laws banning the sale or lease of property to Israel.

To improve the business climate for foreign business, the government recently passed new legislation on investment and taxes. The new investment law permits foreign businesses to invest in Jordan and grants them equal treatment with Jordanian investors, in most industry sectors. The new income tax law is more favorable to corporate business and reduces tax liabilities by at least 15 percent. The sales tax law increases the sales tax from 7 to 10 percent; the increase offsets expected reductions in Jordan's customs tariffs and duties.

The major barriers to exporting to Jordan are increased prices of consumer goods due to U.S. dollar appreciation and high sea freight charges from the United States to Jordan. There are two additional complications to doing business in Jordan: (1) weak enforcement of intellectual property rights, and (2) time-consuming and arbitrary customs clearance procedures. Authority for deciding foreign capital ownership lies with the Council of Ministers, which is committed to loosening restrictions on foreign investment in most sectors.

Jordan imports products of varying quality from around the globe. High-end buyers prefer American consumer products. For 90 percent of the market, however, price is the critical factor in the purchase of most consumer, personal care, and household products. Countries and companies providing generous and flexible export terms and direct support to local business are most capable of sustaining market share and increasing sales in Jordan. Importers buy more products from countries whose companies give their local sales staff access to training and product orientation. The establishment of joint ventures in industry and trade, esepcially in warehousing and distribution, is more attractive in the Jordanian market than direct exporting.

The Prime Minister, in 1997, is determined to continue with Jordan's program of economic reform. However, two key factors will continue to hinder economic development. The first is the continued freezing of international relations with Iraq, a country which previously accounted for one quarter of Jordan's external trade. The second is the deterioration in 1997 with Israel since the election of a new prime minister. Continued uncertainty over the peace process will deter both foreign investors and tourists.

See Abdullah Ibn Hussein (King), Abdullah Ibn Hussein, Amman Middle East/North Africa Economic Summit, boycott, Cairo Agreement, Hussein (King), Israel-Jordan Peace Treaty, Jordanian-Israeli bilateral accords, Jordanian-PLO Accord, Middle East Peace and Stability Fund, Protocol on Economic Relations, Rift Valley.

Jordanian Dinar: The currency of Jordan.

Jordanian-Israeli bilateral accords: The last five Jordanian-Israeli bilateral accords were signed in Aqaba in mid-January 1996, covering cooperation in the fields of science and culture, transportation and aviation, telecommunications, maritime borders, and Aqaba-Eilat developments.

The most important agreements are widely seen to be the ones covering trade and transport, for three reasons: they open the way for ordinary citizens on both sides to develop mutually satisfying working relationships with their counterparts in the other country; they can promote more tangible economic benefits of peace; and they can generate greater understanding of the other on both sides, thereby reducing misconceptions and stereotypes.

The trade accord was actually signed in October 1995, but had not been activated because of the need for the transport agreement, which is now signed. The trade accord grants 20-50 percent exemptions from tariffs for most Jordanian industrial goods and many foodstuffs and consumer items, while Jordan in turn grants a 10 percent reduction of customs for some Israeli exports. The agreement takes into account the large difference in size between the two economies.

The transport agreement establishes daily bus routes linking major cities in both countries, and permits cross-border travel by private cars and vehicles carrying exports and transit trade. It also allows air service between the two countries and calls for work on a future rail link.

Two other important agreements that were signed in late 1995 cover relations in agricultural goods and security affairs. The first comprises joint efforts in research, training, crop spraying, and promotion of bilateral agricultural exports, and the second covers cooperation among the police forces of Jordan and Israel in fighting drugs and crime and jointly investigating cases of murder, fraud, forgery, artifacts thefts, and other illegal activities. This is the first agreement for police cooperation that Israel has ever signed with an Arab country, but it does not cover extradition, which is a judicial rather than a police matter.

See aviation.

Jordanian-Israeli-Palestinian Economic Region (JIPER): See Jordanian-Israeli-Palestinian Economic Zone.

Jordanian-Israeli-Palestinian Economic Zone (JIPEZ): A possible first step of regional cross-border alliances and trade. Not as impactful as the Jordanian-Israeli-Palestinian Trade Agreement, but a less complicated way to begin.

See Jordanian-Israeli-Palestinian Trade Agreement.

Jordanian-Israeli-Palestinian Trade Agreement (JIPTA): Will see the coming together of a Jordan-Israel-Palestinian trade agreement to reduce and ultimately eliminate tariffs and quotas. (JIPTA could await formation or be absorbed by the extended Middle East Free Trade Agreement.) JIPTA identifies the first step in the process of bilateral, extended to multilateral, trade. Commencing a dialogue to materialize into an open trading community of capital, goods, and services between the Jordanians, Israelis, and Palestinians can illustrate to the region and the world that interdependence can be firmed by establishing tariff-lowered and then tariff-free imports and exports.

See Jordanian-Israeli-Palestinian Economic Zone, Middle East Free Trade Agreement.

Jordanian-Palestinian front: See PLO Phased Plan.

Jordanian-PLO Accord: Signed February 11, 1985 and proceeding from the spirit of the Fez summit resolutions, as agreed on by the Arab (world) and the resolutions of the U.N. relating to the Palestine problem; in accordance with international legitimacy, and proceeding from the mutual understanding to establish a special relationship between the Jordanian and Palestinian people; the government of the Hashemite Kingdom of Jordan and the Palestine Liberation Organization have agreed to march together toward the realization of a just and peaceful settlement of the Middle East problem and to put an end to the Israeli occupation of the Arab occupied territories, including Jerusalem, in accordance with the following principles:

1. Land in exchange for peace, as laid down in the resolutions of the United Nations, including the resolutions of the Security Council;
2. The right of the Palestinian people to self-determination. The Palestinians will exercise their inalienable right to self-determination when the Jordanians and the Palestinians will be able to realize this within the framework of an Arab confederal union, which [they] intend to have established between the two states of Jordan and Palestine [dawlatay al-Urdunn wa-Filastin];
3. The solution of the Palestinian refugee problem in accordance with the resolutions of the U.N.;
4. The solution of the Palestinian problem in all its aspects;
5. On this basis, the peace negotiations will be held within the framework of an international conference that will be attended by the five permanent members of the U.N. Security Council and the other parties to the conflict, including the PLO, the sole legitimate representative of the Palestinian people, in a joint delegation [a joint Jordanian-Palestinian delegation].

Jordan-Israel Peace Treaty: See Israel-Jordan Peace Treaty.

Jordan-PLO Accord: See Jordanian-PLO Accord.

Jordan Rift Valley (JRV): An element of a great rift that extends from Syria to the Red Sea and continues through a large portion of Eastern Africa. Riparian population groups in the JRV include Syrians, Palestinians, Israelis, and Jordanians. At its southern extremity, the JRV connects with Egypt and Saudi Arabia.

The area under discussion for integrated development extends from south of Lake Tiberias (Sea of Galilee) to the Gulf of Aqaba/Eilat. It includes a series of distinctive geographic sections:

- the Jordan River Valley south of the Sea of Galilee down to the Dead Sea;
- the Dead Sea and its rugged eastern and western escarpments;
- south of the Dead Sea for a distance of about 40 km; and

- the arid Wadi Arava/Arava Valley further to the south and the Red Sea coastal zone with the border towns of Aqaba in Jordan and Eilat in Israel.

See Rift Valley.

Jordan River: See Israel-Jordan Peace Treaty, Washington Declaration.

Joseph's Tomb: See Interim Agreement (on the West Bank and the Gaza Strip).

JRV: See Jordan Rift Valley.

JSCCC: (Joint Security Coordination and Cooperation Committee): See Gaza Strip and the Jericho Area Agreement.

June War: See Six-Day War.

jurisdiction: See Gaza Strip and the Jericho Area Agreement (Article V).

K

Kammel, Hussein (Lieutenant General): See chronology (August 8, 1995).

Kammel, Saddam: See chronology (August 8, 1995).

Katyusha rockets: See Agreement of Understanding between Israel and Lebanon.

KD: The Kuwaiti dinar.

Khalifa Al-Thani, Shaikh Hamad Bin: See Doha Middle East/North Africa Economic Summit, Shaikh Hamad.

Khatami, Mohammad: See Iran.

Khomeini, Ruholla (Ayatollah): Born in 1902. When the Shah was forced to leave Iran on January 16, 1979 Khomeini returned from exile in Paris. After becoming Supreme Leader in 1979 he imposed strict Islamic puritanism and controlled the country until his death in 1989, one year following the U.N.-brokered ceasefire with Iraq.

See Iran, Iraq; Pahlavi, Mohammed Reza; Rafsanjani, Ali Akbar Hashemi (President).

Kimche, David: See Casablanca Middle East/North Africa Economic Summit.

Kingdom of Saudi Arabia: See Saudi Arabia.

King Hassan: See Hassan II (King).

King Hussein: See Hussein (His Majesty King Hussein Bin Talal).

Kiryat Arba: See Gush Emuium, Hebron, Interim Agreement (the West Bank and the Gaza Strip).

Kissinger, Henry (Secretary of State): See Resolution 338.

Kissufim-Gush Katif road: See Cairo Agreement.

Klinghoffer, Leon: See terrorism.

Knesset: See Beilin-Eitan agreement.

Koran: Synonymous with Quran.

Korei, Ahmed: See Protocol on Economic Relations.

Kozyrev, Andrei: Russian Foreign Minister.

See Amman Middle East/North Africa Economic Summit, Casablanca Middle East/North Africa Economic Summit, Declaration of Principles, Gaza Strip and the Jericho Area Agreement.

kuffar: Used by many Islamists for unbelievers.

Kurds: Found in large numbers in southeastern Turkey, northwest Iran, northern Iraq, and also in Syria. Factions, especially those in Iraq have fought with each other, often taking positions as to their relationship with Iraq.

See Hussein, Saddam; Iraq, minorities.

Kuwait: Officially, the State of Kuwait; ceased being a British protectorate in 1961. This small independent country at the head of the Persian Gulf is ruled by a hereditary emir. Invaded by Iraq on August 2, 1990. The United Nations sent military personnel to evict the Iraqi military in January 1991.

Despite its small size, Kuwait buys large amounts of American products and services and will likely continue to do so. In 1994 Kuwait imported $1.175 billion worth of goods from the United States, a 16.5 percent increase over 1993. (Some 1994 exports are part of much larger multi-year sales, e.g., $4.5 billion in U.S. defense sales to Kuwait since 1991 are exported over several years). Kuwait is the sixth largest market for U.S. exports in the Middle East and North Africa, and the United States is Kuwait's largest trading partner.

Kuwait imports a wide variety of U.S. military, industrial, and consumer products. Leading military imports in the past (1992-1995) have included aircraft and parts, air defense systems, radars, and tanks. Leading industrial imports include all equipment/parts, aircraft parts, and generators. Leading consumer imports include passenger vehicles and trucks.

Kuwait is a highly price-competitive market because it has low tariffs (generally only 4% ad valorem), few import barriers, and no exchange controls. Procurement for large public-sector projects dominates the business scene; there is almost no manufacturing and little non-oil exporting.

Kuwait's oil wealth (10% of world oil reserves) and substantial investments abroad (estimated at $36 billion) have created an affluent population: its 1.83 million people have a per capita income of $16,535. As Kuwait's young population (48% under 15 years of age, 70 percent under 24) grows up, exporters will find opportunities supplying items to them.

Export prospects for industrial products and services are very good as Kuwait government procurement, Kuwait's privatization, and new foreign investments in Kuwait combine to open $15.5 billion in potential export opportunities. Export prospects are tempered somewhat by the $5 billion annual deficits the Kuwait government risks running the rest of the 1990s as it sustains a welfare state for Kuwait nationals, repays Gulf War and reconstruction debts, and pays for large defense purchases.

See Gulf Cooperation Council, Gulf crisis, Gulf nations, intellectual property rights, Jaber, Resolution 687, Shaikh, al-Ahmad al Jaber.

Kuwaiti dinar: The currency of Kuwait and the Neutral Zone (between Saudi Arabia and Iraq).

KWD: The international standard code for the currency of Kuwait, Neutral Zone (between Saudi Arabia and Iraq)—the Kuwaiti dinar.

L

labor: See Protocol on Economic Relations (Article VII).

labor flows: See regional economic integration.

Labor Party (Israel): Established in 1965. More recently, in June 1992 Yitzhak Rabin formed a coalition government, thus becoming Prime Minister. The Labor Party signed a limited self-rule agreement with the PLO in September 1993, and in October 1994 the Israel-Jordan Peace Treaty.

See Netanyau, Binyamih; Peres, Shimon.

labor relations: See Declaration of Principles.

land transport: See transportation, Transportation (Center of the Middle East/North Africa Community).

Larsen, Terje: The Norwegian mediator who piloted the 1992-1993 accord between Israel and the Palestinians. In mid-August 1996 he brought Prime Minister Netanyahu and Chairman Arafat together for a landmark first meeting in September, as the U.N.'s Gaza coordinator.

Latrun Area: See Pentagon Plan.

LBP: The international standard code for the currency of Lebanon—the Lebanese pound.

lead poisoning: See pollution.

lead pollution: See pollution.

League of Arab States, The: See Arab League, Gulf Cooperation Council.

League of Nations: See Hamas Covenant.

League of Nations Mandate: Signed on July 24, 1922 in London as "Mandate for Palestine" by members of the League of Nations, the 28 articles include:

The Council of the League of Nations:

Whereas the Principal Allied Powers have agreed, for the purpose of giving effect to the provisions of Article 22 of the Covenant of the League of Nations, to entrust to a Mandatory selected by the said Powers of the administration of the territory of Palestine, which formerly belonged to the Turkish Empire, within such boundaries as may be fixed by them; and

Whereas the Principal Allied Powers have also agreed that the Mandatory should be responsible for putting into effect the declaration originally made on November 2, 1917, by the Government of His Britannic Majesty, and adopted by the said Powers, in favour of the establishment in Palestine of a national home for the Jewish people, it being clearly understood that nothing should be done which might prejudice the civil and religious rights of existing non-Jewish communities in Palestine, or the rights and political status enjoyed by Jews in any other country; and

Whereas recognition has thereby been given to the historical connection of the Jewish people with Palestine and to the grounds for reconstituting their national home in that country; and....

Article 1—The Mandatory shall have full powers of legislation and of administration, save as they may be limited by the terms of this mandate.

Article 2—The Mandatory shall be responsible for placing the country under such political, administrative, and economic conditions as will secure the establishment of the Jewish national home, as laid down in the preamble, and the development of self-governing institutions, and also for safeguarding the civil and religious rights of all the inhabitants of Palestine, irrespective of race and religion.

Article 3—The Mandatory shall, so far as circumstances permit, encourage local autonomy.

Article 4—An appropriate Jewish agency shall be recognized as a public body for the purpose of advising and cooperating with the Administration of Palestine in such economic, social, and other matters as may affect the establishment of the Jewish national home and the interests of the Jewish population in Palestine, and, subject always to the control of the Administration, to assist and take part in the development of the country.

The Zionist organization, so long as its organization and constitution are in the opinion of the Mandatory appropriate, shall be recognized as such agency. It shall take steps in consultation with His Britannic Majesty's Government to secure the cooperation of all Jews who are willing to assist in the establishment of the Jewish national home.

Article 5—The Mandatory shall be responsible for seeing that no Palestine territory shall be ceded or leased to, or in any way placed under the control of, the government of any foreign power.

Article 6 —The Administration of Palestine, while ensuring that the rights and position of other sections of the population are not prejudiced, shall facilitate Jewish immigration under suitable conditions and shall encourage, in cooperation with the Jewish agency referred to in Article 4, close settlements by Jews on the land, including state lands and waste lands not required for public purposes.

Article 7—The Administration of Palestine shall be responsible for enacting a nationality law. There shall be included in this law provisions framed so as to facilitate the acquisition of Palestinian citizenship by Jews who take up their permanent residence in Palestine.

Article 8—The privileges and immunities of foreigners, including the benefits of consular jurisdiction and protection as formerly enjoyed by Capitulation or usage in the Ottoman Empire, shall not be applicable in Palestine.

Unless the Powers whose nationals enjoyed the aforementioned privileges and immunities on August 1, 1914, shall have previously renounced the right to their reestablishment, or shall have agreed to their non-application for a specified period, these privileges and immunities shall, at the expiration of the mandate, be immediately reestablished in their entirety or with such modifications as may have been agreed upon between the powers concerned.

Article 9—The Mandatory shall be responsible for seeing that the judicial system established in Palestine shall assure to foreigners, as well as to natives, a complete guarantee of their rights.

Respect for the personal status of the various peoples and communities and for their religious interests shall be fully guaranteed. In particular, the control and administration of Wakfs shall be exercised in accordance with religious law and the dispositions of the founders.

Article 10—Pending the making of special extradition agreements relating to Palestine, the extradition treaties in force between the Mandatory and other foreign powers shall apply to Palestine.

Article 11—The Administration of Palestine shall take all necessary measures to safeguard the interests of the community in connection with the development of the country, and subject to any international obligations accepted by the Mandatory, shall have full power to provide for public ownership or control of any of the natural resources of the country or of the public works, services, and utilities established or to be established therein. It shall introduce a land system appropriate to the needs of the country, having regard, among other things, to the desirability of promoting the close settlement and intensive cultivation of the land.

The Administration may arrange with the Jewish agency mentioned in Article 4 to construct or operate, upon fair and equitable terms, any public works, services, and utilities, and to develop any of the natural resources of the country, in so far as these matters are not directly undertaken by the Administration. Any such arrangements shall provide that no profits distributed by such agency, directly or indirectly, shall exceed a reasonable rate of interest on the capital, and any further profits shall be utilized by it for the benefit of the country in a manner approved by the Administration.

Article 12—The Mandatory shall be entrusted with the control of the foreign relations of Palestine and the right to issues exequaturs to consuls appointed by foreign powers. He shall also be entitled to afford diplomatic and consular protection to citizens of Palestine when outside its territorial limits.

Article 13—All responsibility in connection with the Holy Places and religious buildings or sites in Palestine, including that of preserving existing rights and of securing free access to the Holy Places, religious buildings, and sites and the free exercise of worship, while ensuring the requirements of public order and decorum, is assumed by the Mandatory, who shall be responsible solely to the League of Nations in all matters connected herewith, provided that nothing in this article shall prevent the Mandatory from entering into such arrangements as he may deem reasonable with the Administration for the purpose of carrying the provisions of this article into effect; and provided also that nothing in this mandate shall be construed as conferring upon the Mandatory authority to interfere with the fabric or the management of purely Moslem sacred shrines, the immunities of which are guaranteed.

Article 14—A special Commission shall be appointed by the Mandatory to study, define, and determine the rights and claims in connection with the Holy Places and the rights and claims relating to the different religious communities in Palestine. The method of nomination, the composition, and the functions of this Commission shall be submitted to the Council of the League for its approval, and the Commission shall not be appointed or enter upon its functions without the approval of the Council.

Article 15—The Mandatory shall see that complete free of conscience and the free exercise of all forms of worship, subject only to the maintenance of public order and morals, are ensured to all. No discrimination of any kind shall be made between the inhabitants of Palestine on the ground of race, religion, or language. No person shall be excluded from Palestine on the sole ground of his religious belief.

The right of each community to maintain its own schools for the education of its own members in its own language, while conforming to such educational requirements of a general nature as the Administration may impose, shall not be denied or impaired.

Article 16—The Mandatory shall be responsible for exercising such supervision over religious or eleemosynary bodies of all faiths in Palestine as may be required for the maintenance of public order and good government. Subject to such supervision, no measures shall be taken in Palestine to obstruct or interfere with the enterprise of such bodies or to discriminate against any representative or member of them on the ground of his religion or nationality.

Article 17—The Administration of Palestine may organize on a voluntary basis the forces necessary for the preservation of peace and order, and also for the defense of the country, subject, however, to the supervision of the Mandatory, but shall not use them for purposes other than those above specified save with the consent of the Mandatory. Except for such purposes, no military, naval, or air forces shall be raised or maintained by the Administration of Palestine.

Nothing in this article shall preclude the Administration of Palestine from contributing to the cost of the maintenance of the forces of the Mandatory in Palestine.

The Mandatory shall be entitled at all times to use the roads, railways, and ports of Palestine for the movement of armed forces and the carriage of fuel and supplies.

Article 18—The Mandatory shall see that there is no discrimination in Palestine against the nationals of any State Member of the League of Nations (including companies incorporated under its laws) as compared with those of the Mandatory or of any foreign state in matters concerning taxation, commerce or navigation, the exercise of industries or professions, or in the treatment of matching vessels or civil aircraft. Similarly, there shall be no discrimination in Palestine against goods originating in or destined for any of the said states, and there shall be freedom of transit under equitable conditions across the mandated area.

Subject as aforesaid and to the other provisions of this mandate, the Administration of Palestine may, on the advice of the Mandatory, impose such taxes and customs duties as it may consider necessary, and take such steps as it may think best to promote the development of the natural resources of the country and to safeguard the interests of the population. It may also, on the advice of the Mandatory, conclude a special customs agreement with any state the territory of which in 1914 was wholly included in Asiatic Turkey or Arabia.

Article 19—The Mandatory shall adhere on behalf of the Administration of Palestine to any general international conventions already existing, or which may be concluded hereafter with the approval of the League of Nations, respecting the slave traffic, the traffic in arms and ammunition, or the traffic in drugs, or relating to commercial equality, freedom of transit and navigation, aerial navigation and postal, telegraphic, and wireless communication of literary, artistic, or industrial property.

Article 20—The Mandatory shall cooperate on behalf of the Administration of Palestine, so far as religious, social, and other conditions may permit, in the execution of any common policy adopted by the League of Nations for preventing and combating disease, including diseases of plants and animals.

Article 21—The Mandatory shall secure the enactment within 12 months from this date, and shall ensure the execution of a Law of Antiquities based on the following rules. This law shall ensure equality of treatment in the matter of excavations and archaeological research to the nationals of all State Members of the League of Nations; (1) "Antiquity" means any construction or any product of human activity earlier than the year 1700; (2) The law for the protection of antiquities shall proceed by encouragement rather than by threat. Any person who, having discovered an antiquity without being furnished with the authorization referred to in paragraph 5, reports the same to an official of the competent department, shall be rewarded according to the value of the discovery; (3) No antiquity may be disposed of except to the competent department, unless this department renounces the acquisition of any such antiquity. No antiquity may leave the country without an export licence from the said department; (4) Any person who maliciously or negligently destroys or damages an antiquity shall be liable to a penalty to be fixed; (5) No clearing of ground or digging with the object of finding antiquities shall be permitted, under penalty of fine, except to persons authorised by the competent department; (6) Equitable terms shall be fixed for expropriation, temporary or permanent, of lands which might be of historical or archaeological interest; (7) Authorization to excavate shall only be granted to persons who show sufficient guarantees of archaeological experience. The Administration of Palestine shall not, in granting these authorizations, act in such a way as to exclude scholars of any nation without good grounds; (8) The proceeds of excavations may be divided between the excavator and the competent Department in a proportion fixed by that Department. If division seems impossible for scientific reasons, the excavator shall receive a fair indemnity in lieu of a part of the find.

Article 22—English, Arabic, and Hebrew shall be the official languages of Palestine. Any statement or inscription in Arabic on stamps or money in Palestine shall be repeated in Hebrew and any state or inscription in Hebrew shall be repeated in Arabic.

Article 23—The Administration of Palestine shall recognize the holy days of the respective communities in Palestine as legal days of rest for the members of such communities.

Article 24—The Mandatory shall make to the Council of the League of Nations and annual report to the satisfaction of the Council as to the measures taken during the year to carry out the provisions of the mandate. Copies of all laws and regulations promulgated or issued during the year shall be communicated with the report.

Article 25—In the territories lying between the Jordan and the eastern boundary of Palestine as ultimately determined, the Mandatory shall be entitled, with the consent of the Council of the League of Nations, to postpone or withhold application of such provisions of this mandate as he may consider inapplicable to the existing local conditions, and to make such provision for the administration of the territories as he may consider suitable to those conditions, provided that no action shall be taken which is inconsistent with the provisions of Articles 15, 16, and 18.

Article 26—The Mandatory agrees that, if any dispute whatever should arise between the Mandatory and another Member of the League of Nations relating to the interpretation or the application of the provisions of the mandate, such dispute, if it cannot be settled by negotiation, shall be submitted to the Permanent Court of International Justice provided for by Article 14 of the Covenant of the League of Nations.

Article 27—The consent of the Council of the League of Nations is required for any modification of the terms of this mandate.

Article 28—In the event of the termination of the mandate hereby conferred upon the Mandatory, the Council of the League of Nations shall make such arrangements as may be deemed necessary for safeguarding in perpetuity, under the guarantee of the League, the rights secured by Articles 13 and 14, and shall use its influence for securing, under the guarantee of the League, that the Government of Palestine will fully honor the financial obligations legitimately incurred by the Administration of Palestine during the period of the mandate, including the rights of public servants to pensions or gratuities.

The present instrument shall be deposited in original in the archives of the League of Nations and certified copies shall be forwarded by the Secretary-General of the League of Nations to all Members of the League.

Done at London the twenty-fourth day of July, one thousand nine hundred and twenty-two.

SECRETARY-GENERAL

See PLO Charter, San Remo Agreement, United Nations.

Lebanese Pound: The currency of Lebanon.

Lebanon: Officially the Republic of Lebanon. The Israeli-Lebanese peace talks differ from those between Israel and its other Arab neighbors in that neither Israel nor Lebanon has a claim on the other's territory.

For over 35 years Lebanon has been torn by ethnic rivalry and civil war. During her 1975 civil war the Syrian army entered Lebanon and effectively seized control of Lebanese territory and the central government's decision- making powers. No major foreign or domestic policy decision is made by the Lebanese government without consultation with Damascus.

On March 19, 1978 the U.N. Security Council adopted Resolution 425, which calls for "strict respect for the territorial integrity, sovereignty and political independence of Lebanon and for a withdrawal of Israeli forces from Lebanese territory. It reads:

The Security Council,

Taking note of the letters of the Permanent Representative of Lebanon (S/12600 and S/12606) and the Permanent Representative of Israel (S/12607),

Having heard the statements of the Permanent Representatives of Lebanon and Israel,

Gravely concerned at the deterioration of the situation in the Middle East, and its consequences to the maintenance of international peace.

Convinced that the present situation impedes the achievement of a just peace in the Middle East.

1. Calls for strict respect for the territorial integrity, sovereignty, and political independence of Lebanon within its internationally recognized boundaries.
2. Calls upon Israel immediately to cease its military action against Lebanese territorial integrity and withdraw forthwith its forces from all Lebanese territory;
3. Decides, in the light of the request of the government of Lebanon, to establish immediately under its authority a United Nations

interim force for southern Lebanon for the purpose of confirming the withdrawal of Israeli forces, restoring international peace and security, and assisting the government of Lebanon in ensuring the return of its effective authority in the area, the force to be composed of personnel drawn from State Members of the United Nations;

4. Requests the Secretary-General to report to the Council within 24 hours on the implementation of this resolution.

The October 1989 Taif Agreement and the May 1991 "Treaty of Brotherhood, Cooperation and Coordination between Syria and Lebanon" have severely limited Lebanese independence from Syria.

On April 26, 1996 an Agreement of Understanding between Israel and Lebanon terminated 16 days of cross-border fighting.

Prime Minister Rafiq Hariri, who took office in October 1992 after Lebanon emerged from 17 years of civil war, has embarked on an ambitious path toward Lebanon's reconstruction. The government's efforts have fostered a resurgence of investment and commercial activity. Foreign business interest in Lebanon has grown rapidly, particularly since 1993.

The program to rebuild Lebanon's infrastructure provides significant opportunities for medium- and long-term involvement in projects worth more than $10 billion over the next 10 years. Lebanon has a traditionally competitive, open-market commercial orientation and the most liberal banking regime in the Middle East. There are virtually no restrictions on foreign exchange or capital movement. The Lebanese pound has gained value gradually and continuously against the U.S. dollar since mid-1993.

According to Lebanese customs data, Lebanon's imports were worth $4.8 billion in 1993. Italy, Lebanon's most important supplier, has 12.7 percent of the market. The United States is second (10.6%). Lebanon is traditionally a key transit country for goods moving between Eastern Europe and nearby Arab countries.

With Lebanon, in 1997, remaining under Syrian control, it is still vulnerable to the increasing problematic relations between Damascus and Jerusalem. A 45 percent and a current account deficiency of half of GDP has threatened the stability of the Lebanese pound.

In addition, the Beirut stock market remains unable to attract foreign investment. Turnover is low. Too much domestic money is pursuing too few stocks leading to high valuations, and there seems to be little positive potential based on current prices.

It was announced in February 1997 that Hezballah guerrillas in Lebanon received 30 planeloads of arms shipments from Iran over the previous nine months. These arms, delivered through Syria, not only signal the continuing desire of Iran to export its Islamic Revolution, but further dampened the prospects of a resolution to the conflict in Lebanon.

See Agreement of Understanding between Israel and Lebanon, HizbAllah; Hrawi, Elias; Syria, terrorism, water.

Lebanon-Israel Agreement of April 26, 1996: See Agreement of Understanding between Israel and Lebanon.

legal issues: See Interim Agreement (on the West Bank and the Gaza Strip).

Legislation Subcommittee: See Gaza Strip and the Jericho Area Agreement (Article VII).

legislative powers: See Gaza Strip and the Jericho Area Agreement (Article VII).

Lehi: See Arab-Israeli War (I), Hebrew Insurrection Movement.

Letters of Mutual Recognition: In the first steps toward a final settlement of the Palestinian-Israeli conflict, Israel and the PLO signed letters of mutual recognition on September 9, 1993. Official recognition formalized the new Israeli-PLO relationship (which unofficially began during secret negotiations in Norway), laying the foundation for the signing of the Declaration of Principles.

In his letter to Israeli Prime Minister Yitzhak Rabin, PLO Chairman Yasser Arafat recognized Israel's right to exist "in peace and security." Citing the inauguration of "a new epoch of peaceful coexistence, free from violence and all other acts which endanger peace and stability," the PLO renounced "the use of terrorism and other acts of violence" and promised to ensure the compliance of all PLO "elements and personnel." Arafat also pledged to submit the PLO Covenant to the Palestine National Council to invalidate the articles in the Covenant which deny Israel's right to exist. In a supplementary letter addressed to Norwegian Foreign Minister Johan Jorgen Holst, Arafat called upon West Bank and Gaza Palestinians to reject terrorism, and to instead aid in the normalization of the territories by "participating actively in shaping reconstruction, economic development and cooperation."

In his brief response to Arafat's letter, Rabin confirmed that "in light of the PLO commitments included in your letter, the Government of Israel has decided to recognize the PLO as the representative of the Pal-

estinian people and commence negotiations with the PLO within the Middle East peace process."
See Declaration of Principles.

Liaison Bureau: See Cairo Agreement.

Liaison Committee of the Donors: See Ad Hoc Liaison Committee of the Donors.

Libya: In December 1951 Libya became the first independent country created by a resolution of the United Nation's General Assembly. Colonel Muammar al-Qadhafi (Gaddafi) became Libya's leader in September 1969. Lybia joined the Maghreb Union in 1989. Qadhafi also condemned the occupation of Kuwait by Iraq in August 1990.
See Iran and Libya Sanctions Act; Qaddafi, Muammar; terrorism.

Libyan Dinar: The currency of Libya Arab Jamahiriya.

life expectancy: See health.

Lifton, Robert K.: See Council on Foreign Relations.

Likud Party (Israel): Formed in 1973 following the Arab-Israeli War. In 1996 Likud's Benjamin Netanyahu became Prime Minister of Israel vowing to increase Jewish settlements in the West Bank and Gaza, never the separate a unified Jersualem. Prime Minister Netanyahu refused to meet with PLO Chairman Arafat, that meeting finally taking place on September 4, 1996.
See Netanyahu, Binyamin; Peres, Shimon.

Lions Club: See Hamas Covenant.

literacy: Literacy rates vary significantly throughout the MENA region. For example, in 1993 Israel's literacy rate was 95.0 percent, Jordan's 83.9 percent, Syria's 76.7 percent, Saudi Arabia's 60.6 percent, Iraq's 54.6 percent, and Egypt's 49.1 percent.
See Education (Center of the Middle East/North Africa Community).

livestock: See Regional Animal Health Cooperation Program.

living standards: See growth (MENA).

lobbying: Lobbying inevitably will become a major byproduct in the MEC/MENAC. Thousands of lobbyists, representing thousands of firms and trade associations will busily seek access to thousands of bureaucrats, parliamentarians, and commissioners. The petitioners will have a tremendous motive to bend the deliberations toward their views, or to the views of their sponsors, with much to gain or lose, depending on how an economic accord develops.

See dedication, greed, politicians.

logistic trade centers: To be considered in the evolution of MENA, logistic trade centers facilitate trade flow. These centers include loading and unloading facilities, storage and cold storage facilities, packaging and custom brokerage services.

They could be erected in locations with a relatively large flow of goods and people, close to border terminals and main traffic routes. A precondition for the development of these trade centers is easy procedures in border crossing. Some elements of cooperation in this sphere include:

- unification of customs declaration documents.
- unification of working hours of customs and border police officials.
- consolidation of border checkpoints for one-stop passage.
- utilization of local tracks by trains registered in other countries.
- unification of trucking standards.
- reciprocal recognition of insurance policies.

Special attention should be paid to administrative barriers to trade flow with a view to eliminating them, and by so doing pave the way for future economic cooperation. The successful implementation of logistic trade centers should yield the following benefits:

- Strengthen the cooperative spirit between various nations.
- Reduce total travel time, and with it, increase driver and vehicle productivity.
- Reduce required inventory levels resultant from the reduction in turn-around time.
- Enhance fair competition among trucking companies.
- Increase opportunities for greater economic integration, especially for agricultural produce, fish, and other perishable products.
- Increase the flow of tourists.

London Agreement: On April 11, 1987 Shimon Peres of Israel and King Hussein of Jordan met in London and agreed to the following statement which remains subject to approval of the respective governments of Israel and Jordan:

A Three-Part Understanding Between Jordan and Israel

A. Invitation by U.N. Secretary-General.
B. Resolutions of the international conference.
C. The modalities agreed upon by Jordan-Israel.

A. The Secretary-General will issue invitations to the five permanent members of the Security Council and the parties involved in the Arab-Israeli conflict in order to negotiate a peaceful settlement based on Resolutions 242 and 338, with the objects of bringing a comprehensive peace to the area, security to its states, and to respond to the legitimate rights of the Palestinian people.

B. The participants in the conference agree that the purpose of the negotiations is the peaceful solution of the Arab-Israeli conflict based on Resolutions 242 and 338 and a peaceful solution of the Palestinian problem in all its aspects. The conference invites the parties to form geographical bilateral committees to negotiate mutual issues.

C. Jordan and Israel have agreed that:

(1) The international conference will not impose any solution or veto any agreement arrived at between the parties.
(2) The negotiations will be conducted in bilateral committees directly.
(3) The Palestinian issue will be dealt with in the committee of the Jordanian-Palestinian and Israeli delegations.
(4) The Palestinians' representatives will be included in the Jordanian-Palestinian delegation.
(5) Participation in the conference will be based on the parties' acceptance of Resolutions 242 and 338 and the reunification of violence and terrorism.
(6) Each committee will negotiate independently.
(7) Other issues will be decided by mutual agreement between Jordan and Israel.

See Resolution 242, Resolution 338.

lowest park on earth: See parks development.

LYD: The international standard code for the currency of Libyan Arab Jamahiriya—the Libyan dinar.

M

MAD: The international standard code for the currency of Morocco and Western Sahara—the Moroccan dirham.

Madrid Conference: Middle East peace negotiations were carried out within the structure of the Madrid Framework. This framework, described in the Letter of Invitation to the Madrid Conference, was the product of intensive diplomatic efforts carried out in the region immediately following the Gulf War. The framework consisted of three basic elements:

The Opening Conference

The Madrid Invitation called for the convening of an opening conference, in order to inaugurate two separate but parallel negotiating tracks—the bilateral track and the multilateral track. The conference was designed to serve as an opening forum for all the participants, having no power to impose solutions or veto arrangements. The Madrid Conference, cosponsored by the United States and Russia (then the USSR), opening on October 30, 1991 and lasting two days. This conference could only be reconvened with the agreement of all participants.

The Bilateral Track

The bilateral negotiations were meant to resolve the conflicts of the past. The firstever direct talks between Israel, Syria, Lebanon, Jordan, and the Palestinians opened in Madrid on November 3, 1991, immediately following the Madrid Conference. These talks were, in fact, four separate sets of bilateral negotiations. While the talks with the three Arab states were aimed at achieving peace treaties, the negotiations between Israel and the Palestinians were based on a two-stage formula. In the first stage the Israeli-Palestinian talks were to focus on achieving agreement regarding five-year interim self-government arrangements.

In the second phase negotiations were to concentrate on the permanent status issues. Over a dozen formal rounds of bilateral talks were hosted by the U.S. Department of State in Washington.

The Multilateral Track

The multilateral negotiations were meant to build the Middle East of the future while generating confidence among the regional parties. These talks, which opened in Moscow in January 1992, included five separate forums attended by delegations from countries in the region as well as representatives of the international community. The negotiations focused on key issues that concerned the entire Middle East—water, environment, arms control, refugees, and economic development. The talks took place in working groups that met periodically in venues throughout the world, and later, throughout the Middle East.

See Amman Middle East/North Africa Economic Summit, bilateral negotiations, Casablanca Middle East/North Africa Economic Summit, chronology (October 30, 1991), Copenhagen Action Plan, Declaration of Principles, Economic Development, Gaza Strip and the Jericho Area Agreement, Gulf crisis, Hussein (His Majesty King Hussein Bin Talal), Israel-Jordan Common Agenda, multilateral negotiations, Oslo Agreement, security and arms control; Shamir, Yitzhak; Syria, United Nations (Middle East Peace Process), Washington Declaration, water.

Madrid Framework: See Madrid Conference.

Madrid Invitation: See Madrid Conference.

Madrid Peace Conference: See Madrid Conference.

Maghreb: The countries of Morocco, Algeria, and Tunisia. See European Union, Union du Maghreb Arabe, Hassan II (King).

Maghreb Agreement: A trade agreement signed with Algeria, Morocco, and Tunisia in April 1976. Libya and Mauritania would join later.

See Tunisia, Union du Maghreb Arabe.

Maghreb Common Market: See MCM.

Maghreb Union: A trade agreement was signed with Algeria, Morocco, and Tunisia in April 1976 to take effect July 1, 1976. For each nation it meant that there would be duty-free entry of industrial items into the European Union nations except for certain petroleum items and cork.

Some tariff concessions were also given on agricultural goods. Libya and Mauritania joined in February 1989.
See Arab (inter-trade), Hassan II (King).

Maguid, Abdel: See Casablanca Middle East/North Africa Economic Summit.

Mahajir: See PLO Charter.

Mahdi, Sadiq al-: See Omar al Bashir (President).

Majali, Abdul Salam: Prime Minister of Jordan and Israeli Prime Minister Yitzhak Rabin signed a draft Israeli-Jordanian peace treaty in Amman on October 17, 1994.
See Israel-Jordan Peace Treaty, Washington Declaration.

Makhpelah: Hebrew name for the Tomb of the Patriarchs.
See Tomb of the Patriarchs.

Mandate Document: See League of Nations Mandate for Palestine, PLO Charter.

Mandate (League of Nations): See Israel (Declaration of Establishment), League of Nations Mandate for Palestine.

Mandatory for Palestine: See League of Nations Mandate for Palestine.

Mandela, Nelson: See chronology of the Middle East Peace Process (May 10, 1994).

marine transport: See shipping, transport.

Marshall Plan: Officially called the European Recovery Program, a model for U.S. aid to regional areas, has recently been used as a means for promoting peace in the Middle East. It encompassed large-scale American financial aid in the form of grants and loans to the European nations following World War II.
See Casablanca Declaration, Declaration of Principles.

Mashrek (Mashrag) states: Egypt, Jordan, Lebanon, and Syria. These nations signed trade and cooperative agreements in January and February 1977 and thus completed the European Community Mediterranean Policy.
See European Union.

Mauritania: See Ouguiya, Union du Magreb Arabe.

Mazen, Abu: See Beilin-Abu Mazen agreement, Beilin-Eitan agreement.

MCM (Mahgreb Common Market): A multilateral agreement signed in Marakesh, Morocco, in 1989 for the economic trade integration of the Arab nations of the region of North Africa. Member nations are Algeria, Libya, Mauritania, Morocco, and Tunisia.

Synonymous with Arab Mahgreb Union (AMU).

MEC: See Middle East Community.

MED-Invest Program: See European Union.

MEDA-Democracy program: See Euro-Mediterranean Stability Charter.

Mediterranean Basin: As defined by the European Union, the nations of Algeria, Cyprus, Egypt, Jordan, Lebanon, Libya, Morocco, Syria, Tunisia, and ex-Yugoslavia.

Mediterranean Crans Montana Forum: Sponsored The Euro-Mediterranean Space meeting of October 12-15, 1995 in Malta to discuss relationships between nations of the Mediterranean Sea boundaries.

Mediterranean global approach: See Mediterranean Policy.

Mediterranean Noncandidates: See MNCs.

Mediterranean Policy: A European Union policy purporting to create a global relationship with the nations of the Mediterranean basin (Algeria, Morocco, Tunisia, Egypt, Jordan, Libya, Syria, and Israel).

In 1972 the European Community announced a "Mediterranean global approach" that gave them a common framework. Agreements implementing free entry for industrial goods were drawn up with all the Mediterranean nations, except Albania and Libya. The Mediterranean countries were granted duty-free access to the union for their industrial products, various customs concessions for agricultural produce, and European investment bank loans. The Mediterranean countries constitute the union's third largest customer and its fourth largest supplier.

See European Union, Mashrag states, MNCs.

Mediterranean region: See New Middle East, The.

Mediterranean Third Countries (MTCs): A European Union designation for nonmember nations surrounding the Mediterranean Sea.

MEESG: See Middle East/North Africa Economic Strategy Group.

MEFTA: See Middle East Free Trade Agreement.

MENA: See Middle East/North Africa.

MENA I: See Casablanca Middle East/North Africa Economic Summit.

MENA II: See Amman Middle East/North Africa Economic Summit.

MENA III: See Cairo Middle East/North Africa Economic Conference.

MENA IV: See Doha Middle East/North Africa Economic Summit.

MENAC: See Middle East/North Africa Community.

MENAFTA: See Middle East/North Africa Free Trade Agreement.

MENAnet: See Secretariat (Middle East/North Africa Economic Summits).

MENA Series: See Middle East/North Africa Series.

MEPFA: See Middle East Peace Facilitation Act.

Michaelberg Operation: The secret airlift of 100 Iraqi Jews to Israel in the summer of 1947. Problems with legal and illegal immigration from Iraq led to the formulation of several unique plans, one of which was to send a boat to the port of Batzra that would bring new immigrants to Eilat, where they would then cross the desert on camels. Under another plan two American pilots who owned two-engine planes offered to airlift the Iraqi Jews to Israel behind the backs of the British authorities. An improvised takeoff and landing strip was prepared near a Galilee settlement for the mission. Escorted by Israeli intelligence agents, the plane touched down on a landing strip beside the international airport in Bagdad. Both flights were successful, bringing 100 illegal immigrants to the Land of Israel.

Middle East: The traditional description includes: Bahrain, Egypt, Iran, Iraq, Israel, Jordan, Kuwait, Lebanon, Oman, Qatar, Saudi Arabia,

Syria, the United Arab Emirates, and Yemen—an emerging market with a combined gross domestic product of $987 billion in 1996, and a population of 222 million.

A further breakdown forms the Gulf states—Bahrain, Kuwait, Qatar, Oman, Saudi Arabia, and the United Arab Emirates with 26 million people and a collective gross domestic product of $300 billion.

See Gulf states.

Middle East and North Africa (economies): See Middle East/North Africa.

Middle East and North Africa Business International Group (BIG): As stated in the Casablanca Declaration, a Middle East and North Africa Business Interaction Group (BIG) was established by the World Economic Forum. The group's mandate was to propose structures that will encourage interaction among businesses and between business and government in order to capitalize on the summit's ground-breaking work.

A major objective of the summit was, in the words of Shimon Peres, "to privatize the peace...governments in the region have reached out to each other; it is now time for the people to reach out to each other." To this end, BIG purports to facilitate private sector contacts. It focuses international attention on opportunities and issues in the region.

Another major objective of the summit was to develop the partnership between business and government in the Middle East and North Africa. Business, both inside and outside the region, has an important role to play in advancing economic adjustment and structural reform policies. To facilitate this process, BIG provides regular recommendations to the steering committee created by the summit, which is comprised of representatives from governments in the region. The recommendations are based on the advise of board-level executives selected in careful geographical balance from the forum's 1,000 member companies.

Being nonprofit, nongovernmental, and impartial like the forum, BIG will have no other agenda than to encourage the region's long-term economic development and will be a mechanism to help realize the potential of MENA by institutionalizing the new regional cooperation.

In a series of private meetings held during the Cairo Conference in November 1996, BIG put forth suggestions for a blueprint for action:

A. On the political front, the business community wants MENA governments to:
   - accelerate the peace process with a view to achieving a just, comprehensive, and lasting peace in the region;

- involve the private sector much more in decisions that affect investments;
- put forth the necessary conditions to stimulate investment, trade, and business links;
- maintain the search for deeper political stability, democratization, and transparency in government;
- streamline bureaucracies and government decision-making procedures;
- take stronger measures against red tape and investment impediments, factors that influence business investment decisions worldwide;
- work toward accelerating privatization, deregulation, and foreign ownership; and
- concentrate on those regional integrated development projects, for example, Taba-Eilat-Aqaba and the southeast Mediterranean, which have a clear commercial chance of attracting both foreign and local private capital.

B. On the macroeconomic front, the business community pays tribute to:
- the beginning of real change in Egypt, Jordan, Morocco, Tunisia, and other Gulf economies in areas including trade liberalization, fiscal and monetary discipline, privatization, reforming the regulatory framework, and creating a favorable business environment;
- policies adopted in several MENA countries aiming at accelerating growth rates;
- the host government Egypt on its macroeconomic policy reform revolution.

C. On the trade front, the business community urged governments to:
- promote inter- and intra-regional trade and global integration on the basis of free flow of goods, capital, and labor;
- speed up liberaliation, abolish tariff, and nontariff barriers, and open domestic markets to competition;
- create regional infrastructure projects to promote trade;
- support the creation of the MENA development bank;
- adopt internationally commercial standards; and
- enhance trade financing facilities.

D. In the area of investment promotion, additional measures are still needed to attract the type of private capital necessary to generate jobs and boost living standards. Thus, governments should take the following measures, where they have not yet been undertaken:
- eliminate all laws and regulations that ban foreign investment and foreign ownership;

- remove restrictions on the free flow of capital, profits, and dividends;
- promote open and free capital markets with independent regulatory bodies, sounder legal frameworks, and stronger rules of disclosure;
- ensure applying national treatment to foreign concerns and enterprises;
- enforce universally sound and fair business practices;
- require that business transactions be governed by full transparency and accountability;
- create mechanisms to shield business from political risk; and
- lift controls on prices and introduce competition policies.

The business community underlined the extent to which politics and diplomacy have a direct impact on the realization of business opportunities and overcoming investment and business obstacles.

BIG called on both the public and private sector not only to concentrate on large projects, but to take steps as well to foster smaller and more realistic commercial ventures and deals.

BIG pledges to continue to seek new opportunities and ways of supporting the Middle East Peace Process, as well as cementing comprehensive peace by joint ventures and investments in infrastructure, tourism, energy, and communications—sectors that provide the foundation for economic development and which bind people in the region in an interdependent relationship.

See Amman Middle East/North Africa Economic Summit, Cairo Middle/East North Africa Economic Conference, Casablanca Middle East/North Africa Economic Summit, World Economic Forum.

Middle East and North Africa Development Bank: See Amman Middle East/North Africa Economic Summit, Middle East Development Bank.

Middle East Commonwealth: See Hussein (His Majesty King Hussein Bin Talal).

Middle East Community (MEC): Following stages of JIPTA and MEFTA, the Middle East Community will be launched from the mandatory economic relationship (then hopefully, but not necessarily, social and political ties). Although an institutional structure can commence immediately, it is best to await or parallel the establishment of a systematic trade agreement with at least three bodies, such as Jordan-Israel-Palestinians. Once multiple units are involved in creating an effective model for the free movement of capital, goods, and services, a mechanism for governance and regulation is required. The fulfillment of a dream to realize a proper and healthy community must entertain conditions and protections for the fourth freedom—the free movement of people.

See Centers of the Middle East/North Africa Community, Jordan-Israel-Palestinian Trade Agreement, Middle East/North Africa Community; New Middle East, The.

Middle East Development Bank: See Amman Middle East/North Africa Economic Summit, Casablanca Declaration, Casablanca Middle East/North Africa Economic Summit, Declaration of Principles.

Middle East Development Fund: See Declaration of Principles.

Middle East Economic Strategy Group: See Middle East/North Africa Economic Strategy Group.

Middle East Environmental Code of Conduct: See Moscow Multilateral Middle East Conference.

Middle East Free Trade Agreement (MEFTA): Evolving from JIPTA, the ascendence of other contiguous nations to Israel—Lebanon, Syria, and Egypt—will form the Middle East Free Trade Agreement.

To be formed once three or more national bodies within the Middle East region establish trade accords with each other, it would encourage and administer the free movement of capital, goods, and services, but not people, the latter reserved for the creation of a Middle East Community.

See Amman Middle East/North Africa Economic Summit, Casablanca Middle East/North Africa Economic Summit, Jordan-Israel-Palestinian Trade Agreement.

Middle East-Mediterranean Travel and Tourism Association: See Amman Middle East/North Africa Economic Summit.

Middle East/North Africa (MENA): In 1995 The World Bank reported:

With respect to GDP per capita growth, MENA has had two distinct experiences during the past three decades. In the 1960-1985 period it grew faster than all other regions of the developing world with the exception of East Asia. But during the last decade per capital income has actually declined by 2 percent a year, the worst record of any region in the world.

On the positive side, many social indicators improved dramatically. Infant mortality dropped by half and life expectancy increased by 10 years. The number of children attending primary school increased to 98 percent in 1991 from 61 percent at the beginning of the period. Correspondingly, literacy reached 53 percent in 1990, an increase of 56 percent over the 1970 figures. Somewhat surprisingly, MENA was more successful at addressing poverty than other regions. Thus, only 5.6 percent live on less than $1 per day in 1990 as compared to 14.7 percent in East Asia and 28.8 percent in Latin America.

Despite this positive record, there are also underlying themes that led to the economic stagnation of the last decade:

- Rather than being focused on development, political energies were too frequently consumed by regional conflict and rivalries.
- Oil wealth and other investments were frequently expended on activities that had low social and/or economic returns.
- Some investments in health and education helped the rich rather than the poor, for example, disproportionate funds were used for tertiary health care rather than primary care and the environment.
- The two decades ending in 1985 were easier times: rents from natural resources (primarily petroleum) were, while fluctuating, generally increasing, and the industrial world was a less competitive place.

Turning to the present, two objective facts that have severely impacted economic growth stand out: First, the real price of oil declined by about 50 percent during 1980-1985 and has remained essentially stagnant since then. Second, productivity of capital and labor has declined for the following reasons:

a. the inflexibility of enterprises and labor. Whether owned by the state or private individuals, industry that is dependent on high tariffs and state regulation cannot respond to the fiercely competitive world;
b. the old social contract needs to be reassessed. Governments increasingly cannot deliver on the contract, but also the arrangement produces an inflexible and less-productive labor poor;
c. subsidies and transfers, which are quite common, result in the inefficient use of resources.

The World Bank prescription for changing this picture, for stimulating economic growth, is straightforward:

- Maintain macroeconomic stability;
- Promote non-oil exports: determine which private industries and services have comparative advantages, and help them with favorable policy and institutional environments;
- Make the private sector more efficient: improve communications and physical infrastructure, eliminate stultifying regulations, encourage

banks to make credit available at lower rates—the difference between banks' lending and deposit rates are generally 9 percent in MENA as compared to 3-4 percent in Asia and Europe;
- Improve human resources in terms of more skilled and flexible workers: Enhance the quality of primary and secondary schools and make them truly universal; also graduate more technically trained workers demanded by private industry;
- Reduce poverty through faster growth. In addition, rather than subsidizing consumption items for whole populations as is not the practice in some states, target such subsidizes for the truly poor or near poor.

See World Bank.

Middle East/North Africa Business Home Page: See United States (Department of Commerce).

Middle East/North Africa Business Studies (Center for): Established in March 1996 at Rutgers University, Graduate School of Management. The center promotes interdisciplinary organizational research on cross-border activities of trade, investment, foreign direct investment, management education and training, and other business interactions among people from nations of the Middle East and North Africa.

The center's main activities are concerned with:

- publications
- conferences
- lectures and seminars
- research and links with other institutions
- teaching.

The center works closely with agencies established by the Middle East/North Africa Economic summits, for example, the Secretariat in Rabat, Morocco; the Regional Bank in Cairo, Egypt; the Regional Development Working Group (REDWG) in Amman, Jordan.

Middle East/North Africa Community (MENAC): The final stage of evolution, following JIPTA, MEFTA, MEC, MENAFTA will be MENAC. The ultimate goal that will parallel MEC to include the free movement of people. A policy for MENAC means both a policy for peace, and the creation of the community which will simultaneously evolve the centerpiece for a framework of peace in the region that will render any war between participating countries impossible.

Only in the context of MENAC can a stable economic order throughout the region be created, and only through joint community efforts of an international economic policy will the performance of the regional nation economies occur that will also contribute to social justice in the nations of the region. Without internal cohesion, nations of MENAC cannot assert their political and economic independence from the rest of the world, win back influence in the globe, and retrieve a proper place in the brotherhood of nations.

See Casablanca Conference, Centers of the Middle East/North Africa Community, Middle East Community.

Compre New Middle East, The.

Middle East/North Africa Economic Community: See Amman Middle East/North Africa Economic Summit, Cairo Middle East/North Africa Economic Conference, Casablanca Middle East/North Africa Economic Summit, Middle East/North Africa Business Studies (Center for), Middle East/North Africa Community.

Middle East/North Africa Economic Strategy Group (MEESG): One result of the Casablanca Middle East/North Africa Economic Summit was the creation of MEESG. The chair of this group was Paul Volcker, who said:

> From all reports, the Middle East/North Africa Economic Summit represented a historic milestone, bringing ancient rivals and distant neighbors into constructive discussion. The groundwork is being laid for a new partnership between government and business across the region to bring economic prosperity to its peoples. To sustain and build upon the momentum generated by the Summit, the Council on Foreign Relations has created the Middle East/North Africa Economic Strategy Group (MEESG).
>
> The MEESG is a nonprofit and nongovernmental advisory group composed of American, Middle Eastern, European and Japanese experts in international finance, trade and development. Within the context of political and security concerns, this private sector group will recommend strategies for regional economic cooperation and to overcome obstacles to trade and private investment. The MEESG will take a broad perspective in its recommendations for making the Middle East more attractive for investment and for government and private sector cooperation.
>
> Among the issues it will examine are proposals for regional trade regimes, regional financial institutions, and both Western and Middle Eastern governmental policies to facilitate economic development while enhancing the peace process and regional security. The group will assess foreign aid flows and their impact. It will look at the relation between private sector wealth and government debt and the relations between rich and poor states in the region.
>
> MEESG recommendations will be channeled through the Steering Committee and Secretariat created by the Casablanca Economic Summit, as well as through the Steering Committee of the Multilateral Negotiations.
>
> The Group will not represent any particular public or private interest. Consequently, we hope that it can help bridge old rivalries and encourage fresh thinking. Its nongovernmental nature will afford flexibility and encourage thinking about how the resources of government and business can best be utilized. Surely

public-private partnership—fostered by the Casablanca Summit and now continue by MEESG—is one key to firmly anchoring the peace process to sustained economic development.

See Amman Middle East/North Africa Economic Summit, Casablanca Declaration, Casablanca Middle East/North Africa Economic Summit, Centers, Secretariat, Steering Committee, Steering Committee of the Multilateral Negotiations.

Middle East/North Africa Economic Summits: See Amman Middle East/North Africa Economic Summit, Casablanca Middle East/North Africa Economic Summit, Egypt, Middle East/North Africa Business Studies (Center for), Qatar, United States (Department of Commerce).

Compare Cairo Middle East/North Africa Economic Conference.

Middle East/North Africa Free Trade Agreement (MENAFTA): Following stages of JIPTA, MEFTA, and MEC a new level will entitle all Arab nations and Iran to join. Each, when fulfilling requirements of the MEC will apply for membership. In the end, the countries including possibly Turkey (based on her acceptance into the European Union) with a population exceeding 300 million people, will test the forces of the free movement of capital, services, and goods.

As nations beyond the Middle East region agree to participate in trade accords with MEFTA countries, a required change occurs, renaming the authority or secretariat MENAFTA, to recognize inclusion of a non-Middle East member. Each new member will enjoy equal rights of trade, including the free movement of capital, goods, and services, with open movement of people reserved for the creation of a MENAC or joining MEC and signing its respective treaty.

See Amman Middle East/North Africa Economic Summit, Casablanca Middle East/North Africa Economic Summit, Jordan-Israel-Palestinian Trade Agreement, Middle East/North Africa Community.

Middle East/North Africa Series: Publications of the Secretariat of the Middle East/North Africa Economic Summit (Rabat, Morocco). The first book of the series was *The Peace Dividend: Creating a Middle East/North Africa Community*, by Jerry M. Rosenberg.

Middle East Peace and Stability Fund: Announced by President Clinton on June 17, 1997, this U.S. fund will draw on existing allocations of economic assistance to respond to urgent new needs in the region. The 1997, $100 million fund will draw its resources from the redirection of a small percentage of the economic support funds supplied to Israel and Egypt.

Initial focus was on assisting Jordan as it pursues economic modernization and reform. Jordan will receive $50 million annually from Israel to complete a multiyear U.S. program to Jordan of at least $150 million.

Middle East Peace Facilitation Act (MEPFA): MEPFA, a U.S. law allowing American aid to the Palestinian Liberation Organization and its successors, as well as contact with officials of the group, lapsed on August 12, 1997. The expiration of the law meant that the Palestinian Authority's office in Washington, DC, with its staff of eight, was ordered to suspend operations, thus, effectively shutting down the only effective Palestinian presence in the Capital. It also meant that direct U.S. aid to Chairman Arafat's fledging government could end, at least temporarily. However, most of the U.S. $75 million aid goes directly to nongovernmental organizations in Gaza and the West Bank, not the Palestinian Authority.

Middle East Peace Process: Until recently, only Egypt had accepted Israel's offer to negotiate face-to-face. In a bold diplomatic initiative, the late Egyptian President Anwar Sadat accepted Israel's invitation for dialogue, and embarked upon historic bilateral negotiations which led to the 1978 Camp David Accords and the 1979 Israel-Egypt Peace Treaty.

The Gulf War and the breakup of the Soviet Union in the early 1990s reshaped the basic political order of the Middle East, prompting the Arab world to reassess its attitude toward Israel and to enter into negotiations aimed at building a future for the Middle East.

In October 1991 a conference was convened in Madrid to inaugurate direct peace talks. Since then, bilateral negotiations have been conducted between Israel and Syria, Lebanon, Jordan, and the Palestinians. Multilateral talks on key regional issues were also held.

See bilateral negotiations, chronology of the Middle East Peace Process, Madrid negotiations, multilateral negotiations; Sadat, Anwar; United Nations (Middle East Peace Process), United Nations (nuclear-weapon-free zone in the Middle East).

Middle East Peace Process (United Nations): See United Nations (Middle East Peace Process).

Middle East terrorism: See terrorism.

Mideast Free Trade Zone: See Research and Development (Center of the Middle East/North Africa Community).

migrating birds: See birds.

militant Islam: More than any other force, the conflict between militant Islamic groups and the secular authority (dictatorship in most cases) in much of the Arab world may ultimately determine the fate of the Middle East and North Africa. Encouraged by rampant corruption and nepotism, militant groups continue to offer a sense of identity to the frustrated and impoverished, many seeking a cause to fulfill their lives. The community itself will be threatened unless and until it captures the movement in eradicating the seductive forces that bring many into the arms of the fundamentalists. A government with authority is needed, but at the same time we don't want an authoritarian government; balance is not easy to define or find. It may take years to compensate for and overcome the deplorable conditions of the deprived, but the community can only survive after it has satisfied the masses who are hungry, often unemployed, and in general disenfranchised from society.

See fundamentalism, Hamas, Islam, terrorism, United States.

minorities: Each Arab country is a composite of different groups where minorities play a significant role. The largest minority of the total population is as follows:

Country	*1994 Population*	*Largest Minority*	*Size*
Algeria	27.9 mil.	Berber	17.0 %
Egypt	60.8 mil.	Bedouins	n.a.
Iraq	19.9 mil.	Kurds	19.0 %
Jordan	4.0 mil.	Egyptians	n.a.
Lebanon	3.6 mil.	Palestinian	12.0 %
Libya	5.1 mil.	Berber	n.a.
Morocco	28.6 mil.	Berber	30.0 %
Syria	14.9 mil.	Kurds	6.3 %
Saudi Arabia	18.2 mil.	Yemeni	9.6 %
Tunisia	8.7 mil.	Berber	1.2 %

See Christians.

missing persons: See Gaza Strip and the Jericho Area Agreement (Article XIX).

MNCs (Mediterranean Noncandidates): Countries with which the European Union has preferential trade agreements. They are Algeria, Cyprus, Egypt, Israel, Jordan, Lebanon, Malta, Morocco, Syria, Tunisia, Turkey, and Yugoslavia. Albania and Libya have no agreements with the union.

Moderate Territorial Compromise: Put forward by Joseph Alpher of the Jaffee Center for Strategic Studies at Tel Aviv University, Israel would annex 11 percent of the West Bank, but in doing so, would incorporate 70 percent of all settlers. Israel would annex parts of Western Samaria heavily populated with Israelis, and as a result, would move the border eastward. In addition, Israel would annex the Latrun salient near the approaches to Jerusalem, as well as the Greater Jerusalem area. "All remaining territory in the West Bank and Gaza Strip would be turned over to the Palestinian entity. Settlers wishing to remain would be subject to Palestinian authority" (Never implemented).

See settlements (and the Oslo accords). Compare Sharon Plan.

Mohammed, Moulay Hassan Ben: See Hassan II, King.

Monetary Authority: See Protocol on Economic Relations.

monetary issues: See Amman Middle East/North Africa Economic Summit, Protocol on Economic Relations (Article IV).

Monitoring Committee: See Moscow Multilateral Middle East Conference.

Monnet, Jean: Born at Cognac, France in 1888, he was the initiator of the plan which launched the European Communities. At a meeting of the Western heads of state in April 1976 it was decided to confer the title "Honorary Citizen of Europe." He died in 1979. His approach to institutionalizing regional accords has often been used to secure future community models of integration.

See Containment, Conflict, and Common Management of Common Problems.

Moroccan Dirham: The currency of Morocco.

Morocco: Morocco's heavy dependence on agricultural production means that the nation is particularly susceptible to unpredictable desert weather patterns, which can play havoc with its citrus export industry. The world market for phosphates also provides a degree of instability—in general, when phosphate prices are down the economy suffers, when they are up, the nation is prosperous.

The weather shifts so dominate Morocco's farm-weighted economy that the drought in 1995 resulted in a 7.6 percent GDP loss, and the 1996 abundance of rain boosted growth to about 9 percent.

Morocco's largest trading partners are France (23%), the United States (8%), Spain (8%), and Germany (6%). Morocco's trading ties to other Maghreb nations are very formal, but has not had a decided impact on the nation's economy.

Israel and Morocco established diplomatic relations and agreed to open liaison offices in Tel Aviv and Rabat on September 1, 1994. The following month King Hassan II hosted a 61-nation Middle East/North Africa Economic Summit.

The level of trade between Israel and Morocco has increased and in 1994 totalled more than $25 million, with a $100 million trade level expected soon. Israel exports seeds, irrigation systems, and agricultural industry technology to Morocco, as well as computers for banks and travel agencies.

Morocco, a country of 28 million people, enjoys a healthy commercial environment marked by a growing spirit of entrepreneurship and a pragmatic monarchy. The Moroccan government has pursued an economic reform program supported by the International Monetary Fund and the World Bank since the early 1980s.

A cautious but generally successful privatization program of over 100 government entities, including the important energy sector, is helping to kick-start interest in financial services and the stock exchange. Mobile communications and franchising are gaining favor.

Morocco's once healthy foreign exchange reserves headed for a fall in 1995 as food imports increased sharply. Despite the drought and a drop in tourism and workers remittances, many foreign investors view Morocco as a profitable market as investment inflows continue to grow as they did in the early 1990s.

Morocco continues to develop further an already market-oriented infrastructure. Telecommunications, roads, water distribution, and private energy production lead the multilateral bank lending programs to Morocco. Along with creating more jobs and improving healthy delivery services the government also seems pledged to welding a business-friendly network onto industrial growth and agrarian abundance. That determination is further punctuated by the absence of civil and religious unrest.

In anticipation for free trade with the European Union, Morocco's economy and businesses are quickly restructuring. 1997's privatization program expects to result in more than $500 million in receipts.

See Hassan II (King), Maghreb Agreement, Union du Maghreb Arabe.

mortality rate: See health.

Mossad L'Aliyah Bet: The Israeli institution for immigration "B" was established in 1939 to oversee issues related to illegal immigration. The outbreak of World War II caused difficulties for such endeavors. Following the war, over three years, the organization had transported, organized, equipped, and launched 61 ships which made 65 voyages with 70,000 illegal immigrants. Throughout all the years of its operation the institution for immigration "B" brought 75,000 illegal immigrants into Israel, a significant addition to the Jewish population of the Land of Israel.

Moscow Multilateral Middle East Conference: Commenced on January 28-29, 1992 with 36 delegates in all. Five specific working groups were formed to find solutions for key regional problems, while serving as a confidence-building measure (CBM) to promote the development of normalized relations among the nations of the Middle East.

Issues of regional concern were discussed in a forum to foster cooperation and build confidence between parties. Shared infrastructure such as highways and water pipelines, shared development bringing about growth in tourism and trade, and cooperation in preserving the quality of natural resources and the environment are among the realities envisioned for the region.

The Steering Committee

Comprised of representations of the key delegations, coordinates the multilateral talks and sets dates and venues for the various working groups. The committee hears reports of the working groups, confirms their decisions, and sets priorities for the allocation of resources. The steering committee also discusses such broader issues as the overall vision of the future of the Middle East, integrating the work of the individual working groups.

The Water Working Group

The function of this working group is to plan and execute workshops and projects in the realms of water conservation, water-sector training, desalination, and water-date enhancement. Some of its achievements include:

- In an Oman session (April 17-19, 1994), an Israeli proposal was endorsed for the first time in any working group. The proposal was for a project to rehabilitate and make more efficient water systems in small-sized communities in the region.
- An Omani proposal to establish a desalination research and technology center in Muscat was endorsed by the group. This center, whose staff will include Israeli experts, will support regional cooperation in desalination research among all interested parties.
- The United States proposed to develop waste-water treatment and reuse facilities for small communities at several sites throughout the Middle East.

- A joint U.S./European Union regional training program for water personnel was implemented.

The Environment Working Group

The Environment Working Group endeavors to enhance the ability of the regional parties to deal with maritime pollution, waste-water treatment, environmental management, and desertification. Projects already established are:

- Gulf of Eilat/Aqaba Pollution Project—Experts from Egypt, Jordan, Israel, and the European Union met in Nuweiba, Egypt to develop a proposal for emergency preparedness for oil spills in the Upper Gulf of Eilat/Aqaba. The European Union agreed to forward 2.5 million ECUs toward the building of three centers, in Aqaba, Eilat, and Nuweiba, which would stand ready to protect these areas against oil spills and other catastrophes relating to water.
- Environmental Management—Activities include the establishment of a Middle East Environmental Code of Conduct, a workshop on environment monitoring and measurement, a mission to the region to study environmental impact assessments, and the establishment of an environmental data bank in the West Bank and Gaza.
- Desertification—The World Bank is overseeing a project to address natural resource degradation of arid lands in the Middle East, which was launched at a working group meeting in Cairo. This project includes the establishment of grazing lands, wildlife, forestation and planting of orchards, introducing varieties of vegetation appropriate for arid climates, and the purification of brackish water. It would be implemented through five local centers. The participating nations include: Israel, Egypt, Jordan, and Tunisia.

The Arms Control and Regional Security Working Group

This group focuses on security confidence-building measures as well as arms control issues, including information exchange and maritime procedures.

Among the issues discussed were the drafting of a Declaration of Principles on regional disarmament, statements of intent on arms control and regional security, security environment and regional threat perception, and a model safeguard regime for a nuclear weapons-free zone in the Middle East. At the Doha, Qatar, May 1994 meeting further agreements were reached as follows:

- To establish a Crisis Prevention Center and a Regional Security Center.
- Studies to be carried out regarding the establishment of an arms control data bank that would best serve the region.

- There will be a naval exercise on the Turkish coast and a meeting of naval officers from throughout the region in Canada.

The Refugees Working Group

This working group focuses on the sensitive issue of Middle East refugees. Family unification, training and job creation, public health, child welfare, and social and economic infrastructure are among the subjects addressed by this forum. The underlying principle of this group is to improve the living conditions for refugees in their present location.

The members of the Refugees Working Group reached an understanding on the appropriate frameworks for discussing various aspects of the Palestinian refugee problem, in accordance with the terms of the Israel-PLO Declaration of Principles and the Israel-Jordan Common Agenda signed in September 1993, as follows:

- The question of the 1948 refugees will be discussed within the framework of the permanent status negotiations between Israel and the Palestinians.
- Modalities for the admission of persons displaced from the West Bank and Gaza in 1967 will be decided by an Israeli-Palestinian-Jordanian-Egyptian committee.
- The readmission of former Palestinian residents of the West Bank and Gaza who had overstayed their permits abroad will be discussed in the bilateral negotiations with Jordan.

The above understanding, achieved in Tunis at the fourth round of talks, enabled the working group to focus on the humanitarian aspects of the refugee issue—family reunification and improving living conditions. In this context, Israel agreed to approve 2,000 family reunification requests annually, thus according permanent resident status to an additional 6,000 persons who had entered the territories as visitors. In a further humanitarian act, Israel has permitted a number of individuals deported in the early 1970s due to prior terrorist activities to return to the territories, together with their families.

At another round which took place in Cairo, plans were developed to ensure a real and positive impact on the lives of the refugees. A substantial portion of the World Bank's emergency assistance program will be directed toward this effort. The following activities were agreed upon:

- Human Resources and Development—Several countries including Israel, the United States, the Netherlands, Germany, Turkey, and China will be conducting courses for the refugees as follows: paramedic training, a drylands agricultural seminar, a public-sector management training course, an animal health workshop, an "employment guaranteed" skills training program (to develop small- and medium-scale enterprises in Gaza and the West Bank), agricultural training, and public health, community planning, and education seminars.

- Child Welfare—Sweden has agreed to make available $2 million for child welfare programs, with a scheduled conference on assistance to Palestinian children in the West Bank, Gaza, and neighboring countries.
- Social and Economic Infrastructure—The United States is providing aid to develop housing facilities and rehabilitation for refugees in Syria and Lebanon.
- Public Health—A regional laboratory is to be established in one of the West Bank cities in order to complement a medical center already in place.

The Regional Economic Development Working Group

This group addresses the issues of infrastructure, training, and tourism development in the region, including the West Bank and Gaza. In November 1993 the working group adopted the "Copenhagen Action Plan," comprising 35 projects in various fields: communications, transportation, energy, tourism, agriculture, financial markets and investment, trade, training, and regional networks.

The round of talks held at Rabat in June 1994 centered around two major subjects: advancing the implementation of the Copenhagen Action Plan, and establishing a Monitoring Committee as a tool to formulate a regional agenda and set priorities for the working group.

The talks of the Monitoring Committee were defined as follows:

- To monitor the implementation of the Copenhagen Action Plan;
- To organize expert meetings and subcommittees, as required, in the various fields of regional economic cooperation;
- To promote the region's integration in global markets;
- To fully exploit respective advantages by promoting regional trade, facilitating investment, and developing infrastructure.

The Monitoring Committee represents a qualitative leap in the institutionalization of the regional dialogue. With the establishment of this committee, the principle of regional cooperation has been anchored in specific subcommittees in the different fields of economic activity.

The working group heard progress on specific projects in the areas of transportation (highway infrastructure, railways and ports, motor transportation, and civil aviation); energy (linking the electricity grids of Israel, the autonomous areas, Egypt, and Jordan, and the alternatives of a Mediterranean-Dead Sea or Red Sea-Dead Sea hydroelectric project); finance (financial markets and stock exchanges, training of banking personnel in the autonomy); regional trade; and training programs.

The following intersessional activities were scheduled:

- The establishment of a regional tourist center in Egypt, probably in Ismaliya, in September 1994;
- A business conference to be held in Marrakech (it was held in Casablanca) in October/November 1994;
- Meetings of tourist agents to be held in Cairo and Amman in October and December 1994.

Parallel with these developments, the World Bank is considering holding a regional workshop to study ways to integrate the private sector in infrastructure projects.

See desertification, multilateral negotiations.

most-favored nation: See Free Trade Agreement (European Union and Israel).

Mouwad, Rene: See Hrawi, Elias (President).

MRO: The international standard code for the currency of Mauritania and Western Sahara—the ouguiya.

MTCs: See Mediterranean Third Countries.

Mubarak, Hosni (President): Born in 1928, Mubarak had a distinguished career in the air force before the late Anwar Sadat elevated him from relative political obscurity to become vice president in June 1975. He became president in October 1981 when Sadat was assassinated and has ruled ever since under emergency laws. Mubarak has himself survived assassination attempts. Against much criticism he deployed Egyptian troops against Iraq in the Gulf crisis in 1991.

Reconfirmed as president for a third six-year term by referendum in October 1993, Mubarak tends to chart a very cautious course, although he has frozen out opposition and occasionally used iron-fist tactics. The head of state believes in long-term alliances; for example, the regime's relationship with the United States has maintained multibillion dollar aid flows and promoted economic reform, although the changes do not affect the regime's political interests. Stolid and hard working as ever, Mubarak is entering a period of doubt; that Egypt may lose the advantages (such as U.S. aid) associated with its role as the region's leading pro-Western state, and that the violence of Islamist radicals may undermine the regime.

President Mubarak hosted the March 1996 "Summit of Peacemakers."

See Cairo Declaration, Cairo Middle East/North Africa Economic Conference, Egypt, Gaza Strip and the Jericho Area Agreement, New Middle East; Sadat, Anwar.

Mujahideen: Foreign Islamic volunteer fighters who tend to be radical Islamic in orientation, some belonging to Islamic opposition movements in their countries of origin. They tend to operate in small, cohesive units.

multilateral negotiations: Multilateral talks purport to build the Middle East future. As outlined in the Madrid invitation, the multilateral track deals with a number of issues that are of common interest to peoples of the region; water, environment, arms control and regional security, refugees, and regional economic development. Specific multilateral negotiations commenced on January 28, 1992 at the Moscow Multilateral Middle East Conference.

See bilateral negotiations, Casablanca Declaration, Israel-Jordan Common Agenda, Israel-Jordan Peace Treaty, Madrid Conference, Moscow Multilateral Middle East Conference.

Multilateral Working Group on Arms Control and Regional Security: See United Nations (nuclear-weapon-free zone in the region of the Middle East).

Multilateral Working Group on Refugees (MWGR): See Israel-Jordan Peace Treaty, refugees.

Munich Olympics: See Black September, terrorism.

Museums without Frontiers: See tourism.

mutual trust: See vision, the.

MWGR (Multilateral Working Group on Refugees): See refugees.

N

Nahal Oz-Karni Nezzarim road: See Cairo Agreement.

Naharayim area: See Israel-Jordan Peace Treaty.

Nasser, Gamal Abd al-: Born in 1918, a charismatic leader who helped to oust King Faruq, Egyptian President Nasser nationalized the Suez Canal on July 26, 1956, which led to war with Israel (the Sinai Campaign). Ten years later the Six-Day War with Israel followed with the resignation of Nasser on June 9. Probably his greatest achievement was the construction of the Aswan Dam. He died of a heart attack on September 28, 1970 and was succeeded by Anwar Sadat as President.

See Egypt, Gaza Strip.

nation: See powerless state.

National Agreement Regarding the Negotiations on the Permanent Settlement with the Palestinians: See Beilin-Eitan agreement.

National Animal Health Code: See Protocol on Economic Relations (Article VIII).

National Council of the Palestine Liberation Organization: See Palestine National Council.

National Insurance Law: See Protocol on Economic Relations (Article VII).

nationalism: As defined by the major theoretician of modern nationalism, Ernest Gellner: "is a theory of political legitimacy, which requires that ethnic boundaries should not cut across political ones, and, in particular, that ethnic boundaries within a given state—a contingency already formally excluded by the principle in its general formulation—should not separate the power-holders from the rest."

Nations, especially those in the region, who have been denied freedom, legitimately want to proclaim themselves. Nationalism can be a valid source of energy, but when it becomes egotistical and is directed against each other and against neighbors and minorities, then it is unacceptable and dangerous.

Nationalism, when carried to an extreme, can poison relations between countries with suspicions and fears, which may become a permanent and potentially dangerous obsession among leaders. It can make people of the MEC/MENAC believe in the inevitable and salutary nature of aggression. Nationalism can break the dream of harmony between community countries where each nation marches to its own drumbeat, leading to conflicts and hatred between borders.

Nationalists when agitated will oppose unification of any sort. Only in moderation can nationalism be acceptable; at that point it is more properly called patriotism. This self-pride is a fulfilling objective.

National Palestinian Covenant: See Hebron, PLO Phased Plan.

National Salvation Front: Alliance of anti-Arafat factions sponsored by Syria. Formed in 1983 to oppose U.S.-brokered PLO withdrawal from Beirut.

natural gas: See Energy and Environment (Center of the Middle East/North Africa Community), gas, Qatar.

natural resources: See Israel-Jordan Common Agenda.

navigation: See Israel-Jordan Peace Treaty.

Nazism: See Hamas Covenant.

Near East: Originally coined to distinguish the region from the Middle East and the Far East. Extending from the Mediterranean Sea to the Persian Gulf, the Near East was defined as possessing the boundaries of the Ottoman Empire. Considered obsolete by most countries, but still retained by some governments, including the United States.

Nebi Mousa: See Cairo Agreement.

negotiations: See borders, non-paper.

Netanyahu, Binyamin (Benjamin) (Bibi): Born in Tel-Aviv in 1950; elected Prime Minister of Israel on May 29, 1996 by less than one percentage point, defeating Shimon Peres. Nicknamed Bibi, this 46 year old became the first prime minister directly elected by the people.

His family moved to a suburb of Philadelphia in 1963. Bibi returned to Israel to enlist in the army on the eve of the Six-Day War, but wasn't accepted until a few months afterward, when he turned 18. Afterward he became an officer in the most exalted of the military's elite units, Sayeret Matcal.

He then earned a bachelor's degree in architecture and a master's in business at the Massachusetts Institute of Technology.

In the 12th Knesset he was Deputy Minister of Foreign Affairs and a member on the Knesset Immigration and Absorption Committee. In the 13th Knesset he was a member of the Foreign Affairs and Defense Committees as well as the Constitution, Law, and Justice Committee.

Netanyahu organized the International Conference on Terror in Jerusalem in 1979 and in Washington in 1984.

In 1982 the Israeli Ambassador to the United States asked Netanyahu to be his deputy in the Washington embassy. Then, in 1984, as the Israeli delegate to the United Nations, he led a campaign against the General Assembly's infamous resolution equating Zionism with racism.

In 1987 he initiated the opening of the U.N. Archives on Nazi War Crimes.

Once Prime Minister he was cautious in choosing his ministers and even more reluctant to take a firm position on the future of the peace process, including a meeting with Arafat which finally took place in early September 1996; the "dark channel," so named because these meetings of opposite sides were shadowed in gray on a flow chart presented to Netanyahu. A set of understandings, called a "non-paper" in diplomatic jargon, included:

- increasing the number of Palestinian laborers permitted in Israel from 37,000 to at least 50,000;
- Israel will permit the opening of a Palestinian airport in Gaza, with the understanding that Israel will retain security inspection controls;
- Israel and the PA will meet to discuss mutual changes in the Hebron redeployment agreement;
- Following resolution of the Hebron issue, Israel pledged to negotiate further redeployments in the territories that are mandated by the Oslo 2.

See Albright, Madeleine K., Allon-Plus Plan, Arab Summit, Cairo Middle East/North Africa Economic Conference, Hebron, Israel (economy); Larsen, Terje; Likud Party; Peres, Shimon; reciprocity, Third Way Plan.

Netivot Shalom: See Beilin-Abu Mazen Plan.

Network of Ancient Theaters for the Performing Arts: See tourism.

Network of Major Popular and Traditional Mediterranean Festivals: See tourism.

Neutral Zone (between Saudi Arabia and Iraq): See Iraqi Dinar, Kuwaiti Dinar, Saudi Riyal.

New Declaration of Principles: See Council on Foreign Relations.

New Israeli Sheqel (NIS): See Protocol on Economic Relations (Article IV).

New Middle East, The: A vision for the Middle East region developed in a 1993 book by Shimon Peres, while Minister of Foreign Affairs for Israel. His model calls for regional economic integration as the key to peace and security, to promote stability and economic development, national growth, and individual prosperity. Peres calls for a three-tiered pyramid program of cooperation.

Stage 1 will include binational or multinational projects, such as a joint research institute for desert management or cooperative desalination plants. The ongoing productive cooperation between Israel and Egypt in agriculture is a good example of this approach. Projects may even be put into effect before a permanent peace is established. Israel already has mutual economic and research programs with various countries with which it has not as yet signed peace treaties.

Stage 2 will involve international consortiums that will carry out projects requiring large capital investment under the supervision of relevant countries in the region and perhaps other interested parties as well. Examples of such projects are a Red Sea-Dead Sea canal coupled with development of free trade and tourism along its length; a joint Israeli-Jordanian-Saudi Arabian port; development of hydroelectric power for electricity and desalination and well-planned, rapid development of Dead Sea industries.

Stage 3 will include regional community policy, with gradual development of official institutions. Our era has witnessed the emergence of two contradictory trends: particularist nationalism and ultranational development of regional communities. In every area in which the first has staked a claim, the social order has been subverted and hostility and violence have taken root. In contrast, everywhere the ultranational trend predominated there is sensitivity to human needs, opportunities and desires leading to a more lasting international order that strives for prosperity, development, and human rights.

To a great extent, The New Middle East is already upon the region. With the visit of King Hussein and Egypt's President Mubarak at the Rabin's November 1995 funeral with eulogizes for the man who engineered their national defeat in 1967, along with the 13 Arab leaders who attended the Sharm al-Sheikh summit on terrorism and showed a special solidarity with Israel, along with the already standing seven Arab diplomatic missions in Israel (which remained opened during the winter 1996 Grapes of Wrath operation in Lebanon), it is already a fact of life that The New Middle East exists.

In addition, Israel exports technology to the Gulf nations, offers tourist packages with Jordan and Egypt, and construes the containment of Islamic terrorism with Arafat and Mubarak. The New Middle East is different today, it is a region with a greater commitment to peace.

See MENA Series.

Compare Middle East/North Africa Community.

new trade: See input sharing.

NIS: New Israeli Sheqel. See Protocol on Economic Relations.

Night of the Bridges: See Hebrew Insurrection Movement.

Nobel Peace Prize laureates: See Arafat, Yasir; Peres, Shimon; Rabin, Yitzhak.

nondiscrimination: See Treaty of the Middle East/North Africa Community (Article 3).

non-Mediterranean: As defined by the European Union in negotiating trade agreements, the countries of Cyprus, Malta, and Turkey.

non-paper: Popularized during Israel-Syrian negotiations. It is a document which does not officially exist, because either party retains the right to deny it was ever drawn up.

nonproliferation (of weapons): See unconventional weapons.

North Africa: Nations on the southern rim of the Mediterranean Sea. Includes Morocco, Algeria, Tunisia, Libya, and Egypt. The Sudan is at times included. Egypt has a dual geographic identity, also being a nation of the Middle East.

Norway: See Letters of Mutual Recognition.

Note for the Record: See Hebron.

nuclear-weapon-free zone: See United Nations (nuclear-weapon-free zone in the region of the Middle East).

O

OAPEC: See Organization of Arab Petroleum Exporting Countries.

Occupied Territories: Coined in 1967 for the areas of Palestine under the British mandate that were occupied by Israel following the June 1967 Arab-Israeli War. Basically, it is the Gaza Strip and the West Bank, including East Jerusalem.

See Gaza Strip, West Bank.

October 1973 Arab-Israel War: See Yom Kippur War. Synonymous with Arab-Israeli War (IV), Ramadan War.

October War: See Nasser, Gamal Abd al-; Yom Kippur War. Synonymous with Arab-Israeli War (IV), Ramadan War.

OECD: See Organization for Economic Cooperation and Development.

OIE: See Protocol on Economic Relations (Article VIII).

oil: Several nations of MENA have major sources of oil, housing two-thirds of the world's crude oil reserves. Most of the oil resources are concentrated in a few deposits in Saudi Arabia, Kuwait, the Emirates, Iran, and Iraq. Syria, Egypt, and Libya also boast reserves and are currently oil exporters.

At present, most exports to other nations around the world are transported by sea. Political conflicts have greatly precluded export via pipelines to the Mediterranean.

However, the cost of transporting oil from MENA nations to other parts of the world could be substantially reduced if it could be conveyed via pipeline to a port along the Mediterranean, and from there by tankers.

According to the International Petroleum Encyclopedia approximately 10 percent of the total world production of oil is occurring in MENA nations that are either currently involved in an ethnic conflict or are at a high risk for future conflicts. Production in the thousand barrels/day, they are:

Algeria	771.3
Iran	3,415.3
Iraq	417.3
Libya	1,468.7

In addition, MENA nations that are a moderate risk for future ethnic conflicts include:

Egypt	870.7
Kuwait	845.3
Oman	729.0
Saudi Arabia	8,206.7
United Arab Emirates	2,337.2

The MENA nations provide 75 percent of the oil available to importing countries, including the United States (the United States consumes 25% of world oil consumption), and only the Middle East with its vast reserves is capable of meeting the future growth of world demand.

Economic sanctions imposed on Iran, Iraq, and Libya, together with widening polarization between the United States and some other Muslim nations, encourage radical action against American interests.

Oil export revenues have risen in 1996. For example, in billions of U.S. dollars:

Country	*1995*	*1996*
Saudi Arabia	40.5	47.8
Iran	14.4	15.3
Kuwait	11.9	14.5
UAE	12.1	14.4
Libya	7.7	8.8
Algeria	6.7	8.7
Oman	4.7	5.6
Bahrain	2.5	3.3
Qatar	2.5	3.3

See Amman Middle East/North Africa Economic Summit, energy, Cairo Middle East/North Africa Economic Conference, Energy and Environment (Center of the Middle East/ North Africa Community), Egyptian-Israeli Peace Treaty, growth (MENA), Middle East/North Africa (MENA), Qatar, transport.

oil spill: See Moscow Multilateral Middle East Conference, Upper Gulf of Aqaba Oil Spill Contingency Project.

Oman: Officially the Sultanate of Oman. During the nineteenth century the most powerful of the Arabia states. The policy of Sultan Qaboos has been to encourage greater ties with the West.

In the 1960s two developments inside Oman brought the nation to the attention of the world. First, was the discovery of large quantities of oil. Second, an outbreak in 1964 of a major insurrection against the rule of Sultan Said. For 25 years Sultan Qaboos has transformed Oman, with a basic infrastructure of superior schools, hospitals, housing, electricity, piped water, telecommunications, roads, and ports.

Oman has played a constructive role in the Middle East peace process. Oman became the first Gulf Arab state to host a session of a working group under the multilateral track of the peace process and thus the first to host an official Israeli delegation. Israeli Deputy Foreign Minister Yossi Beilin led an Israeli delegation on a groundbreaking trip to Oman to participate in the multilateral talks on water. On December 26, 1994 Sultan Qaboos hosted a visit by Israeli Prime Minister Yitzhak Rabin, who had the honor of being the first Israeli prime minister to set foot on the Arabian Peninsula.

The Sultanate of Oman is a small nation of 2 million people living in the arid mountains and desert plains of the southeastern part of the Arabian Peninsula. The country controls one side of the strategic strait of Hormuz.

Although the economy is driven by Oman's oil resources, these resources are modest. The oil revenues provide a comfortable cushion but yield a per capita income of only US$5-6,000.

In 1993 the United States exported US$265 million worth of goods to Oman. The government, which operates under five-year development plans, is trying to diversify its economy away from oil. In May 1994 the government announced its intention to give priority to development of a US$9 billion liquefied natural gas (LNG) project. The government hopes to have its first LNG exports by the year 2000. In addition, a number of industrial estates are being set up throughout the country to provide subsidized services for investors in plants for light manufacturing. Other development initiatives include a focus on developing the infrastructure of Oman's interior in order to provide services and employment for Oman's growing population, much of that growth coming in those rural, interior areas.

The government's emphasis on income diversification has opened the country to foreign participation in the economy, particularly in the form of joint ventures. Oman is actively seeking foreign investors, particularly in the industrial field. Although bureaucratic processing of commercial licenses can be cumbersome and time consuming, the government does promote a free market economy. There are no restrictions on the flow of capital, whether in terms of salaries or the repatriation of corporate profits. Nationalization of foreign enterprises is unknown in Oman and the government is beginning to privatize some of its state-owned companies.

In addition to the oil and gas sectors, the areas with the greatest commercial opportunities, there are other sectors in Oman with potential for exporters. These include water-saving technologies for agriculture, equipment for the treatment of waste water, medical equipment, telecommunications, joint ventures in light industry, training and vocational education, and development of an infrastructure for increasing tourism to the Sultanate.

See Gulf Cooperation Council, Gulf nations, Moscow Multilateral Middle East Conference.

Oman Development Bank: See Arab (inter-trade).

Omar al-Bashir (President): Sudan's head of a military regime in the firing line for human rights violations. Took power in a 1989 coup, subsequently banning all political opposition. Imprisoned former Prime minister Sadiq al-Mahdi. Allegations about his regime's activities pushed it onto the U.S. list of terrorist states in 1993.

See Sudan.

OMR: The international standard code for the currency of Oman—the rial omani.

On the Wings of Eagles: See Operation Magic Carpet.

Operation Grapes of Wrath: See Agreement of Understanding between Israel and Lebanon.

Operation Magic Carpet: The informal name for the airlift of Yemenite Jews to Israel in 1949 and 1950. The operation was officially called "On the Wings of Eagles," after the passage in the Book of Exodus, "and I will carry you forth on the wings of eagles." Yemenite Jews were gathered in Aden, Yemen. Many of them came long distances—some crossed the desert, others walked. Because the Suez Canal was closed to Israeli ships, the immigrants were flown directly from Yemen. Approximately 50,000 Yemenite Jews came to Israel on a total of 500 flights.

opportunity database: See Energizing Business Actors.

Organisation de Cooperation et de Developpement Economiques: See Organization for Economic Cooperation and Development.

Organization for Economic Cooperation and Development (OECD): An organization created by the Paris Convention of 1960 by 20 founder states (Austria, Belgium, Denmark, France, Germany, Greece, Iceland, Ireland, Italy, Luxembourg, the Netherlands, Norway, Portugal, Spain, Sweden, Switzerland, Turkey, the United Kingdom, the United States, and Canada). It was later joined by Japan, Finland, Australia, and New Zealand. Its objective is promoting economic and social welfare throughout the OECD area by assisting member governments in the formulation and coordination of policies; stimulating and harmonizing members' aid efforts in favor of developing nations; and contributing to the expansion of world trade.

See Amman Middle East/North Africa Economic Summit, Declaration of Principles, Middle East/North Africa Community (MENAC), regional economic integration.

Compare International Monetary Fund.

Synonymous with Organisation de Cooperation et de Developpement Economiques.

Organization of Arab Petroleum Exporting Countries (OAPEC): Created in January 1968 in Kuwait, following the Arab loss in the June 1967 Arab-Israeli War. Includes members of OPEC less Ecuador, Gabon, Indonesia, Iran, Nigeria, and Venezuela, plus Bahrain, Egypt, and Syria; aids in setting international oil prices and acts as liaison between the Arab oil states and other Arab nations.

During the Iraqi occupation of Kuwait in 1991, headquarters were moved from Kuwait to Cairo.

See Organization of Petroleum Exporting Countries.

Organization of Petroleum Exporting Countries (OPEC): Following a conference in Bagdad in September 1960 OPEC was formally launched in January 1961 (Geneva) and in 1965 set up its permanent headquarters in Vienna. The group's 12 present (Ecuador left in 1992) members are concentrated in the Middle East but also include countries in Africa, South America, and the Far East. By virtue of their large exports, Saudi Arabia and Iran have been the most powerful influences.

See Organization of Arab Petroleum Exporting Countries.

Oslo I: See Oslo Agreement.

Oslo II: See Interim Agreement.

Oslo accords: See Oslo Agreement.

Oslo Agreement: On September 9, 1993, after almost two years of bilateral talks, a breakthrough occurred in Israeli-Palestinian relations. As a result of intense behind-the-scenes contacts between Israeli and Palestinian negotiators in Oslo, Norway, PLO Chairman Arafat sent a letter to Prime Minister Rabin of Israel in which he stated unequivocally that the PLO:

- Recognizes the right of Israel to exist in peace and security;
- Accepts U.N. Security Council Resolutions 242 and 338;
- Commits itself to a peaceful resolution of the conflict;
- Renounces the use of terrorism and other acts of violence;
- Assumes responsibility over all PLO elements to ensure their compliance, prevent violations, and discipline violators;
- Affirms that those articles of the PLO Covenant which deny Israel's right to exist are now inoperative and no longer valid;
- Undertakes to submit to the Palestinian National Council for formal approval the necessary changes to the covenant.

On September 13, 1995 a joint Israeli-Palestinian Declaration of Principles was signed by the two parties in Washington, outlining the proposed interim self-government arrangements as envisioned and agreed by both sides.

See Council on Foreign Relations, Declaration of Principles, Hebron, Interim Agreement, Madrid Conference, PLO Charter, settlements (and the Oslo accords).

Ouguiya: The currency of Mauritania and Western Sahara.

P

PADICO: The Palestinian Development and Investment Company develops a marketing strategy and detailed promotion plan for the industrial estate. Its initial focus is placed on local Gaza industries, selected Israeli manufacturers, and overseas Palestinian business groups.

PADICO signed a contract in August, 1996 to construct the first PA industrial zone, following a Palestinian Authority decision to proceed with the scheme unilaterally and position it completely outside Israeli-controlled territory.

The 500-hectare park (the former airport of Gaza) at Karni in Gaza, will be entirely on the PA-controlled side and will be supported by the United States.

The first operating phase of the estate is aimed at the development of local industry, while the second will open the estate to foreign investors in the heavy industry sector. French, British, and German companies are interested in setting up there. Seventeen thousand-five hundred employees will be needed on the estate which will also generate 35,000 indirect jobs.

See Palestinian Stock Exchange.

Pahlavi, Mohammed Reza: The Shah of Iran from 1941 to 1979. Rioting throughout the last several months of 1978 led the Shah to leave Iran in January 1979, never to return. He was disposed two months later, with the Ayatollah Khomeini becoming the de facto ruler. He lived briefly in Morocco, Mexico, the United States and Panama, and finally died as a guest of President Sadat in Cairo on July 27, 1980.

Palestine: The southern part of Syria was called "Palestine," a corruption of "Philistia," the coastal strip occupied by the Philistines more than a thousand years before Christ. While no country had ever called itself

"Palestine," it was a geographic term current in the Christian Western world to describe the Holy Land.

Palestine became a British mandated territory from 1920 to 1948. In November 1947 the United Nations urged partition, and the British mandate ended on May 14, 1948. The Arabs would not recognize the new State of Israel, nor a larger Jordan. War broke out from May 1948 until January 1949. Not to be confused with the Palestine Authority.

See Abdullah Ibn Hussein, Balfour Declaration, East Jerusalem, Egyptian-Israeli Peace Treaty, League of Nations Mandate, Palestinian (entries), PLO Charter.

Palestine Authority: (Also referred to as the Palestine Liberation Authority, Palestinian Authority, Palestinian Liberation Authority) The name given to the organization headed by Chairman Yasir Arafat.

With the signing of the Gaza-Jericho Agreement in May 1994, and ongoing negotiations to reach an interim agreement between Israel and the Palestinian Authority (PA), the PA is gradually assuming more control over many economic functions. The PA is making slow but steady progress in developing transparent legal, regulatory, and commercial codes.

Synonymous with Palestinian National Authority, the name of the Palestine Authority used by the Palestinians.

See Arafat, Yasir; Cairo Agreement, Casablanca Declaration, Declaration of Principles, European Union, Free Trade Agreement (United States and the West Bank and the Gaza Strip), Gaza Strip and the Jericho Area Agreement, Israel (economy), Japan, PADICO, Palestine National Council, Paris Economic Agreement, Protocol on Economic Relations.

Palestine Liberation Army: The military wing of the Palestinian Liberation Organization; formed in 1964. After the signing of the PLO-Israeli Accord in September 1993, some 10,000 became policemen within the Palestinian Authority. As many as 8,000 Palestinian troops are stationed in other Arab nations.

Palestine Liberation Authority: See Palestine Authority.

Palestine Liberation Organization: See Palestinian Liberation Organization, PLO Charter.

Palestine Liberation Organization Charter: See Palestinian Liberation Organization, PLO Charter.

Palestine Liberation Organization Covenant: See PLO Charter.

Palestine National Congress (PNC): Parliament-in-exile of all Palestinians inside and outside occupied territories. PLO gets its authority from PNC which meets annually to elect an executive committee and approve PLO policies.

Palestine National Council: Official title, The National Council of the Palestine Liberation Organization. Chairman Yasir Arafat was sworn in as the first president of the Palestine National Council on February 12, 1996 in Gaza. The 88-member council held its first meeting on February 20, 1996.

On April 24, 1996 the Palestine National Council altered its original covenant, which had called for the destruction of Israel.

Seven of the 18 member executive committee resigned following the signing of the Israeli-Arab Accord in September 1993.

See Arafat, Yasir (Letter to Y. Rabin recognizing Israel's right to exist in peace, September 9, 1993), Letters of Mutual Recognition, Oslo Agreement, PLO Charter.

Palestine National Fund: The unit of the PLO created to raise funds for its parent organization.

Palestine Securities Exchange (PSE): See stock exchanges.

Palestine War: Synonymous with Arab-Israeli War.

Palestinian: Any person who is a resident within the Israeli-occupied territories of the West Bank (including East Jerusalem) and the Gaza Strip, as well as those in exile outside these territories, and Israel itself. It also reflects the concerns and interests of those Palestinian residents of Israel itself, although they have Israeli citizenship. These issues of definition are important because a tendency, chiefly encouraged by Israel, has developed which considers the Palestinian problem as being confined only to the inhabitants of the West Bank and the Gaza Strip. Yet, just as world Jewry has often viewed Israel as a home for all Jews, so Palestinians regard the whole of Palestine as a home for all Palestinians. By the early 1990s the Palestinian populations throughout the Middle East totalled 5,700,000.

See Camp David Accords, Fatah, Palestine (entries), PLO Charter, refugees.

Palestinian Airlines: In July 1997 Palestinian Airlines flew its inaugural commercial flight from El-Arish Airport in Egypt to Amman. Officials said they hoped the airline would soon be flying from its home soil in the Gaza Strip rather than its temporary base in Egypt.

Palestinian Authority: See Palestine Authority.

Palestinian Council: See Interim Agreement.

Palestinian Council Elections: See Japan.

Palestinian Covenant: See Hebron, Interim Agreement, PLO Phased Plan.

Palestinian Delegation: See Declaration of Principles.

Palestinian Development and Investment Company: See PADICO.

Palestinian Development Bank: See Declaration of Principles.

Palestinian Economic Council for Development and Reconstruction (PECDAR): See Paris Economic Agreement.

Palestinian economy: Often referred to as a peace factor, the economic policy of the Palestinian Authority is built on free enterprise, private investment, and minimal government intervention. In spite of the influence of substantial recession, the Palestinian economy retains considerable assets.

The first of these is an important human resource, that is the level of training and education reached by the population. The territories have no shortage of qualified manpower or business people.

This wealth of human resource is compounded by the monetary wealth of the Palestinian diaspora.

The 200,000 active Palestinians dispersed throughout the Gulf and industrialized countries meet a high level of qualification.

The Palestinian economy also possesses a considerable geographic asset: its location and businessmen's astuteness enable the Palestinian economy to serve as an intermediary between the Western and the Arab world. The Palestinian's per capita annual income is $1,100, compared to $17,000 in Israel (1997).

The Palestinian economy inherited a debt-free country, an exceptional feat in today's developing world.

Palestinian Electricity Authority: See Declaration of Principles.

Palestinian Environmental Authority: See Declaration of Principles.

Palestinian Export Promotion Board: See Declaration of Principles.

Palestinian Fund: See Protocol on Economic Relations (Article XI).

Palestinian-Israeli Coordination and Cooperation Committee: See Declaration of Principles.

Palestinian-Israel Joint Economic Committee: See Protocol on Economic Relations.

Palestinian-Israel peace accord: See Interim Agreement (on the West Bank and the Gaza Strip).

Palestinian Land Authority: See Declaration of Principles, Palestine Authority.

Palestinian Liberation Front (PLF): Iraqi-backed radical PLO faction formed in 1977. Originally worked closely with Syrian intelligence. Reunited in 1987 with opposing PLF faction. Responsible for 1985 Achille Lauro hijacking and 1990 attack on Tel Aviv beach.

Palestinian Liberation Organization (PLO): Created in May 1964 by Egyptian President Gamal Abdel Nasser and other Arab states to gain support of Palestinian nationalists. Arafat became its leader in 1969. Umbrella organization for Palestine liberation movement. Named sole legitimate representative of Palestinian people at 1974 Arab summit in Rabat. Governed by the Center Committee and ruling 15-member Executive Committee. Comprises eight main organizations.

Following attempts to restrain extremists of the PLO, King Hussein of Jordan forced the PLO guerrilla groups to shift to Syria and Lebanon.

Yasir Arafat has led the PLO in its struggle against Israel, moving his headquarters to Gaza following the signed peace treaty with Israel.

See Camp David Accords, Fatah, Gaza Strip and the Jericho Area Agreement, Jordanian-PLO Accord, Palestine Authority, PLO, Rabin, Yitzhak (Letter to Arafat recognizing the PLO, September 9, 1993), terrorism.

Palestinian Liberation Organization Charter: See PLO Charter.

Palestinian Liberation Organization Covenant: See PLO Charter.

Palestinian Monetary Authority (PMA): See Paris Economic Agreement, Protocol on Economic Relations (Article IV).

Palestinian National Authority: Synonymous with Palestine Authority.

Palestinian National Authority Plan (border): A border plan separating Israel and Palestine. The plan assumes that there is no real border between Israel and the Palestinian entity, which controls pieces of the West Bank area. The official Palestinian position is a map in which the border is the Israel-West Bank boundary as it stood before the 1967 war. Under this map the entire West Bank and the eastern half of Jerusalem are part of Palestine. There are also several proposals that would allow Jewish settlers to remain in the West Bank, although under Palestinian sovereignty.

Compare Alpher Plan, Beilin-Abu Mazen Plan, Sharon Plan, Third Way Plan.

Palestinian National Charter: See Hebron, PLO Charter.

Palestinian National Congress: See Palestine National Congress.

Palestinian National Council: See Palestine National Council.

Palestinian National Covenant: See PLO Charter.

Palestinian National Liberation Movement: See Fatah.

Palestinian police force: See Declaration of Principles, Gaza Strip and the Jericho Area Agreement, Interim Agreement.

Palestinian refugees: See Moscow Multilateral Middle East Conference.

Palestinian Stock Market: Inaugurated in the West Bank town of Nablus in November 1996. PADICO is the driving force for the exchange.

At 10:01 a.m., in February 1997, the Palestinian stock market officially opened. Initially the exchange was opened for only one trading session a week.

Arab Insurance Company was the first transaction—100 shares.

See PADICO.

Palestinian Tax Administration: See Paris Economic Agreement.

Palestinian Tourism Authority: See Paris Economic Agreement, Protocol on Economic Relations (Article X).

Palestinian Tourist Authority: See Paris Economic Agreement, Protocol on Economic Relations (Article X).

Palestinian Water Administration Authority: See Declaration of Principles.

Palestinian Water Data Unit: See Water Data Banks Project.

Palestinian Wing: See Cairo Agreement.

Palmach: See Arab-Israeli War (I), Hebrew Insurrection Movement.

Pan Am 103: See Popular Front for the Liberation of Palestine-General Command.

Paris, Jonathan: Fellow at the Council on Foreign Relations.
See Council on Foreign Relations.

Paris Club: Synonymous with Group of Ten.

Paris Economic Agreement: Negotiated over six months of arduous Israeli-Palestinian economic talks the agreement was signed on April 29, 1994. Eight rounds were needed. The accord grants Palestinians economic independence and glues the economies of Israel and the Palestinians. Israel agreed to open up Israeli markets to Palestinians' agricultural products. The final agreement ensures the free movement of goods into Israel except for five items on which quotas were imposed for five years. In addition, Palestinians have no restrictions on exporting to other markets.

This protocol allows Arab businessmen and others previously denied entry into Israel's markets the unprecedented opportunity to access them. More importantly for the Palestinians, this agreement brings about a substantial lowering in their longstanding trade deficit with Israel, by providing them with virtually free access to Israel's markets, and by allowing them to purchase goods from other cheaper suppliers outside of Israel.

The economic protocol provides for free trade and the establishment of a limited form of customs union, initially between Israel and the autonomous areas of Gaza and Jericho, to be extended to the rest of the West Bank. Exceptions include industrial inputs, durable goods and basic food products from Arab countries, basic food products from other Islamic nations, and basic industrial and agricultural machinery from all countries.

As hoped, should the Israeli-Palestinian Trade Accord help the Palestinian economy develop and raise the standard of living in the West Bank and Gaza Strip, this could boost demand in these markets, and Israel could be one of the beneficiaries of increased demand. Likewise, it should provide Israeli businesses with greater access to other Arab markets.

A Palestinian Tax Administration would be created to conduct its own tax policy. A Palestinian Monetary Authority will soon be regulating banks in the territories and manage its own foreign currency. A Palestinian Authority will set import duties on a list of goods based on the GATT 1994 Uruguay Round.

The agreement established a Palestinian Tourist Agency, maintaining "the normality of movement of labor" between Israel and the territories and the free movement of manufactured goods, transfers insurance licensing and supervising to the Palestinian Authority, and ensures that gasoline in the territories not be sold for more than 15 percent less than in Israel.

The accord established a Joint Economic Committee composed of Israeli and Palestinian officials who review economic issues on an ongoing basis.

The conclusion of the economic agreement paved the way for the World Bank to unveil, on May 2, 1994, its $1.2 billion, three-year, economic development plan for the West Bank and Gaza. Based on financial pledges of 40 donor countries, the blueprint offered specific strategies for implementing the economic arrangements agreed upon between Israel and the Palestinians. The plan is carried out by the PLO's Economic Council for Development and Reconstruction (PECDAR), in cooperation with the World Bank, Israel, and the donor nations.

See chronology (April 29, 1994), General Agreement on Tariffs and Trade, World Bank.

parks development: Individual nations of MENA take responsibility for park creation and maintenance. However, the birth and development of joint international parks is a domain which demands extensive involvement on the part of the participating governments. The construction, development, and administration of these parks require the involvement of a myriad of sectors, including tourism, transportation, communication, energy, environment, to mention only a few. The establishment of parks in cross-border regions will facilitate joint action plans on environmental protection and the preservation of scenery and nature, which are related to cultural heritage.

For example, the "lowest park on earth" could be developed more fully. The Dead Sea is a unique lake given its geographical distinction as the lowest point on earth, its mineral composition, seclusion, and climatic conditions.

The entire MENA region has great potential for attracting tourists and other visitors to their natural parks.

See tourism.

particularist nationalism: See New Middle East, The.

Partition Plan of the United Nations: See United Nations (Partition Plan on Palestine).

partner database: See Energizing Business Actors.

Party of God: See HizbAllah.

Pasadaran: The Islamic Revolutionary Guards Corps of Iran.

passages: See Gaza Strip and the Jericho Area Agreement (Article X).

passengers customs procedures: See Protocol on Economic Relations.

patents: See intellectual property rights.

patriotism: See nationalism.

peace: See investing in peace; vision, the.

peace factor: See Palestinian economy.

Peace Kites: A kite-flying contest held annually in the Jezreel Valley in northern Israel (renamed The Eighth Annual Peace Kites International Show). Children from Israel, Morocco, Jericho (in the Palestinian Autonomy area), and Lebanon participated. To celebrate the Washington Declaration, Jordanian children were also invited in 1995 to fly kites for the occasion across the Jordan River border, as Israeli children launched their kites in the area. The organizers hope to assemble participants from 60 countries in future annual competition.

Peace Process: See chronology of the Middle East Peace Process, Council of Foreign Relations, Middle East Peace Process, private sector.

Peace Pulse: An index intended to gauge the peace process based on economic, diplomatic, and other trends that is a joint project of the Israel

Policy Forum and the Center for International Relations at the University of California at Los Angeles.

The index combines 14 qualitative and quantitative indicators to rate the overall peace process at given historical moments on a scale of 0-100, where 0 is "total war" and 100 is "total peace."

Virtually all progress made since the 1993 Rabin-Arafat handshake on the White House lawn was erased during the first quarter of 1997.

According to the index, 1996 was a year of "precipitious decline" that followed five years of steady progress. There was some progress in 1997, due to the Hebron accord and the Palestinian redeployment, according to the index. At the end of the first quarter of 1997, however, the Peace Pulse score was 39.05, the lowest level since 1992. The score for 1992 was 38.75; in October 1995, just before the assassination of Prime Minister Yitzhak Rabin, the score was 69.4

See Israel Policy Forum.

PECDAR (Palestinian Economic Council for Development and Reconstruction): See Paris Economic Agreement.

Peel Commission: The delegation sent by the British government go Palestine to determine the factors underlying the Arab riots of 1936. The commission was requested to investigate means of fulfilling the British Mandate, and to determine whether Arab or Jewish residents of Palestine had justified complaints regarding the way in which the mandate was being implemented. The panel, which took the name of its chairman, Lord William Robert Peel, heard testimony for two months. In June 1937 the commission released its report, which concluded that the national ideals of the Jewish and the Arab residents would not enable them to combine to form one state. The commission proposed canceling the mandate and dividing the territory between the two peoples. The British government accepted the recommendations, the Arabs rejected them, and serious disagreements broke out among the Jews on the issue.

See Choma U'migdal, League of Nations Mandate for Palestine.

Peel, William Robert (Lord): See Peel Commission.

pensions: See Protocol on Economic Relations (Article VII).

Pentagon Plan: Signed on June 29, 1967 it is a "Memorandum for the Secretary of Defense (U.S.) on the Subject of Middle East Boundaries":

1. Reference is made to your memorandum dated June 19, 1967, subject as above, which requested the views of the Joint Chiefs of Staff, without regard to political factors, on the minimum territory in addition to that held on June 4, 1967, Israel might be justified in retaining in order

to permit a more effective defense against possible conventional Arab attack and terrorist raids.

2. From a strictly military point of view Israel would require the retention of some captured Arab territory in order to provide militarily defensible borders. Determination of territory to be retained should be based on accepted tactical principles such as control of commanding terrain, use of natural obstacles, elimination of enemy-held salients, and provision of defense in-depth for important facilities and installations. More detailed discussions of the key border areas mentioned in the reference are contained in the Appendix hereto. In summary, the views of the Joint Chiefs of Staff regarding these areas are as follows:

A. The Jordanian West Bank

Control of the prominent high ground running north-south through the middle of West Jordan generally east of the main north-south highway along the axis Jenin-Nablus-Bira-Jerusalem and the southwest to a junction with the Dead Sea at the Wadi el Daraja would provide Israel with a militarily defensible border. The envisioned defensive line would run just east of Jerusalem; however, provision could be made for internationalization of the city without significant detriment to Israel's defensive posture.

B. Syrian Territory Contiguous to Israel

Israel is particularly sensitive to the prevalence of terrorist raids and border incidents in this area. The presently occupied territory, the high ground running generally north-south on a line with Qnaitra about 15 miles inside the Syrian border, would give Israel control of the terrain which Syria has used effectively in harassing the border area.

C. The Jerusalem-Latrun Area

See Subparagraph 2A above.

D. The Gaza Strip

By occupying the Gaza Strip, Israel would trade approximately 45 miles of hostile border for eight. Configured as it is, the strip serves as a salient for introduction of Arab subversion and terrorism, and its retention would be to Israel's military advantage.

E. The Negev-Sinai Border

Except for retention of the demilitarized zone around Al Awja and some territory for the protection of the port of Eilat, discussed below, continued occupation of the Sinai would present Israel with problems outweighing any military gains.

F. The Negev-Jordan-Aqaba-Strait of Tiran Area

Israel's objectives here would be innocent passage through the Gulf of Aqaba and protection of its ports at Eilat. Israel could occupy Sharm ash-Shaykh with considerable inconvenience but could rely on some form of internationalization to secure free access to the Gulf. Failing this, Israel would require key terrain in the Sinai to protect its use of the Strait of Tiran. Eilat, situated at the apex of Israel's narrow southern tip, is vulnerable to direct ground action for Egyptian territory. Israel would lessen the threat by retention of a portion of the Sinai Peninsula south and east of the Wadi el Gerafi then east to an intersection with the Gulf of Aqaba at approximately 29 degrees 20' north latitude.

3. It is emphasized that the above conclusions, in accordance with your terms of reference are based solely on military considerations from the Israeli point of view.

FOR THE JOINT CHIEFS OF STAFF; SIGNED: Earle G. Wheeler
Chairman, Joint Chiefs of Staff

See PLO Charter, Syria, terrorism.

per capita income (MENA): See Middle East/North Africa (MENA).

perceptions: What various constituents believe, what attitudes they project, will determine the destiny of MENA. Perceptions of all participants are critical, not only from those in the region, but from around the world. For example, a summer 1996 survey (Israel Policy Forum) of 1,260 American Jewish attitudes toward Israel and the peace process showed:

- Would you say that you strongly approved, somewhat approved, somewhat disapproved, or strongly disapproved of the Oslo Accords initiated by Yitzhak Rabin three years ago, the set of agreements between Israel and the Palestinians?

strongly approve	36%	somewhat approve	34%
somewhat disapprove	7%	strongly disapprove	8%
don't know	16%		

- Do you strongly approve, somewhat approve, somewhat disapprove, or strongly disapprove of there being an Independent Palestinian State in the Middle East?

strongly approve	25%	somewhat approve	38%
somewhat disapprove	13%	strongly disapprove	16%
don't know	8%		

- Do you strongly support, somewhat support, somewhat oppose, or strongly oppose building new settlements in the West Bank and Gaza?

strongly support	19%	somewhat support	22%
somewhat oppose	23%	strongly oppose	26%
don't know	11%		

- What is your opinion of Yasser Arafat, the elected head of the Palestinian Authority? Is it...?

very favorable	5%	somewhat favorable	30%
somewhat unfavorable	33%	very unfavorable	26%
don't know	6%		

- Do you believe that Yasser Arafat is committed to peace with Israel, or do you not believe this?

do believe	51%	do not believe	38%
don't know	11%		

- The United States is currently providing economic assistance to the Palestinian Authority in the West Bank and Gaza. Should the United States continue this practice?

yes	67%	no	25%
don't know	8%		

- Do you strongly support, somewhat support, somewhat oppose, or strongly oppose Israel continuing to pursue peace with the Syrians?

strongly support	57%	somewhat support	30%
somewhat oppose	5%	strongly oppose	5%
don't know	3%		

- Do you think peace with Syria would be possible without Israel giving up all or part of the Golan Heights, or would this not be possible?

possible	38%	not possible	48%
don't know	14%		

- Do you strongly support, somewhat support, somewhat oppose, or strongly oppose the idea of exchanging all or part of the Golan Heights for peace with Syria?

strongly in favor	10%	somewhat in favor	30%
somewhat oppose	18%	strongly oppose	36%
don't know	7%		

- Do you believe the continuation of the peace process is important to the United States' interests and to a strong U.S./Israel relationship, or do you not believe this?

do believe	94%	do not believe	5%
don't know	1%		

- Given the budget constraints that America now faces, and the drive to balance the budget in a way that is fair to all, do you strongly support, somewhat support, somewhat oppose, or strongly oppose cutting American aid to Israel?

strongly support	12%	somewhat suppory	15%
somewhat oppose	21%	strongly oppose	47%
don't know	5%		

Peres, Shimon: Born in Poland in 1923, Peres is considered a persistent dreamer and innovator, and many of his dreams have become a reality. In 1947, at only 24, he became the manpower chief of the Haganah, the fledgling Jewish military. He initiated the acquisition of weapons for national defense, bolstered the morale of the Israel Defense Forces, creatively originated negotiating ideas and options, and instigated diplomatic forays abroad. In the 1950s and 1960s he bound Israel's arms acquisition to France, Germany, and the United States; his secret negotiations with King Hussein of Jordan began in the 1970s and culminated in the 1994 Jordanian-Israeli Peace Treaty.

In 1974 he lost the contest for party leadership to Yitzhak Rabin and became Rabin's Defense Minister and was given the task to rebuild the military. Then, in 1977 Peres temporarily became prime minister when Rabin resigned over the issue of his wife's illegal bank account, but lost the election to Menachem Begin of the Likud Party. Again, in 1981 he lost the national election to Begin.

After a deadlocked election in 1984 Peres became the Prime Minister in the Labor-Likud unity government and was widely praised for bringing troops out of Lebanon and slashing runaway inflation.

At the United Nations in 1985 Peres outlined an eight-point plan for structuring an international conference; those ideas were the bedrock for the procedures eventually used in holding the 1991 Madrid Middle East Peace Conference.

After Labor's electoral victory in 1992, Shimon Peres began his second tenure as Israel's Foreign Minister, during which time he played a central role in the negotiations surrounding the Middle East Peace Process. In December 1994, he was awarded the Nobel Peace Prize with Yitzhak Rabin and Yasir Arafat for his tireless efforts to achieve peace with Israel's neighbors.

Peres was instrumental in forging forward with the economic conferences in Casablanca in 1994 and in Amman in 1995. He, more than anyone else is the visionary behind the New Middle East.

At the Amman Economic Summit he said on October 30, 1995 "There are almost 300 million people in the Middle East. The choice before them is to remain poor and bitter, wearing the cloak of protest and the mantle of fundamentalism, or to become producers and consumers, investing in prosperity—to enter the twenty-first century as equals with the most developed nations."

Following the assassination of Prime Minister Rabin in November 1995, Shimon Peres formed the 26th Government of Israel, and became its Prime Minister and Minister of Defense on November 22, 1995.

He then lost a close election for prime minister to Binyamin Netanyahu on May 29, 1996 and lost the leadership of the Labor Party in 1997. He continues to promote his vision of a "New Middle East."

See Amman Economic Summit, Beilin-Eitan agreement, Casablanca Middle East/North Africa Economic Summit, fences, free processing zone, Hamas, Hebron, Interim Agreement, Israel, London Agreement, Middle East and North Africa Business Interaction Group, Middle East/North Africa Community; Netanyahu, Binyamin; New Middle East, PLO Charter; Rabin, Yitzhak; terrorism, Trilateral Israel-Jordan-U.S. Economic Committee, Washington Declaration.

permanent settlement with the Palestinians: See Beilin-Eitan agreement.

Permit Palestinians: See refugees.

Persia: The ancient empire that in 1935 became Iran.

PFLP: See Popular Front for the Liberation of Palestine.

PFLP-GC: See Popular Front for the Liberation of Palestine-General Command.

PG: Persian Gulf.

pharmaceuticals: See intellectual property rights.

Phased Plan (of the PLO): See PLO Phased Plan.

Philistia: See Palestine.

philosophy of the Community: See MENA Series; vision, the.

Pisgat Shmu'el: See chronology (March 18, 1997).

pitfalls: See forewarning.

plant health controls: See Agriculture (Center of the Middle East/North Africa Community), Health and Housing (Center of the Middle East/North Africa Community).

PLF: See Palestinian Liberation Front.

PLO: See PLO Charter, Jordanian-PLO Accord, Palestinian Liberation Organization, Tunisia.

PLO (Palestinian Liberation Organization) Charter (Covenant): The PLO Charter was first adopted in 1964 and focused more on a pan-Arab approach and less on the need for armed struggle. The latest version was made in 1968. Major excerpts of 33 articles include:

This covenant will be called "The Palestinian National Covenant" (Al-Mihaq Al-Watani Al-Filastini).

Article 1—Palestine is the homeland of the Palestine Arab people and an integral part of the great Arab homeland, and the people of Palestine are a part of the Arab Nation.

Article 2—Palestine with its boundaries that existed at the time of the British Mandate is an integral regional unit.

Article 3—The Palestinian Arab people possess the legal right to its homeland, and when the liberation of its homeland is completed it will exercise self-determination solely according to its own will and choice.

Article 4—The Palestinian personality is an innate, persistent characteristic that does not disappear, and it is transferred from fathers to sons. The Zionist occupation, and the dispersal of the Palestinian Arab people as a result of the disasters which came over it, do not deprive it of its Palestinian personality and affiliation and do not nullify them.

Article 5—The Palestinians are the Arab citizens who were living permanently in Palestine until 1947, whether they were expelled from there or remained. Whoever is born to a Palestinian Arab father after this date, within Palestine or outside it, is a Palestinian.

Article 6—Jews who were living permanently in Palestine until the beginning of the Zionist invasion will be considered Palestinians.

Article 7—The Palestinian affiliation and the material, spiritual, and historical tie with Palestine are permanent realities. The upbringing of the Palestinian individual in an Arab and revolutionary fashion, the undertaking of all means of forging consciousness and training, the Palestinian, in order to acquaint him profoundly with his homeland, spiritually and materially, and preparing him for the conflict and the armed struggle, as well as for the sacrifice of his property to his life to restore his homeland, until the liberation—all this is a national duty.

Article 8—The phase in which the people of Palestine are living is that of the national (Watani) struggle for the liberation of Palestine. Therefore, the contradictions among the Palestinian national forces are of a secondary order which must be suspended in the interest of the fundamental contradiction between Zionism and colonialism on the one side and the Palestinian Arab people on the other. On this basis, the Palestinian masses, whether in the homeland or in places of exile (Mahajir), organizations and individuals, comprise one national front which acts to restore Palestine and liberate it through armed struggle.

Article 9—Armed struggle is the only way to liberate Palestine and is therefore a strategy and not tactics. The Palestinian Arab people affirm its absolute resolution and abiding determination to pursue the armed struggle and to march forward toward the armed popular revolution, to liberate its homeland and return to it, (to maintain) its right to a natural life in it, and to exercise its right of self-determination in it and sovereignty over it.

Article 10—Fedayeen action forms the nucleus of the popular Palestinian war of liberation. This demands its promotion, extension, and protection, and the mobilization of all the mass and scientific capacities of the Palestinians, their organization and involvement in the armed Palestinian revolution, and cohesion in the national (Watani) struggle among the various groups of the people of Palestine, and between them and the Arab masses, to guarantee the continuation of the revolution, its advancement, and victory.

Article 11—The Palestinians will have three mottoes: national (Wataniyya) unity, national (Qawmiyya) mobilization, and liberation.

Article 12—The Palestinian Arab people believe in Arab unity. In order to fulfill its role in realizing this, it must preserve, in this phase of its national (Watani) struggle, its Palestinian personality and the constituents thereof increase consciousness of its existence and resist any plan that tends to disintegrate or weaken it.

Article 13—Arab unity and the liberation of Palestine are two complementary aims. Each one paves the way for realization of the other. Arab unity leads to the liberation of Palestine, and the liberation of Palestine leads to Arab unity. Working for both goes hand in hand.

Article 14—The destiny of the Arab nation, indeed the very Arab existence, depends upon the destiny of the Palestine issue. The endeavor and effort of the Arab nation to liberate Palestine follows from this connection. The people of Palestine assume its vanguard role in realizing this sacred national (Qawmi).

Article 15—The liberation of Palestine, from an Arab viewpoint, is a national (Qawmi) duty to repulse the Zionist, imperialist invasion from the great Arab homeland and to purge the Zionist presence from Palestine. Its full responsibilities fall upon the Arab nation, peoples and governments, with the Palestinian Arab people at their head.

For this purpose, the Arab nation must mobilize its military, human, material, and spiritual capabilities to participate actively with the people of Palestine. They must, especially in the present stage of armed Palestinian revolution, grant and offer the people of Palestine all possible help and every material and human support, and afford its every sure means and opportunity enabling it to continue to assume its vanguard role in pursuing its armed revolution until the liberation of its homeland.

Article 16—The liberation of Palestine, from a spiritual viewpoint, will prepare an atmosphere of tranquility and peace for the Holy Land, in the shade of which all the holy places will be safeguarded, and freedom of worship and visitation to all will be guaranteed, without distinction or discrimination of race, color, language, or religion. For this reason, the people of Palestine look to the support of all the spiritual forces in the world.

Article 17—The liberation of Palestine, from a human viewpoint, will restore to the Palestinian man his dignity, glory, and freedom. For this, the Palestinian Arab people look to the support of those in the world who believe in the dignity and freedom of man.

Article 18—The liberation of Palestine, from an international viewpoint, is a defensive act necessitated by the requirements of self-defense. For this reason, the people of Palestine, desiring to befriend all peoples, look to the support of the states that love freedom, justice, and peace in restoring the legal situation to Palestine, establishing security and peace in its territory, and enabling its people to exercise national (Wataniyya) sovereignty and national (Qawmiyya) freedom.

Article 19—The partitioning of Palestine in 1947 and the establishment of Israel is fundamentally null and void, whatever time has elapsed, because it was contrary to the wish of the people of Palestine and its natural right to its homeland, and contradicts the principles embodied in the Charter of the United Nations, the first of which is the right of self-determination.

Article 20—The Balfour Declaration, the Mandate Document, and what has been based upon them are considered null and void. The claim of a historical or spiritual tie between Jews and Palestine does not tally with historical realities nor with the constituents of statehood in their true sense. Judaism, in its character as a religion of revelation, is not a nationality with an independent existence. Likewise, the Jews are not one people with an independent personality. They are rather citizens of the states to which they belong.

Article 21—The Palestinian Arab people, in expressing itself through the armed Palestinian revolution, rejects every solution that is a substitute for a complete liberation of Palestine, and rejects all plans that aim at the settlement of the Palestine issue or its internationalization.

Article 22—Zionism is a political movement organizationally related to world imperialism and hostile to all movements of liberation and progress in the world. It is a racist and fanatical movement in its formation; aggressive, expansionist and colonialist in its aims; and Fascist and Nazi in its means. Israel is the tool of the Zionist movement and a human and geographical base for world imperialism. It is a concentration and jumping-off point for imperialism in the heart of the Arab homeland, to strike at the hope of the Arab nation for liberation, unity and progress.

Israel is a constant threat to peace in the Middle East and the entire world. Since the liberation of Palestine will liquidate the Zionist and imperialist presence and bring about the stabilization of peace in the Middle East, the people of Palestine look to the support of all liberal men of the world and all the forces of good progress and peace; and implores all of them, regardless of their different leanings and orientations, to offer all help and support to the people of Palestine in its just and legal struggle to liberate its homeland.

Article 23—The demands of security and peace and the requirements of truth and justice oblige all states that preserve friendly relations among peoples and maintain the loyalty of citizens to their homelands to consider Zionism an illegitimate movement and to prohibit its existence and activity.

Article 24—The Palestinian Arab people believe in the principles of justice, freedom, sovereignty, self-determination, human dignity, and the right of peoples to exercise them.

Article 25—To realize the aims of this covenant and its principles the Palestine Liberation Organization will undertake its full role in liberating Palestine.

Article 26—The Palestine Liberation Organization, which represents the forces of the Palestinian revolution, is responsible for the movement of the Palestinian Arab people in its struggle to restore its homeland, liberate it, return to it, and exercise the right of self-determination in it. This responsibility extends to all military, political, and financial matters, and all else that the Palestine issue requires in the Arab and international spheres.

Article 27—The Palestine Liberation Organization will cooperate with all Arab states, each according to its capacities, and will maintain neutrality in their mutual relations in the light of, and on the basis of, the requirements of the battle of liberation, and will not interfere in the internal affairs of any Arab state.

Article 28—The Palestinian Arab people insist upon the originality and independence of its national (Wataniyya) revolution and rejects every manner of interference, guardianship and subordination.

Article 29—The Palestinian Arab people possess the prior and original right in liberating and restoring its homeland and will define its position with reference to all states and powers on the basis of their positions with reference to the issue (of Palestine) and the extent of their support for (the Palestinian Arab people) in its revolution to realize its aims.

Article 30—The fighters and bearers of arms in the battle of liberation are the nucleus of the Popular Army, which will be the protecting arm of the Palestinian Arab people.

Article 31—This organization shall have a flag, oath, and anthem, all of which will be determined in accordance with a special system.

Article 32—To this covenant is attached a law known as the Fundamental Law of the Palestine Liberation Organization, in which is determined the manner of the organization's formation, its committees, institutions, the special functions in every one of them and all the requisite duties associated with them in accordance with the covenant.

Article 33—This covenant cannot be amended except by a two-thirds majority of all the members of the National Council of the Palestine Liberation Organization in a special session called for this purpose.

The 33-clause Palestinian National Covenant includes some of the most anti-Israeli policy ever adopted by an official Arab institution, not even recognizing the possibility of Jewish nationalism—only religion. The covenant's bottom line is not so much a call for the destruction of Israel, as a complete negation of Israel's existence.

Over the years, Israeli officials have demanded that it either be scrapped or that 29 of its 33 clauses be amended. In fact, the chief objections are found in six or seven clauses.

Article 2 states that "Palestine, with the boundaries it had during the British Mandate, is an indivisible territorial unit." The 1947 partition and the establishment of the State of Israel are "entirely illegal." The Balfour Declaration, the mandate for Palestine, and everything that has been based on them "are deemed null and void." Israel is the "Zionist occupation" and a "geographical base for world imperialism placed strategically in the midst of the Arab homeland." Zionism itself is "racist and fanatic in its nature." The primary objective is the "elimination of Zionism in Palestine." The strategy is the armed struggle.

"The Jews who resided in Palestine until the beginning of Zionist invasion will be considered Palestinians," notes Article 6, but Jews who arrived after 1917 or were born to Jews who arrived after 1917 cannot live here.

In 1968 Article 7 was amended to say that every Palestinian "must be prepared for the armed struggle and ready to sacrifice his wealth and his life in order to win back his homeland and bring about its liberation."

Although the 1968 amendments emphasized certain elements of Palestinian distinctiveness, the document retains a highly pan-Arab tone, focusing on the liberation of all "Palestine." It notes that "the Arab nation must mobilize all its military, human, moral, and spiritual capabilities to participate actively with the Palestinian people in the liberation of Palestine."

Articles 3 and 21 state that only Palestinian Arabs have the right of self-determination, and that all the land belongs to them.

Prime Minister Shimon Peres believes that in the wake of the January 1996 Palestinian elections, Arafat now has the legitimacy to make the desired changes. His legal advisor states "If [two years ago when I was in Oslo it was being negotiated] we have insisted on immediately amending the covenant, we would have blown up the DOP (Declaration of Principles)."

In late April 1996 the PNC (Palestine National Council) met to consider amending the charter. Officially, it would have 669 members, including 483 from abroad with 186 reserved for the West Bank, Gaza Strip, and East Jerusalem. Amending the charter required a two-thirds majority, or at least 446 yes votes.

On April 24, 1996, bowing to the insistent demands of Yasir Arafat, the main assembly of the Palestine Liberation Organization voted to revoke the clauses in its 32-year-old charter that called for an armed struggle to destroy the Jewish state. The vote was 504 in favor of amending the document and 54 against. Fourteen members abstained, and 97 of the 669 members of the council were absent.

Formally the resolution consisted of two simple clauses. The first declared that the council "decides to amend the Palestinian National Covenant by canceling clauses which contradict the letters exchanged between the PLO and the Israeli government." The second ordered a new charter to be drafted within six months.

The April 24 Palestine National Council meeting led Shimon Peres to hail the announcement as "the most important development in our region in 100 years." As it turned out, the text of the covenant was not changed at all; instead, the council had authorized a subcommittee to explore ways of amending the covenant and report back to the council within six months.

By the end of October 1996 the six-month deadline came and went with no action whatsoever taken regarding amending or revoking any or all of the covenant.

See Hamas Covenant, Hebron, Interim Agreement (on the West Bank and the Gaza Strip), Oslo Agreement, Palestine Authority, Pentagon Plan, PLO Phased Plan.

Synonymous with PLO Covenant.

PLO Council: See PLO Phased Plan.

PLO Covenant: Synonymous with PLO Charter.

PLO Phased Plan: The PLO Council approved a political plan, the Phased Plan, on June 8, 1974, initially broadcast on Egyptian radio. It reads:

On the basis of the National Palestinian Covenant and the PLO's political plan as approved at the 11th session (January 6-12, 1973); and in the belief that a just and lasting peace in the region is impossible without restoration of the full national rights of the Palestinian nation, and first and foremost the right of return and self-determination on the homeland's soil entire; and after having studied the political circumstances as they developed during the period between its previous and its present session, the council resolves as follows:

1. Emphasis of the PLO's position with relation to Security Council Resolution 242, which overlooks the national rights of our nation and approaches the Palestinian issues as a refugee problem.

   The council therefore rejects any action on that basis on any level of Arab and international operation, including the Geneva Conference.
2. The PLO is fighting by every means, and primarily by the armed struggle, to free the Palestinian land and establish a national, independent, and fighting government over every part of the soil of Palestine to be freed. This calls for a considerable change in the balance of forces, for the good of our nation and its struggle.
3. The PLO objects to any plan for a Palestinian entity at the price of recognition, peace (sulh), secure boundaries, surrender of national rights, and deprivation of our nation's prerogative of return and of self-determination in its homeland.
4. Any steps of liberation is a link in realizing the strategy of the PLO for the establishment of a Palestinian-democratic state, as resolved by the previous councils.
5. It is necessary to struggle alongside the national Jordanian forces for the establishment of a national Jordanian-Palestinian front, with the purpose of forming a national democratic government in Jordan that will safeguard solidarity.
6. The PLO struggles to unify the efforts of the two nations and all the forces of the Arab liberation movements which subscribe to this plan.
7. In view of this plan, the PLO fights to strengthen national unity and raise it to a height where it will be capable of fulfilling its national missions.

8. After its establishment, the national Palestinian government will fight for the unity of the countries of confrontation, to complete the liberation of all the Palestinian land and as a step in the direction of overall Arab unity.
9. The PLO fights to strengthen its solidarity with the socialist countries and the forces of liberation and progress, to frustrate all the Zionist reactionary and imperialist plans.
10. On the basis of this plan, the leadership of the revolution will formulate tactics that will enable these objectives to be realized.

See Pentagon Plan, PLO Charter.

PLO's right to exist: See Rabin, Yitzhak (Letter to Arafat recognizing the PLO (September 9, 1993).

PMA (Palestinian Monetary Authority): See Paris Economic Agreement, Protocol on Economic Relations (Article IV).

PNC: See Palestine National Congress.

police agreements: See Jordanian-Israeli bilateral accords.

policy issues (MENA): According to the International Monetary Fund, four key policy issues face most, if not all, MENA economies:

1. Higher and more efficient investment is needed.
2. The improvement in public savings must continue.
3. Financial reforms will enhance private savings and resource allocation.
4. Trade regimes require further liberalization.

See Middle East/North Africa (MENA).

political plan (of the PLO): See PLO Phased Plan.

politicians: Politicians stray and often vote not by conscience, but instead are swayed by their ability to influence, thus they are by nature influenced by others. In the evolution of the MEC/MENAC, leaders (since they will be selected by the people they are properly, though perhaps regrettably, also politicians) must be accountable each step of the way, and instead of pretending to know what they are doing, be reminded of their obligation that they indeed do know what they are doing, and what the implications of their actions can be. The collective leadership of the community should be wary of agreeing to things they don't fully understand, or of taking actions the long-term consequences of which they cannot foresee.

Every action in institutional building has a political sweep. Each decision can be argued politically. As behavior and recommendations cross borders and enter the global scene, politics becomes ever more pronounced (and usually, more influential). The reality of a democratic or pseudo-democratic style of government encourages the few to control the destiny of the masses. Benevolent as those in power may be, political pressure can betray ideals.

MEC/MENAC elected politicians must be held accountable and have their own watchdog organization. When their conclusions are unacceptable, the public shall have the right to demand further accountability by forming an independent commission (ad-hoc) to resolve perceptions and identify any misdeeds.

Politics will also be found among nonelected people. They too, should be kept at bay by the sound scrutiny of their conduct, by subjecting each person, when appropriate, to defend his or her behavior, and if necessary to suffer the consequences of any misconduct. When MEC/MENAC politics leaves the arena of greed and self-serving goals, and enters the world of purpose and ideals, everyone will come out ahead with heads proudly raised.

See dedication, forewarning, greed, lobbying, trade reform (MENA).

politics: See politicians, trade reform (MENA).

Pollack, Lester: See Council on Foreign Relations.

pollution: Several cities in the MENA region have the worst air pollution in terms of suspended particulates and lead. For example, the most serious of Egypt's pollution is lead pollution, leading to 25,000 deaths or more each year. In addition, children raised in Cairo are in danger of losing on average an estimated 4.25 IQ points resulting in an economic cost of this IQ loss estimated at $13 billion.

See Amman Middle East/North Africa Economic Summit, environment, Moscow Multilateral Middle East Conference.

Popular Front for the Liberation of Palestine (PFLP): Formed in 1967 as an outgrowth of the Arab Nationalist Movement, it is responsible for terrorist operations, chiefly airplane hijackings in early 1970s. PFLP is Fatah's main rival within the PLO but conditionally supports Arafat's peace initiative.

See Popular Front for the Liberation of Palestine-General Command.

Popular Front for the Liberation of Palestine-General Command (PFLP-GC): Radical Syrian-backed and anti-Arafat faction formed in 1969. Responsible for terrorist operations both inside and outside Israel, including hangglider attacks, and allegedly was involved in the bombing of Pan Am flight 103.

Popular Struggle Front (PSF): Radical anti-Arafat faction formed in 1960, based in Damascus.

population: See vision, the.

population growth: See Middle East/North Africa (MENA), Water & Desertification (Center of the Middle East/North Africa Community).

Porter, Michael: See Middle East Competitive Advantage project.

postal services: See Israel-Jordan Peace Treaty.

potential (of MENA nations): See foreign direct investment.

poultry: See Regional Animal Health Cooperation Program.

poverty: See Social Rights (Center of the Middle East/North Africa Community).

powerless state: Assuredly, regional and national economies have changed profoundly over the past years. International trade has grown rapidly and capital roams freely. What impact will a MEC or MENAC have on economic national sovereignty? Will cross-border trade emasculate the role of the government? Will governments be able to do important things or will they be relegated to dealing with minor matters as MEC and MENAC unfolds?

Certainly, with or without regional integration the role of the nation-state is undermined. The ability of governments to decide their exchange rate, interest rate, trade flows, investment, and output has been altered.

However, governments will retain wide discretion over the extent to which they control national resources. Economic integration will increase the opportunities for MEC/MENAC governments to attract inward flows of capital, especially foreign direct investment. Nations can and will continue to realize that barriers against global integration will do more harm than good. State expenditures for public spending will continue and the nation-state will continue to play its major and historic role.

See sovereignty.

power stations: See Energy and Environment (Center of the Middle East/North Africa Community).

preferential trading arrangements: See regional economic integration.

Preparatory Transfer of Powers and Responsibilities: On August 29, 1994, the Agreement on Preparatory Transfer of Powers and Responsibilities was signed by Israel and the Palestinians. The agreement put into effect the next phase (early empowerment) of the Declaration of Principles, as follows:

Spheres of Authority:

The Agreement on Preparatory Transfer of Powers and Responsibilities, in accordance with the DOP, provides for the transfer of powers to the Palestinian Authority within specified spheres:

a. Education and Culture: responsibility over higher education, special education, cultural and educational training activities, institutions and programs, and private, public nongovernmental, or other educational or cultural activities or institutions.
b. Health: authority over all health institutions, whether private, public, nongovernmental, or other.
c. Social Welfare: authority over governmental and nongovernmental organizations and institutions, including charitable societies and institutions and voluntary and nonprofit organizations.
d. Tourism: regulating, licensing, grading, supervising, and developing the tourist industry.
e. Direct Taxation: The DOP provides for a transfer of authority regarding direct taxation. Accordingly, the Palestinian Authority will establish a revenue collection system to collect income tax. Property tax will be collected by Israel and transferred to the Palestinian Authority.

   In addition, it was agreed to transfer the sphere of indirect taxation (VAT) to the Palestinian Authority as an attempt to ensure a balanced budget.
f. Indirect Taxation: This is an additional sphere, agreed upon on August 29, 1994. VAT will be collected on local production in the West Bank—outside the settlements and the military locations. This sphere was added to help fill the shortfall that will occur while the tax system is being set up.

The transfer of power in the sphere of education and culture was carried out on August 29, 1994 to enable the Palestinian Authority to implement its plans at the beginning of the school year.

The transfer of power in the spheres of social welfare and tourism was carried out on November 13-14, 1994. Transfer to the Palestinian Authority of the remaining two spheres, health and taxation, took place on December 1, 1994.

Jurisdiction:

In accordance with the DOP, the jurisdiction of the Palestinian Authority with regard to this agreement does not include Jerusalem, settlements, military locations, and unless otherwise provided for in this agreement, Israelis.

Smooth Transfer of Authority:

The Israeli authorities will assist and support the Palestinian Authority in promoting the effective implementation of the transfer, including turning over all information on budgets and providing the services of Israeli experts. The Palestinian Authority in turn will prevent activities with military orientation, and will coordinate with the Israeli authorities regarding any large-scale events.

Legislation:

The Palestinian Authority may promulgate secondary legislation, such as regulations within the spheres transferred to it. The procedure for enacting legislation gives the Israeli authorities an effective veto over such legislation when it

- exceeds the powers and responsibilities transferred to the Palestinian Authority;
- is inconsistent with the agreement; or
- affects legislation, power, and/or responsibilities which was not transferred to the Palestinian Authority.

Enforcement:

The Palestinian Authority may authorize employees in each of the spheres to act as civilian inspectors to monitor compliance with the laws and regulations. These inspectors are not police, and cannot carry arms, wear uniforms, or organize into a central unit. These names will be submitted to Israel, and those afforded special privileges must be approved by Israel.

Funding:

The Palestinian Authority and the Israeli authorities jointly approached the donor countries for funding designed specifically to offset the shortfall that may be created in the collection of taxes during the initial period while the Palestinian Authority establishes its own revenue collection system.

On August 27, 1995, a Protocol on Further Transfer of Powers and Responsibilities was signed, transferring the following additional spheres to the Palestinian Authority: labor, trade and industry, gas and gasoline, insurance, postal services, statistics, agriculture, and local government.

primary boycott: See boycott.

Principal Allied Powers: See League of Nations Mandate for Palestine.

prisoners: See Interim Agreement (on the West Bank and the Gaza Strip).

private investment: See private sector, privatization.

private participation in infrastructure financing and management: See private sector, privatization.

private sector: Peace and economic development are complementary goals. Economic cooperation bolsters the peace process by rendering it stability and reconfirming the mutual commitment held by all parties. At the same time, peace is essential for sustained economic development in the region. Private participation in infrastructure financing and management stems, according to Avraham Shohat, Minister of Finance of Israel, from several factors:

a. There is an increasing body of evidence supporting the theory that in general, the private sector out performs the public sector in terms of reducing construction costs and time, providing consumer oriented services, and improving operational efficiency.
b. Given diminishing sources of both foreign and domestic public funds available to developing nations, the need for private financing is becoming increasingly apparent.
c. Technological developments in some infrastructure subsectors have changed their structural nature. Once natural monopolies, such as port authorities, telephone companies, and power suppliers are now open to private sector competition in niche markets.
d. Financial innovation, in conjunction with global financial market integration, have combined to create a market for emerging market securities. The number of private and institutional investors seeking to diversify portfolios by investing in these markets has steadily

increased over the past decade. Hence, the ability to raise private capital is currently greater than it has been in the past.

Between 1984 and 1995, according to the World Bank, some 90 countries privatized nearly 550 infrastructure companies and more than 80 countries had nearly 600 active new private infrastructure projects. But barely 1 percent of this investment flowed to the Middle East and North Africa. The region had fewer than 10 infrastructure projects involving private investment—mainly gas pipelines and telecommunications projects. Fortunately, today most MENA nations are starting to experiment with private participation in infrastrucutre in a variety of sectors, including telecommunications, power, transport, and water and sanitation.

See investment, privatization.

privatization: Until recently, privatization in MENA nations has lagged behind regions such as Eastern Europe, Latin America, and Asia. The share of public enterprises is higher in the nations of the MENA region than in other middle-income countries, there have been very few privatization actions, and the revenues from privatization has been small.

Then in the early 1990s, most MENA countries launched privatization programs of some form. For example, privatization of state-owned enterprises in sectors such as tourism, manufacturing, real estate, and banking has accelerated. In short, privatization is now gaining momentum.

Nevertheless, to reap the efficiency and productivity gains that privatization can bring, several key issues will have to be addressed in MENA nations:

a. Transactions will have to be transparent, which requires clear procedures, competition among buyers, and an incorruptible privatization agency.
b. Regulations must provide incentives for private suppliers of infrastructure services to cut costs and provide better services to consumers.
c. Foreign capital and expertise are much needed in the region, and allowing foreigners to participate in privatizations is important. Political fears of foreign controls over strategic sectors can be mitigated through effective regulation.
d. Popular participation in privatization is an important means of overcoming public resistance. Voucher schemes, underpricing of public offerings, and employee share ownership schemes are all mechanisms for encouraging participation and garnering political support.
e. Fears of layoff have thwarted privatization in many MENA nations. The solution to overmanning is to deal fairly with unnecessary workers by giving them a choice of generous severance payments,

retraining, relocation grants, and/or participation in small business promotion programs.

As important as paving the way for meaningful privatization is the process of institutional building that is doing remarkably well in MENA nations, where there is a growing recognition, both officially and among the general public, of the importance of introducing more participatory democracy and legal reforms.

By 1997, in most countries of MENA, some form of economic reform is underway. More often than not it involves privatization which goes hand in hand with the development of the local stock market. A major movement on trade barriers was initiated in early 1997.

See Amman Middle East/North Africa Economic Summit, Cairo Middle East/North Africa Economic Conference, Egypt, Jordan, Lebanon, Morocco.

production structure: See trade reform (MENA).

protectionism: See Middle East/North Africa (trade), trade reform (MENA).

protocol (Gaza Strip and Jericho): See Declaration of Principles, Protocols on Economic Relations.

Protocol Concerning the Redeployment in Hebron: See Hebron.

Protocol on Economic Relations: Approved by officials from Israel (Finance Minister Avraham Shohat) and the PLO (Abu Ala [Ahmed Korei]) on April 29, 1994, it deals only with the Gaza Strip and the Jericho Area. Primary excerpts are:

Preamble

The two parties view the economic domain as one of the cornerstones in their mutual relations with a view to enhance their interest in the achievement of a just, lasting, and comprehensive peace. Both parties shall cooperate in this field in order to establish a sound economic base for these relations, which will be governed in various economic spheres by the principles of mutual respect of each other's economic interests, reciprocity, equity, and fairness.

This protocol lays the groundwork for strengthening the economic base of the Palestinian side and for exercising its right of economic decision making in accordance with its own development plan and priorities. The two parties recognize each other's economic ties with other markets and the need to create a better economic environment for their peoples and individuals.

Article I—Framework and Scope of this Proposal

1. This Protocol establishes the contractual agreement that will govern the economic relations between the two sides and will cover the West Bank and the Gaza Strip during the interim period. The implementation will be according to the stages envisaged in the Declaration of Principles on Interim- Self-Government Arrangements signed in Washington, DC on September 13, 1993 and the agreed minutes thereto. It will therefore begin in the Gaza Strip and the Jericho Area and at a later stage will also apply to the rest of the West Bank, according to the provisions of the Interim Agreement and to any other agreed arrangements between the two sides.
2. This Protocol, including its Appendices, will be incorporated into the Agreement on the Gaza Strip and the Jericho Area (in this Protocol, the Agreement), will be an integral part thereof and interpreted accordingly. This paragraph refers solely to the Gaza Strip and the Jericho Area.
3. This Protocol will come into force upon the signing of the agreement.
4. For the purpose of this Protocol, the term "Areas" means the areas under the jurisdiction of the Palestinian Authority, according to the provisions of the agreement regarding territorial jurisdiction.

Article II—The Joint Economic Committee

1. Both parties will establish a Palestinian-Israel Joint Economic Committee (hereinafter, the JEC) to follow up the implementation of this Protocol and to decide on problems related to it that may arise from time to time. Each side may request the review of any issue related to this Agreement by the JEC.
2. The JEC will serve as the continuing committee for economic cooperation envisaged in Annex III of the Declaration of Principles.
3. The JEC will consist of an equal number of members from each side and may establish subcommittees specified in this Protocol. A subcommittee may include experts as necessary.
4. The JEC and its subcommittees shall reach their decisions by agreement and shall determine their rules of procedure and operation, including the frequency and place or places of their meetings.

Article III—Import Taxes and Import Policy

1. The import and customs policies of both sides will be according to the principles and arrangements detailed in this Article.

2.
   a. the Palestinian Authority will have all powers and responsibilities in the sphere of import and customs policy and procedures with regard to the following:
      (1) Goods on List A1, attached hereto to Appendix I locally produced in Jordan and in Egypt particularly and in the other Arab countries, which the Palestinians will be able to import in quantities agreed upon by the two sides up to the Palestinian market needs as estimated according to paragraph 3 below.
      (2) Goods on List A2, attached hereto as Appendix II, from the Arab, Islamic, and other countries, which the Palestinians will be able to import in quantities agreed upon by the two sides up to the Palestinian market needs as estimated according to paragraph 3 below.
   b. The import policy of the Palestinian Authority for Lists A1 and A2 will include independently determining and changing from time to time the rates of customs, purchase tax, levies, excises, and other charges, the regulation of licensing requirements and procedures and of standard requirements. The valuation for customs purposes will be based upon the GATT 1994 agreement as of the date it will be introduced in Israel, and until then on the Brussels Definition of Valuation (BDV) system.
3. For the purposes of paragraph 2(a) above, the Palestinian market needs for 1994 will be estimated by a subcommittee of experts. These estimates will be based on the best available data regarding past consumption, production, investment, and external trade of the Areas.
4. The Palestinian Authority will have all powers and responsibilities to independently determine and change from time to time the rates of customs, purchase taxes, levies, excises, and other charges on the goods on List B, attached hereto as Appendix III, of basic food items and other goods for the Palestinian economic development program, imported by the Palestinians to the Areas.
6. Each side will notify the other side immediately of changes made in rates and in other matters of import policy, regulations, and procedures determined by it within its respective powers and responsibilities as detailed in this Article. With regard to changes which do not require immediate application upon decision, there will be a process of advance notifications and mutual consultations which will take into consideration all aspects and economic implications.
7. The Palestinian Authority will levy VAT at one rate on both locally produced goods and services and on imports by the Palestinians

(whether covered by the three lists mentioned above or not), and may fix it at the level of 15 to 16 percent.

8. Goods imported from Jordan, Egypt, and other Arab countries according to paragraph 2(a)(1) above (List A1) will comply with rules of origin agreed upon by a joint subcommittee within three months of the date of the signing of the Agreement.
9. Each side will issue import licenses to its own importers, subject to the principles of this Article and will be responsible for the implementation of the licensing requirements and procedures prevailing at the time of the issuance of the licenses. Mutual arrangements will be made for the exchange of information relevant to licensing matters.
10. Except for the goods on Lists A1 and A2 and their Quantities—in which the Palestinian Authority has all powers and responsibilities, both sides will maintain the same import policy (except for rates of import taxes and other charges for goods in List B) and regulations including classification, valuation, and other customs procedures, which are based on the principles governing international codes, and the same policies of import licensing and of standards for imported goods, all as applied by Israel with respect to its importation. Israel may from time to time introduce changes in any of the above, provided that changes in standard requirements will not constitute a non-tariff barrier and will be based on consideration of health, safety, and the protection of the environment in conformity with Article 2.2 of the Agreement on Technical Barriers to trade of the Final Act of the Uruguay Round of Trade Negotiations.
11.
    a. The Palestinian Authority will determine its own rates of customs and purchase tax on motor vehicles imported as such, to be registered with the Palestinian Authority. However, the Palestinian Authority may request, through the subcommittee on transportation, that in special cases different standards will apply.
    b. Each side may determine the terms and conditions for the transfer of motor vehicles registered in the other side to the ownership or use of a resident of its own side, including the payment of the difference of import taxes, if any, and the vehicle having been tested and found compatible with the standards required at that time by its own registration administration, and may prohibit transfer of vehicles.
12.
    a. Jordanian standards, as specified in the attached Appendix I, will be acceptable in importing petroleum products into the

Areas, once they meet the average of the standards existing in the European Union countries, or the U.S. standards, which parameters have been set at the values prescribed for the geographical conditions of Israel, the Gaza Strip, and the West Bank. Cases of petroleum products which do not meet these specifications will be referred to a joint experts' committee for a suitable solution.

b. The difference in the final price of gasoline to consumers in Israel and to consumers in the Areas, will not exceed 15 percent of the official final consumer price in Israel. The Palestinian Authority has the right to determine the prices of petroleum products, other than gasoline, for consumption in the Areas.

14. In the entry points of the Jordan River and the Gaza Strip:
    a. Freight shipment: The Palestinian Authority will have full responsibility and powers in the Palestinian customs points (freight area) for the implementation of the agreed upon customs and importation policy as specified in this Protocol, including the inspection and the collection of taxes and other charges, when due. Israeli customs officials will be present and will receive from the Palestinian customs officials a copy of the necessary relevant documents related to the specific shipment and will be entitled to ask for inspection in their presence of both goods and tax collection. The Palestinian customs officials will be responsible for the handling of the customs procedure including the inspection and collection of due taxes.
    b. Passengers customs lane: Each side will administer its own passengers customs procedures, including inspection and tax collection. The inspection and collection of taxes due in the Palestinian customs lane will be conducted by customs officials of the Palestinian Authority. Israel customs officials will be invisibly present in the Palestinian customs lane and entitled to request inspection of goods and collection of taxes when due. In the case of suspicion, the inspection will be carried out by the Palestinian official in a separate room in the presence of the Israeli customs official.

15. The clearance of revenues from all import taxes and levies, between Israel and the Palestinian Authority, will be based on the principle of the place of final destination. In addition, these tax revenues will be allocated to the Palestinian Authority even if the importation was carried out by Israeli importers when the final destination explicitly stated in the import documentation is a corporation registered by the Palestinian Authority and conducting business activity in the Areas.

17. The Palestinian Authority will have the right to exempt the Palestinian returnees who will be granted permanent residency in the Areas from import taxes on personal belonging including house appliances and passenger cars as long as they are for personal use.
19. Donations in kind to the Palestinian Authority will be exempted from customs and other import taxes if destined and used for defined development projects or noncommercial humanitarian purposes. The Palestinian Authority will be responsible exclusively for planning and management of the donors' assistance to the Palestinian people.

Article IV—Monetary and Financial Issues

1. The Palestinian Authority will establish a Monetary Authority in the Areas. The PMA will have the powers and responsibilities for the regulation and implementation of the monetary policies within the functions described in this Article.
2. The PMA will act as the Palestinian Authority's official economic and financial advisor.
3. The PMA will act as the Palestinian Authority's and the public-sector entities' sole financial agent, locally and internationally.
4. The foreign currency reserves (including gold) of the Palestinian Authority and all Palestinian public-sector entities will be deposited solely with the PMA and managed by it.
5. The PMA will act as the lender of last resort for the banking system in the Areas.
6. The PMA will authorize foreign exchange dealers in the Areas and will exercise control (regulation and supervision) over foreign exchange transactions within the Areas and with the rest of the world.
7.
    a. The PMA will have a banking supervision department that will be responsible for the proper functioning, stability, solvency, and liquidity of the banks operating in the Areas.
    c. The supervision department will be charged with the general supervision of every such bank, including:
        - the regulation of all kinds of banking activities, including their foreign activities;
        - the licensing of banks formed locally and of branches, subsidiaries, joint ventures, and representative offices of foreign banks and the approval of controlling shareholders;
        - the supervision and inspection of banks.
8. The PMA will relicense each of the five branches of the Israeli banks operating at present in the Gaza Strip and the West Bank, as soon as

its location or the authorities regarding it come under the jurisdiction of the Palestinian Authority. These branches will be required to comply with the general rules and regulations of the PMA concerning foreign banks.

9.

a. Any other Israeli bank wishing to open a branch or a subsidiary in the Areas will apply for a license to the PMA and will be treated equally to other foreign banks, provided that the same will apply to the Palestinian banks wishing to open a branch or a subsidiary in Israel.

e. The host authority will regularly examine the activities of branches and subsidiaries in the area under its jurisdiction. The home authority will have the right to conduct on-site examinations in the branches and subsidiaries in the host area. However, the supervision responsibilities of the home authority concerning subsidiaries will be according to the "Basle Concordat."

f. The BOI and the PMA will establish a mechanism for cooperation and for the exchange of information on issues of mutual interest.

10.

a. The New Israeli Sheqel (NIS) will be one of the circulating currencies in the Areas and will legally serve there as means of payment for all purposes including official transactions. Any circulating currency, including the NIS, will be accepted by the Palestinian Authority and by all its institutions, local authorities, and banks, when offered as a means of payment for any transaction.

b. Both sides will continue to discuss, through the JEC, the possibility of introducing mutually agreed Palestinian currency or temporary alternative currency arrangements for the Palestinian Authority.

11.

a. The liquidity requirements on all deposits in banks operating in the Areas will be determined and announced by the PMA.

b. Banks in the Areas will accept NIS deposits.

c. The supervision and inspection of the implementation of all liquidity requirements will be carried out by the PMA.

15. The PMA will have the right to convert at the BOI excess NIS received from banks operating in the Areas into foreign currency, in which the BOI trades in the domestic interbank market, up to the amounts determined per period, according to the arrangements detailed in paragraph 16 below.

17. The PMA and the BOI will meet annually to discuss and determine the annual amount of convertible NIS during the following calendar year and will meet semiannually to adjust the said amount.
19. There will be no ceiling on the annual foreign currency conversions by the PMA into NIS. However, in order to avoid undesirable fluctuations in the foreign exchange market, monthly ceilings of such conversions will be agreed upon in the annual and semiannual meetings referred to in paragraph 17.
20. Banks in the Areas will convert NIS into other circulating currencies and vice versa.
21. The Palestinian Authority will have the authorities, powers, and responsibilities regarding the regulation and supervision of capital activities in the Areas, including the licensing of capital market institutions, finance companies, and investment funds.

Article V—Direct Taxation

1. Israel and the Palestinian Authority will each determine and regulate independently its own tax policy in matters of direct taxation, including income tax on individuals and corporations, property taxes, municipal taxes, and fees.
2. Each tax administration will have the right to levy the direct taxes generated by economic activities within its area.
3. Each tax administration may impose additional taxes on residents within its area on (individuals and corporations) who conduct economic activities in the other side's area.
4. Israel will transfer to the Palestinian Authority a sum equal to:
   a. 75 percent of the income taxes collected from Palestinians from the Gaza Strip and the Jericho Area employed in Israel.
   b. The full amount of income taxes collected from Palestinians from the Gaza Strip and Jericho Area employed in the settlements.
5. The two sides will agree on a set of procedures that will address all issues concerning double taxation.

Article VI—Indirect Taxes on Local Production

1. The Israel and the Palestinian tax administrations will levy and collect VAT and purchase taxes on local production, as well as any other indirect taxes, in their respective areas.
2. The purchase tax rates within the jurisdiction of each tax administration will be identical as regards locally produced and imported goods.
3. The present Israeli VAT rate is 17 percent. The Palestinian VAT rate will be 15 to 16 percent.

4. The Palestinian Authority will decide on the maximum annual turnover for businesses under its jurisdiction to be exempt from VAT, within an upper limit of US$12,000.
5. The VAT on purchases by businesses registered for VAT purposes will accrue to the tax administration with which the respective business is registered. Businesses will register for VAT purposes with the tax administration of the side of their residence, or on the side of their ongoing operation.
6. VAT paid by not-for-profit Palestinian organizations and institutions, registered by the Palestinian Authority, on transactions in Israel, will accrue to the Palestinian tax administration.

Article VII—Labor

1. Both sides will attempt to maintain the normality of movement of labor between them, subject to each side's right to determine from time to time the extent and conditions of the labor movement into its area. If the normal movement is suspended temporarily by either side, it will give the other side immediate notification, and the other side may request that the matter be discussed in the Joint Economic Committee. The placement and employment of workers from one side in the area of the other side will be through the employment service of the other side and in accordance with the other sides' legislation. The Palestinian side has the right to regulate the employment of Palestinian labor in Israel through the Palestinian employment service, and the Israel Employment Service will cooperate and coordinate in this regard.
2.
   a. Palestinians employed in Israel will be insured in the Israeli social insurance system according to the National Insurance Law for employment injuries that occur in Israel, bankruptcy of employers and maternity leave allowance.
   c. Implementation procedures relating thereto will be agreed upon between the Israeli National Insurance Institute and the Palestinian Authority or the appropriate Palestinian social insurance institution.
3. Israel will transfer to the Palestinian Authority, on a monthly basis, the equalization deductions as defined by Israeli legislation, if imposed and to the extent levied by Israel. The sums so transferred will be used for social benefits and health services, decided upon by the Palestinian Authority, for Palestinians employed in Israel and for their families.
4. Israel will transfer, on a monthly basis, to a relevant pension insurance institution to be established by the Palestinian Authority, pen-

sion insurance deductions collected after the establishment of the above institution and the completion of the documents mentioned in paragraph 6.

5. Upon the receipt of the deductions, the Palestinian Authority and its relevant social institutions will assume full responsibility in accordance with the Palestinian legislation and arrangements, for pension rights and other social benefits of Palestinians employed in Israel, that accrue from the transferred deductions related to these rights and benefits.
7. The above arrangements concerning equalization deductions and/ or pension deductions may be reviewed and changed by Israel if an authorized court in Israel will determine that the deductions or any part thereof must be paid to individuals, or used for individual social benefits or insurance in Israel, or that it is otherwise unlawful.
9.
   a. The Palestinian Authority may integrate the existing health insurance scheme for Palestinians employed in Israel and their families in its health insurance services.
   b. The Palestinian Authority may integrate the existing health insurance scheme for Palestinians who were employed in Israel and are receiving pension payments through the Israeli Employment Service, in its health insurance services.
10. The JEC will meet upon the request of either side and review the implementation of this Article and other issues concerning labor, social insurance, and social rights.
12. Palestinians employed in Israel will have the right to bring disputes arising out of employee-employer relationships and other issues before the Israeli Labor Courts, within these courts' jurisdiction.
13. This Article governs the future labor relations between the two sides and will not impair any labor rights to the date of signing of the Agreement.

Article VIII—Agriculture

1. There will be free movement of agricultural produce, free of customs and import taxes, between the two sides, subject to the following exceptions and arrangements.
2. The official veterinary and plant protection services of each side will be responsible, within the limits of their respective jurisdiction, for controlling animal health, animal products and biological products, and plants and parts thereof, as well as their importation and exportation.

3. The relations between the official veterinary and plant protection services of both sides will be based on mutuality in accordance with the following principles, which will be applied in all the areas under their respective jurisdiction.
4. Trade between the two sides in animals, animal products, and biological products will be in keeping with the principles and definitions set out in the current edition of the OIE National Animal Health Code as updated from time to time (hereinafter, IAHC).
5. Transit of livestock, animal products, and biological products from one side through the area under the jurisdiction of the other side, should be conducted in a manner aimed at the prevention of diseases spreading to or from the consignment during its movement.
6. The official veterinary services of each side have the authority to issue veterinary import permits for import of animals, animal products, and biological products to the areas under its jurisdiction.
8. Transportation of livestock and poultry, animal products, and biological products destined for Israel from the Areas and vice versa will be subject to veterinary permits issued by the official veterinary services of the recipient side, in keeping with the OIE standards used in international traffic in this field.
10. The agricultural produce of both sides will have free and unrestricted access to each other's markets, with the temporary exception of sales from one side to the other side of the following items only: poultry, eggs, potatoes, cucumbers, tomatoes, and melons. The temporary restrictions on these items will be gradually removed on an increasing scale until they are finally eliminated by 1998.
11. The Palestinians will have the right to export their agricultural produce to external markets without restrictions, on the basis of certificates of origin issued by the Palestinian Authority.
12. Without prejudice to obligations arising out of existing international agreements, the two sides will refrain from importing agricultural products from third parties which may adversely affect the interests of each other's farmers.
13. Each side will take the necessary measures in the area under its jurisdiction to prevent damage which may be caused by its agriculture to the environment of the other side.
14. The two sides will establish a subcommittee of their respective official veterinary and plant protection services, which will update the information and review issues, policies, and procedures in these fields. Any changes in the provisions of this Article will be agreed upon by both sides.
15. The two sides will establish a subcommittee of experts in the dairy sector in order to exchange information, discuss and coordinate

their production in this sector so as to protect the interests of both sides. In principle, each side will produce according to its domestic consumption.

Article IX—Industry

1. There will be free movement of industrial goods free of any restrictions including customs and import taxes between the two sides, subject to each side's legislation.
2.
   a. The Palestinian side has the right to employ various methods in encouraging and promoting the development of the Palestinian industry by way of providing grants, loans, research and development assistance, and direct-tax benefits. The Palestinian side has also the right to employ other methods of encouraging industry resorted to in Israel.
   b. Both sides will exchange information about the methods employed by them in the encouragement of their respective industries.
3. Each side will do its best to avoid damage to the industry of the other side and will take into consideration the concerns of the other side in its industrial policy.
4. Both sides will cooperate in the prevention of deceptive practices, trade in goods which may endanger health, safety, and the environment and in goods of expired validity.
5. Each side will take the necessary measures in the area under its jurisdiction to prevent damage which may be caused by its industry to the environment of the other side.
6. The Palestinians will have the right to export their industrial produce to external markets without restrictions, on the basis of certificates of origin issued by the Palestinian Authority.
7. The JEC will meet and review issues pertaining to this Article.

Article X—Tourism

1. The Palestinian Authority will establish a Palestinian Tourism Authority which will exercise, inter alia, the following powers in the Areas.
   a. Regulating, licensing, classifying, and supervising tourist services, sites, and industries.
   b. Promoting foreign and domestic tourism and developing the Palestinian tourist resources and sites.
   c. Supervising the marketing, promotion, and information activities related to foreign and domestic tourism.

2. Each side shall, under its respective jurisdiction, protect, guard, and ensure the maintenance and good upkeep of historical, archaeological, cultural and religious sites, and all other tourist sites, to fit their status as well as their purpose as destination for visitors.
3. Each side will determine reasonable visiting hours and days for all tourist sites in order to facilitate visits at a wide variety of days and hours, taking into consideration religious and national holidays. Each side shall publicize such opening times.
4. Tourist buses or any other form of tourist transport authorized by either side, and operated by companies registered and licensed by it, will be allowed to enter and proceed on their tour within the area under the jurisdiction of the other side, provided that such buses or other vehicles conform with the EEC technical specifications.
6. Tourism companies and agencies licensed by either side shall enjoy equal access to tourism-related facilities and amenities in border points of exit and entry according to the regulations of the authority operating them.
8. Each side will make its own arrangement for compensation of tourists for bodily injury and property damages caused by political violence in the areas under its respective jurisdiction.
9. The JEC or a tourism subcommittee established by it shall meet upon the request of either side in order to discuss the implementation of the provisions of this Article and resolve problems that may arise.

Article XI—Insurance Issues

1. The authorities, powers, and responsibilities in the insurance sphere in the Areas, including inter alia the licensing of insurers, insurance agents, and the supervision of their activities, will be transferred to the Palestinian Authority.
2.
   a. The Palestinian Authority will maintain a compulsory absolute liability system for road accident victims with a ceiling on the amount of compensation.
   c. Any change by either side in the rules and regulations regarding the implementation of the abovementioned principles will require prior notice to the other side. A change which might substantially affect the other side will require prior notice of at least three months.
3.
   a. Upon the signing of the Agreement the Palestinian Authority will establish a fund to the Areas. The Palestinian Fund will assume the responsibilities of the Statutory Road Accident Vic-

tims Compensation Fund in the West Bank and the Gaza Strip (hereinafter, the Existing Fund) regarding the Areas, according to the prevailing law at that time.

b. The Existing Fund will transfer to the Palestinian Fund, after the assumption of the above mentioned responsibilities by it, the premiums paid to the Existing Fund by the insurers for vehicles registered in the Areas, pro-rata to the unexpired period of each insurance policy.

4.

a. Compulsory motor vehicles insurance policies issued by insurers licensed by either side will be valid in the territories of both sides. Accordingly, a vehicle registered in one side covered by such a policy will not be required to have an additional insurance coverage for travel in the areas under the other side's jurisdiction. These insurance policies will cover all the liabilities according to the legislation of the place of the accident.

b. In order to cover page of the liabilities which may incur due to road accidents in Israel by uninsured vehicles registered in the Palestinian Authority, the Palestinian Fund will transfer to the Israeli Fund, on a monthly basis, for each insured vehicle, an amount equal to 30 percent of the amount paid to the Israeli Fund by an insurer registered in Israel, for the same type of vehicle, for the same period of insurance.

7. The fund of each side will be responsible toward the victims of the other side for any liability of the insurers of its side regarding the compulsory insurance and will guarantee their liabilities.

12. Insurers from both sides may apply for a license to the relevant authorities of the other side, according to the rules and regulations regarding foreign insurers in the latter side. The two sides agree not to discriminate against such applicants.

Protocol of the Elders of Zion: See Hamas, Hamas Covenant.

Protocol on Further Transfer of Powers and Responsibility: See Preparatory Transfer of Powers and Responsibility.

PSE: See Palestine Securities Exchange.

PSF: See Popular Struggle Front.

public enterprises: See privatization.

public health: See health.

public procurement: See European Union, Industrial Cooperation and Trade (Center of the Middle East/North Africa Community).

public sector: See investment, private sector.

Q

Qaboos, Sultan Bin Said: See Oman.

Qaddafi, Muammar Colonel: The son of nomads, born in 1942 in the desert at Sirete, Qaddafi took a university diploma in history. His army career began in 1963. Six years later he joined the September coup d'etat which overthrew King Idriss Al-Sanoussi and propelled Qaddafi to the presidency of the revolutionary command council in Libya. His leadership has been reinforced by unyielding security and punctuated by conflict with the West.

Since the 1980s Colonel Qaddafi has been in constant friction with the West and the United States over issues of terrorism.

See Libya.

QAR: The international standard code for the currency of Qatar—the Qatari rial.

Qassam Bridges: See Hamas.

Qatar: Officially the State of Qatar. An independent Arab sheikhdom in the Persian Gulf. In 1971 it was declared independence and charted a constitution. Sheik Hamad Bin Khalifa Al-Thani ousted his father as emir in June 1995.

By October 1995 a Memorandum of Understanding was agreed upon between Qatar and Israel regarding the supply of gas. The deal involves the conversion of natural gas to liquid form while in Qatar, the transfer of the liquid to Israel in special ships, and the reconversion of the liquid into gas in Israel.

Qatar plans to host the Middle East/North Africa Economic Summit in November 1997.

See Cairo Declaration, Doha Middle East/North Africa Economic Summit, Gulf Cooperation Council, Gulf nations, intellectual property rights, U.S.-Qatar Business Council.

Qatar Middle East Conference: See Doha Middle East/North Africa Economic Summit.

Qatar Middle East/North Africa Economic Summit: See Doha Middle East/North Africa Economic Summit.

Qatari Rial: The currency of Qatar.

Qawmi: See PLO Charter.

Qawmiyya: See PLO Charter.

quantitative restrictions: See Free Trade Agreement (European Free Trade Association and Israel), Treaty of the Middle East/North Africa Community (Article 6).

quaternary boycott: See boycott.

quotas: See Paris Economic Agreement.

Quran: (Arabic for Recitation). Regarded by muslims as the heavenly book. Composed of the divine revelations received by the Prophet Muhammad (570-632 AD) over the last 20 years of his life. The Quran is the ultimate authority for the Muslim people; is the ultimate source and continual inspiration of Islam.

Synonymous in English with Koran.

R

Rabat Arab Summit: See Hassan II (King).

Rabin, Yitzhak: Born in Jerusalem in 1923, Rabin was the first Prime Minister of Israel to be native-born. A much-decorated army commander in the 1967 Six-Day War, he entered politics in 1968. A Nobel Peace Prize laureate (along with Peres and Arafat), he appeared to be the only leader with sufficient clout among Israeli hawks to sell the peace plan to a skeptical public.

To the end Yitzhak Rabin urged "a fabric of coexistence and cooperation." His last international conference took place at the Amman Middle East/North Africa Economic Summit (October 29, 1995). Less than one week before his death, Rabin announced a turning point between investing in war and investing in peace, pivotal points of balance throughout his life.

Addressing more than 1,700 global leaders in business and government from 63 nations he drew attention to shared difficulties noting that the same region which had four million more mouths to feed than it did last year (1994) had only raised 3 percent of the international funds earmarked for its development.

He asserted Israel's commitment to playing a constructive and cooperative role in developing the hidden economic potential of the region and referred to the dozens of development projects already initiated in such fields as water, transportation, agriculture, environment, and energy.

"We must overcome, on both sides, psychological barriers and generations of hostilities....But in the drowsy Middle East...we must hurry to make up lost time," reminding MENA participants that greater efforts must be directed toward the economy and improving the lives of people.

He called on the leaders of the region to concentrate on economic development and improve the quality of life: "But the Middle East needs extra regional help to achieve it...." Applause was heard when he stated that "No one will come here because of our winning smiles....They all want to make money."

At the opening plenary session, in an apparent rejoiner to Chairman Arafat, the consummate politician Rabin—both Prime Minister and Defense Minister—noted that he had arrived one hour earlier from Jerusalem, which he described as "Israel's capital."

He concluded his speech by expressing the hope that the Middle East would become one of the most viable and prosperous regions in the world. As he urged all participants at the Amman Summit to stand up to the challenges facing them, he set himself as the foremost model.

Returning from the Amman Economic Summit, Prime Minister Rabin was assassinated by Yigal Amir, a fanatic Jewish Israeli, while at a peace rally in Jerusalem on November 4, 1995.

See Amman Middle East/North Africa Economic Summit, Beilin-Eitan agreement, Casablanca Middle East/North Africa Economic Summit, chronology (November 5, 1995), Declaration of Principles, Ein Avrona, Gaza Strip and the Jericho Area Agreement, Interim Agreement (on the West Bank and the Gaza Strip), Israel, Israel-Jordan Peace Treaty, Jordan, Labor Party (Israel), Letters of Mutual Recognition, New Middle East, Oman, Oslo Agreement, perceptions; Peres, Shimon; Rabin, Yitzhak (Letter to Arafat recognizing the PLO, September 9, 1993), Washington Declaration.

Rabin, Yitzhak (Letter to Arafat recognizing the PLO, September 9, 1993):

Yasser Araft
Chairman
The Palestinian Liberation Organization
Mr. Chairman:

In response to your letter of September 9, 1993, I wish to confirm to you that, in light of the PLO commitments included in your letter, the government of Israel has decided to recognize the PLO as the representative of the Palestinian people and commence negotiations with the PLO within the Middle East Peace Process.

Sincerely,

Yitzhak Rabin
Prime Minister of Israel

See Arafat, Yasir (Letter to Y. Rabin recognizing Israel's right to exist in peace, September 9, 1993).

Rachel's Tomb: See Interim Agreement (on the West Bank and the Gaza Strip).

Ra'ees Yasser Arafat: See Arafat, Yasser.

Rafah crossing: See Cairo Agreement, Gulf Strip.

Rafsanjani, Ali Akbar Hashemi (President): Born in 1933, Rafsanjani took over as president of Iran in August 1989 and was reelected in June 1993, during which time he worked to institutionalize the Islamic revolution and engineer a cautious opening to the West in the post-Khomeini period. By temperament a free marketeer, Rafsanjani wears the robe and turban of a cleric, but favors rational management. This has brought him into conflict with hardliners. He denies claims that his regime sponsors international terrorism and has attempted to remove the U.S. investment and trade embargo with the message that the door is open to better bilateral relations. The effects of isolation and especially of payments problems in 1993, means that Rafsanjani's reforms have not been a great success.

See Iran.

rail transport: See transport.

Ramadan War: See Yom Kippur War.

Ras Muhammad: See Camp David Accords.

RCD (Regional Cooperation for Development): A regional international institution, formed in 1964 by Iran, Pakistan, and Turkey, with headquarters in Tehran, Iran. Purports to develop and promote technical cooperation and economic progress to further integrate members' economies and trade.

reciprocity: A policy innovation by Prime Minister Netanyahu whereby Israel refrains from doing what it is supposed to do under the terms of the Oslo accords until the Palestinians, for their part, live up to the requiremens Oslo imposed on them.

See Banking and Financial Services (Center of the Middle East/North Africa Community).

recognition of the State of Israel: See Stockholm PLO-American delegation statement.

redeployment: Neither the DOP, nor the Interim Agreement contains a definition of redeployment. However, the word is to be distinugished from "withdrawal," as used for example in the context of the withdrawal of Israeli forces from the Gaza Strip. Unlike "withdrawal," which required the removal of the majority of forces from the areas in question, "redeployment" relates only to the location of the forces; it places no restriction on the number of forces and military equipment or the possibility of introducing further forces and equipment if necessary.

See further redeployments.

Redeployment Accord: See Beilin-Eitan agreement, Declaration of Principles, further redeployments, Hebron, Interim Agreement (on the West Bank and the Gaza Strip), redeployment.

Redirected Mediterranean Policy: See European Union.

red tape: See informal economy.

REDWG (Regional Economic Development Working Group): See Amman Middle East/North Africa Economic Summit, Casablanca Declaration, Euro-Mediterranean Partnership, Moscow Multilateral Middle East Conference, Swiss Trade Initiative for the Middle East and North Africa.

reforms: See Amman Middle East/North Africa Economic Summit.

refugees: As dealing with Palestinians, refugees are people displaced during war, and as a result of the War of Independence in 1948. They include Palestinians who lived within the Green Line, and Jews ejected from Arab countries. The Palestinians demand the right of return for all Palestinians displaced in 1948. The Israelis reject this claim and view it as a direct threat to Israeli statehood.

Displaced persons are Palestinians displaced from the West Bank and Gaza Strip as a result of the 1967 War.

Expired "Permit Palestinians" are former Palestinian residents who went abroad and overstayed their permits. As a result, they have not been allowed to reenter the territories. There are approximately 50,000 such cases (Jordan claims 90,000).

A Multilateral Working Group on Refugees (MWGR) was created by the Madrid Conference to conduct studies and initiate projects to improve the living conditions of Palestinian refugees wherever they reside. MWGR purports to improve the overall welfare of Palestinians living in the territories and in refugee camps in Jordan, Lebanon, and Syria. Strong emphasis is placed on improving housing and economic conditions, including the generation of employment and income.

See Beilin-Eitan agreement, Israel-Jordan Common Agenda, Israel-Jordan Peace Treaty, Jordanian-PLO Accord, Moscow Multilateral Middle East Conference, Stockholm PLO-American delegation statement.

Refugees Working Group: See Moscow Multilateral Middle East Conference.

region: A region is defined by the most basic characteristics of geography and culture. However, there is also potential for welfare-enhancing regional economic interaction. These are also determined by economic and political factors. The more similar the economic and political systems, and the more similar the political goals among countries, the easier it has been to promote effective regional economic integration.
See regional economic integration.

Regional Business Council: See Amman Middle East/North Africa Economic Summit.

Regional Cooperation for Development: See RCD.

Regional Cooperative Network in Science and Technology: See Secretariat (Middle East/North Africa Economic Summits).

regional development centers: Establishing regional development centers along borders with member states of a community not only gives the message of economic commitment but also serves as a confidence-building measure. For example, in May 1995 the Israel-Morocco Chamber of Commerce opened in Tel Aviv. The chamber is the first binational chamber of commerce established between Israel and an Arab state.

These mini-community locals should result in economies of scale and scope, and in the conservation of precious land resources. Each one will have its own relative advantages, for example, activities of agriculture, industry, tourism, transportation, energy, protection of the environment, which are dependent on climatic, topographic, panoramic, availability of historic sites, and so on.

Projects resulting from these operations will:

- receive inputs from the periphery, including energy, water, roads, and raw materials;
- have a density of activities as compact as possible so as to save land resources and energy; high-rise and deep underground structures may be preferred;

- be able to benefit from services such as banking, insurance, communications, advertising, management advisory services, and so on—concentration of the services should help to lower their cost;
- become an integral part of their immediate surroundings and the activities carried out in them will be integrated with communal life in the area;
- take all precautions to ensure that damage to the environment will be controlled and minimal, if at all.

See investing in peace.

regional economic alliances: See regional economic integration (MENA).

regional economic arrangements: See regional economic integration (MENA).

Regional Economic Development Working Group: See Moscow Multilateral Middle East Conference, REDWG.

regional economic integration (MENA): The Organization for Economic and Cooperation and Development (OECD) identifies three reasons countries seek greater regional integration:

a. economic welfare gains;
b. increasing the region's collective political bargaining power in extra-regional issues; and
c. achieving other noneconomic national goals, such as meeting security concerns and preventing future conflict.

According to the International Monetary Fund report of April 1996:

> Despite favorable geographical and cultural elements, the MENA region remains remarkably unintegrated economically. Intraregional trade is small, tourist and labor flows are skewed, and private capital transactions remain limited. Moreover, there has been very little effective regional economic policy coordination.
>
> Given this low level of interaction, the paper finds that there are gains to be achieved from increased regional interaction in MENA, including the stimulation of growth and employment, which is the major challenge facing the region. The scope for such gains is limited, however, by some of the economic characteristics of the economies (particularly, the similarity of their resource endowments) and political factors (including the legacy of the prolonged Arab-Israeli conflict).
>
> Realizing gains from increased regional interaction requires policy changes, most of which are also needed if these countries are to benefit from the broader process of globalization of the international economic system. Indeed, the policy implications of the analysis in this paper is that, rather than set as their first economic policy priority the goal of regional integration, MENA countries should focus on domestic policy reforms and the associated process of greater integration into the world economy. Successful liberalization will promote regional interaction.

In addition, the IMF characterizes the possible forms of regional economic arrangements as:

- preferential trading arrangements, in which regional trading partners enjoy more favorable trading conditions, including lower tariffs, than other countries;
- a free trade area (FTA), in which trade within the area is not subject to taxation but external tariffs on third parties may differ among member countries;
- a customs union, in which trade in goods and capital flows within the region is free, where market conditions (including standards) have been made uniform so that producers in one country face no non-market restrictions in selling in other countries in the common market, and where restrictions on labor flows are more limited; and
- an economic union, where members eventually share a single currency, and the concept of national residency is eliminated in all economic relations through two key principles: mutual recognition and minimum harmonization.

See Amman Middle East/North Africa Economic Summit, Cairo Economic Conference, Casablanca Middle East/North Africa Economic Summit, free trade (theory), investing in peace, Middle East/North Africa Community, New Middle East, The; region.

regional economic interaction: See regional economic integration.

regional investment: See Amman Middle East/North Africa Economic Summit.

Regional Investment Promotion Program: See Secretariat (Middle East/North Africa Economic Summits).

Regional Investment Promotion Resource Center: See Secretariat (Middle East/North Africa Economic Summits).

regionalism: Goals of regionalism within the Middle East and North Africa will be to:

- present the concrete meaning of peace;
- create confidence building measures;
- establish commitments that will reduce the likelihood of hostile activities;
- promote regional and national economies;

- create additional sources of income and raise the standard of living; and
- present an order of priorities for development of joint projects and for allocation of resources for their implementation.

See Amman Middle East/North Africa Economic Summit, benefits (of Middle East/North Africa regionalization), Casablanca Middle East/North Africa Economic Summit; containment, conflict and common management of common problems; regional economic integration.

Regional Security and Arms Control Working Group (RSAC): See security and arms control.

Regional Security Center: See Moscow Multilateral Middle East Conference.

regional subgroupings: See Middle East/North Africa (MENA).

Regional Tourism Board: See Amman Middle East/North Africa Economic Summit.

religion: See Social Rights (Center of the Middle East/North Africa Community, Tourism (Center of the Middle East/North Africa Community).

religious places: See Interim Agreement (on the West Bank and the Gaza Strip), Israel-Jordan Peace Treaty.

Renovated Mediterranean Policy (RMP): In 1992 the European Union implemented the Renovated Mediterranean Policy. RMP introduced several innovations:

- the financial effort for the period 1992-1996 increased by 50 percent compared with the resources contained in the previous financial protocols;
- a major qualitative leap forward was achieved with the launch of a regional cooperation policy;
- by means of programs of decentralized cooperation which go to the very heart of civil society, the path to a genuine partnership relationship was opened up.

See Euro-Mediterranean Partnership, European Union.

Republic of Iraq: See Iraq.

Republic of Lebanon: See Lebanon.

Republic of Yemen: See Yemen.

Research and Development (Center of the Middle East/North Africa Community): After the signing of the interim agreement between Israel and the PLO, discussions began to study joint ventures. A Mideast Free Trade Zone with a centralized administration would provide revenues that could be returned to participating nations for joint research projects.

Research projects can multiply across all participating nations. Broken into categories, billions of dollars could be well spent on already commenced researched areas, for example:

- regional development projects;
- joint economic and business ventures;
- intercountries economic cooperation.

No one can doubt the importance and urgency of cooperation in research among nations. By implementing a substantial R&D policy, the community can complement and integrate the efforts being made at national levels. The community should ensure that all significant intellectual and financial resources available within the boundaries of participating nations are used in a rational manner, avoiding unnecessary duplication. All findings and applications require an open and full sharing with participants.

Future community R&D efforts should be linked to the following areas:

- quality of life;
- information technology and telecommunications;
- industrial technologies;
- energy;
- biotechnologies and agriculture.

The community can encourage R&D efforts by

- stimulating cooperation and exchanges;
- providing development aid;
- linking international programs to community efforts.

The Middle East/North Africa Community's involvement in research and technology has two beneficial effects: valuable results in the form of advances in scientific knowledge or technological processes, and the development of a genuine tradition of collaborative research involving close, frequent, and long-term cooperation between research centers, universities, and firms in each of the member nations. Individually, nations of the community have made significant contributions. Combined efforts via the region will provide the encouragement needed to continue these, and other expanded tasks which, though progressing well, will prosper while building a genuine research and technology community.

See Centers of the Middle East/North Africa Community, Middle East/ North Africa Community.

Resolution 181 (United Nations): See Arab-Israeli War (I); Arafat, Yasir (Press Statement of December 14, 1988).

Resolution 194 (United Nations): ESTABLISHING A U.N. CONCILIATION COMMISSION, RESOLVING THAT JERUSALEM SHOULD BE PLACED UNDER A PERMANENT INTERNATIONAL REGIME, AND RESOLVING THAT THE REFUGEES SHOULD BE PERMITTED TO RETURN TO THEIR HOMES. The General Assembly having considered further the situation in Palestine,

1. Expresses its deep appreciation of the progress achieved through the good offices of the United Nations Mediator in promoting a peaceful adjustment of the future situation of Palestine, for which cause he sacrificed his life; and Extends its thanks to the Acting Mediator and his staff for their continued efforts and devotion to duty in Palestine;
2. Establishes a Conciliation Commission consisting of three States Members of the United Nations which shall have the following functions:
    a. To assume, in so far as it considers necessary in existing circumstances, the functions given to the United Nations Mediator on Palestine by resolution 186 (S-2) of the General Assembly of May 14, 1948:
    b. To carry out the specific function and directives given to it by the present resolution and such additional functions and directives as may be given to it by the General Assembly or by the Security Council;
    c. To undertake, upon the request of the Security Council, any of the functions now assigned to the United Nations Mediator on Palestine or to the United Nations Truce Commission by resolutions of the Security Council; upon such request to the Conciliation Commission by the Security Council with respect to all the remaining functions of the United Nations Mediator on Palestine under Security Council resolutions, the office of the Mediator shall be terminated;
3. Decides that a Committee of the Assembly, consisting of China, France, the Union of the Soviet Socialist Republic, the United King-

dom, and the United States of America, shall present, before the end of the first part of the present session of the General Assembly, for the approval of the Assembly, a proposal concerning the names of the three states which will constitute the Conciliation Commission;

4. Requests the commission to begin its functions at once, with a view to the establishment of contact between the parties themselves and the commission at the earliest possible date;
5. Calls upon the governments and authorities concerned to extend the scope of negotiations conducted either with the Conciliation Commission or directly, with a view to the final settlement of all questions outstanding between them;
6. Instructs the Conciliation Commission to take steps to assist the governments and authorities concerned to achieve a final settlement of all questions outstanding between them;
7. Resolves that the Holy Places—including Nazareth—religious buildings and sites in Palestine should be protected and free access to them assured, in accordance with existing rights and historical practice; that arrangements to this end should be under effective United Nations supervision that the United Nations Conciliation Commission, in presenting to the fourth regular session of the General Assembly its detailed proposals for a permanent international regime for the territory of Jerusalem, should include recommendations concerning the Holy Places in that territory; that with regard to the Holy Places in the rest of Palestine the Commission should call upon the political authorities of the area concerned to give appropriate formal guarantees as to the protection of the Holy Places and access to them; and that these undertakings should be presented to the General Assembly for approval;
8. Resolves that, in view of its association with three world religions, the Jerusalem area, including the present municipality of Jerusalem plus the surrounding villages and towns, the most eastern of which shall be Abu Dis; the most southern, Bethlehem; the most western, Ein Karim (including also the built-up area of Motsa); and the most northern Shu'fat, should be accorded special and separate treatment from the rest of Palestine and should be placed under effective United Nations control;

Requests the Security Council to take further steps to ensure the demilitarization of Jerusalem at the earliest possible date;

Instructs the Commission to present to the fourth regular session of the General Assembly detailed proposals for permanent international regime for the Jerusalem area which will provide for the maximum local autonomy for distinctive groups consistent with the special international status of the Jerusalem area:

The Conciliation Commission is authorized to appoint a United Nations representative, who shall cooperate with the local authorities with respect to the interim administration of the Jerusalem area;

9. Resolves that, pending agreement on more details arrangements among the governments and authorities concerned, the freest possible access to Jerusalem by road, rail, or air should be accorded to all inhabitants of Palestine.

   Instructs the Conciliation Commission to report immediately to the Security Council, for appropriate action by the organ, any attempt by any party to impede such access;

10. Instructs the Conciliation Commission to seed arrangement among the governments and authorities concerned which will facilitate the economic developments of the area, including arrangements for access to ports and airfields and the use of transportation and communication facilities;

11. Resolves that the refugees wishing to return to the homes and live at peace with their neighbors should be permitted to do so at the earliest practicable date, and the compensation should be paid of the property of the choosing not to return and for loss of or damage to property which, under principles of international law or in equity, should be made good by the governments or authorities responsible;

    Instructs the Conciliation Commission to facilitate the repatriation, resettlement, and economic and social rehabilitation of the refugees and the payment of compensation and to maintain close relations with the Director of the United Nations Relief for Palestine Refugees and, through him, with the appropriate organs and agencies of the United Nations;

12. Authorizes the Conciliation Commission to appoint such subsidiary bodies and to employ such technical experts, acting under this authority, as it may find necessary for the effective discharge of its functions and responsibilities under the present resolutions;

    The Conciliation Commission will have its official headquarters at Jerusalem. The authorities responsible for maintaining order in Jerusalem will be responsible for taking all measures necessary to ensure the security of the Commission. The Secretary General will provide a limited number of guards for the protection of the staff and premises of the Commission;

13. Instructs the Conciliation Commission to render progress reports periodically to the Secretary General for transmission to the Security Council and to the Members of the United Nations.

14. Calls upon all governments and authorities concerned to cooperate with the Conciliation Commission and to take all possible steps to assist in the implementation of the present solution;
15. Requests the Secretary General to provide the necessary staff and facilities and to make appropriate arrangements to provide the necessary funds required in carrying out the terms of the present resolution.

ADOPTED AT THE 186TH PLENARY MEETING.

See Jordanian-PLO Accord, Resolution 242 (United Nations), Resolution 338 (United Nations).

Resolution 242 (United Nations): STATING THE PRINCIPLES OF A JUST AND LASTING PEACE IN THE MIDDLE EAST. Having failed to take any decision, the General Assembly on July 21, 1967 voted to adjourn and to refer the discussion on the Middle East crisis back to the Security Council. Although the Middle East was also the subject of many of the speeches heard during the regular session of the General Assembly in September 1967, the Security Council continued its debates and was presented with draft resolutions by Britain, the Soviet Union, and the United States. On November 22, 1967 it unanimously adopted Resolution 242, which formed the terms of reference for the mission of Ambassador Gunnar Jarring. The resolution was formulated by the British Ambassador to the United Nations, Lord Caradon; it established provisions and principles which, it was hoped, would lead to an agreement.

The Security Council

Expressing its continuing concern with the grave situation in the Middle East, Emphasizing the inadmissibility of the acquisition of territory by war and the need to work for a just and lasting peace in which every state in the area can live in security, Emphasizing further that all member states in their acceptance of the Charter of the United Nations have undertaken a commitment to act in accordance with Article 2 of the Charter.

1. Affirms that the fulfillment of Charter principles requires the establishment of a just and last peace in the Middle East which should include the application of both the following principles:
   (i) Withdrawal of Israeli armed forces from territories occupied in the recent conflict;
   (ii) Termination of all claims or states of belligerency and respect for and acknowledgment of the sovereignty, territorial integrity, and political independence of every state in the area and their right to live in peace within secure and recognized boundaries free from threats or acts of force;

2. Affirms further the necessity
   (a) for guaranteeing freedom of navigation through international waterways in the area;
   (b) for achieving a just settlement of the refugee problem;
   (c) for guaranteeing the territorial inviolability and political independence of every state in the area, through measures including the establishment of demilitarized zones;
3. Requests the Secretary-General to designate a special representative to proceed to the Middle East to establish and maintain contacts with the states concerned in order to promote agreement and assist efforts to achieve a peaceful and accepted settlement in accordance with the provisions and principles in this resolution;
4. Requests the Secretary-General to report to the Security Council on the progress of the efforts of the special representative as soon as possible.

See Arafat, Yasir (Letter to Y. Rabin recognizing Israel's right to exist in peace, September 9, 1993); Arafat, Yasir (Press Statement of December 14, 1988), Camp David Accords, Casablanca Declaration, Declaration of Principles, Egyptian-Israeli Peace Treaty, Framework for the Conclusion of a Peace Treaty between Egypt and Israel, Gaza Strip and the Jericho Area Agreement, Israel-Jordan Common Agenda, London Agreement, PLO Phased Plan, Resolution 338 (United Nations), Security Council, United Nations, Stockholm PLO-American delegation statement, Washington Declaration.

Resolution 338 (United Nations): CALLING FOR A CEASE-FIRE AND FOR THE IMPLEMENTATION OF RESOLUTION 242 IN ALL OF ITS PARTS. Following the recovery of the Golan Heights and the occupation of a bulge inside Syria, and the extension of the Israeli presence in the west bank of the Suez Canal, international efforts to stop the war were intensified. United States Secretary of State Dr. Henry Kissinger flew to Moscow on October 20, 1973 and, together with the Soviet government, the United States proposed a cease-fire resolution. The Security Council met on October 21 at the urgent request of the United States and the USSR. By 14 votes to none and no abstentions (China did not participate in the vote), the council adopted the following resolution on October 22, 1973:

The Security Council

1. Calls upon all parties to the present fighting to cease all firing and terminate all military activity immediately, no later than 12 hours

after the moment of the adoption of this decision, in the positions they now occupy;

2. Calls upon the parties concerned to start immediately after the cease-fire the implementation of Security Council 242 (1967) in all of its parts;
3. Decides that immediately and concurrently with the cease-fire, negotiations start between the parties concerned under appropriate auspices aimed at establishing a just and durable peace in the Middle East.

See Arafat, Yasir (Letter to Y. Rabin recognizing Israel's right to exist in peace, September 9, 1993); Arafat, Yasir (Press Statement of December 14, 1988), Camp David Accords, Casablanca Declaration, Declaration of Principles, Egyptian-Israeli Peace Treaty, Gaza Strip and the Jericho Area Agreement, Israel-Jordan Common Agenda, Jordanian-PLO Accord, London Agreement, Oslo Agreement, Resolution 242 (United Nations), Security Council, Stockholm PLO-American delegation statement, United Nations, Washington Declaration.

Resolution 425 (United Nations): See Lebanon.

Resolution 598 (United Nations): Adopted unanimously on July 20, 1987 by the UN Security Council. The 10-clause resolution included the following operative clauses:

The Security Council

Acting under Articles 39 and 40 of the Charter of the United Nations,

1. Demands that, as a first step toward a negotiated settlement, Iran and Iraq observe an immediate ceasefire, discontinue all military actions on land, at sea and in the air, and withdraw all forces to the internationally recognized boundaries without delay;
2. Requests the Secretary-General to dispatch a team of United Nations observers to verify, confirm, and supervise the cease-fire and withdrawal and further requests the Secretary-General to make necessary arrangements in consultation with the parties and to submit a report thereon to the Security Council;
3. Urges that prisoners of war be released and repatriated without delay after cessation of active hostilities in accordance with the Third Geneva Convention of August 12, 1949;
4. Calls upon Iran and Iraq to cooperate with the Secretary-General in implementing this resolution and in mediation efforts to achieve a comprehensive, just and honorable settlement, acceptable to both sides, of all outstanding issues in accordance with the principles contained in the Charter of the United Nations;

6. Requests the Secretary-General to explore, in consultation with Iran and Iraq, the question of entrusting an impartial body with inquiring into responsibility for the conflict and to report to the Security Council as soon as possible;
7. Recognizes the magnitude of the damage inflicted and the need for reconstruction efforts with appropriate international assistance once the conflict is ended and in this regard requests the Secretary-General to assign a team of experts to study the question of reconstruction and to report to the Security Council.

Resolution 660 (United Nations): See Gulf crisis.

Resolution 661 (United Nations): See Gulf crisis.

Resolution 687 (United Nations): See Gulf crisis.

Resolution 904 (United Nations): See Agreement on Security Arrangements for Hebron.

Resolution 3236 (General Assembly, United Nations): On November 22, 1974 the United Nations General Assembly reaffirmed the Palestinian people's right to self-determination, independence, and sovereignty, and their right to return to their homes and properties.

retaliation: See forewarning.

retirement: See Social Rights (Center of the Middle East/North Africa Community).

revenues: See Interim Agreement.

Revolutionary Palestinian Community Party (RPCP): Moscow-backed PLO faction. Early advocate of two-state solution and supporter of Arafat peace initiative.

Reza Pahlavi, Mohammed: See Pahlavi, Mohamed Reza.

rial: The monetary unit of Iran, Oman, Qatar, and Yemen.
Compare riyal.

Rift Valley: See Israel-Jordan Peace Treaty, Jordan Rift Valley, Washington Declaration.

right of establishment: See Euro-Mediterranean agreements, Treaty of the Middle East/North Africa Community (Article 9).

right of return: See Stockholm PLO-American delegation statement.

riyal: The monetary unit of Oman, Saudi Arabia, and Yemen.
Cf. rial.

RMP: See Renovated Mediterranean Policy.

Road Accident Victims Compensation Fund: See Protocol on Economic Relations (Article XI).

roads: See Israel-Jordan Peace Treaty.

road transport: See Amman Middle East/North Africa Economic Summit, Transportation (Center of the Middle East/ North Africa Community).

Rogers Plan: A U.S. 1969 peace proposal made by U.S. Secretary of State William Rogers; called for Israel to withdraw to its 1967 borders, subject to minor modifications for mutual security, except for the Gaza Strip which was to be negotiated between Israel, Jordan, and Egypt. Both Egypt and Israel rejected the proposal.

Rosenberg, Jerry M.: See Middle East/North Africa Series.

Ross, Dennis: U.S. Special Middle East Coordinator.
See chronology (July 24, 1996) (August 11, 1997), Hebron.

Rotary Clubs: See Hamas Covenant.

Rothchild, Lord: See Balfour Declaration.

Royal Jordanian: See aviation.

RPCP: See Revolutionary Palestinian Community Party.

RSAC (Regional Security and Arms Control Working Group): See regional security and arms control.

rule of law: See Gaza Strip and the Jericho Area Agreement (Article XIV).

rules of competition: See Treaty of the Middle East/North Africa Community (Article ll).

Rutgers University: See Center for Middle East/North Africa Business Studies.

S

Sadat, Anwar: Born in 1918; President of Egypt from 1970-1981. Faced with severe economic problems he went to the Knesset in Israel, pleading the cause for a Middle East settlement. Israel's Prime Minister Begin hosted Sadat in November 1977, suggesting the first tacit recognition of Israel's sovereignty by an Arab leader, which all led to the historic Camp David Accords.

In March 1979 Sadat and Begin signed the bilateral Egyptian-Israeli Peace Treaty at the White House. Egypt was immediately suspended from the Arab League. During a military parade (marking the anniversary of the October 1973 Arab-Israeli War), Islamic militants assassinated Sadat on October 6, 1981.

See Begin, Menachim; Boumedienne, Camp David Accords, Egypt, Egyptian-Israeli Peace Treaty, Framework for the Conclusion of a Peace Treaty between Egypt and Israel, Islam; Nasser, Gamal Abd al-; Middle East Peace Process; Mubarak, Hosni; Pahlavi, Mohammed Reza, Suez Canal.

Sadat Initiative: See Sadat, Anwar.

Saddam Hussain: See Hussain, Saddam.

Sahyuniun: Used by many Arab nationalists for Zionists.

Said, Qaboos Bib Taymour Bin: A one-man band who is in all, prime minister, minister of foreign affairs, finance, and defense. Born in 1940, he was educated privately in Britain and graduated from the British Royal Military Academy at Sandhurst, returning to Oman after brief service with the British armed forces. He became sultan in 1970 after his father's forced abdication and has been steadily implementing a reform program that has strengthened Oman.

See Oman.

Saiqa: Acronym in Arabic for Vanguard of the Popular War of Liberation. Created by Syrian Ba'athist regime in 1967 as a Palestinian military unit, it is an anti-Arafat faction based in Damascus which split from the PLO in 1983. Supports struggle to regain all of Palestine.

salinity: See water, Water & Desertification (Center of the Middle East/ North Africa Community).

San Remo Agreement (1920): The Supreme Council of the League of Nations met in San Remo, Italy in April 1920 making some critical decisions regarding the Middle East. It concluded that France and Britain would decide the nature of the mandates for the region to be voted upon by the League. The mandates of Syria and Lebanon were ultimately given to France while Mesopotamia (later Iraq) and Palestine would be given to Britain. The British mandate for Palestine includes the Balfour Declaration.

See Balfour Declaration.

SAR: The international standard code for the currencies of the neutral zone (between Saudi Arabia and Iraq) and Saudi Arabia—the Saudi riyal.

satellite television: See television.

Satloff, Robert: See Council on Foreign Relations.

Saudi Arabia: Officially the Kingdom of Saudi Arabia, it is the largest Middle East kingdom, formed in September 1932. In 1981 King Fahd tried to mediate the continuing crises over the existence of the State of Israel but his attempts were rejected because of close ties to the United States and by President Assad of Syria.

In 1990, during the Gulf crisis, U.N. forces protected the nation from possible Iraqi aggression.

Although Saudi gross domestic product (GDP) recorded near-zero growth in 1994, the Saudi economy is the largest in the region of the Middle East and North Africa, with output topping $120 billion in 1994. In fact, Saudi per capita gross national product surpasses every African and Latin American state and is exceeded in Southeast Asia only by Japan, Singapore, and Hong Kong.

Despite declines in public-sector expenditures, the kingdom's private sector grew an estimated 4 percent in 1994. The Saudi private sector is assuming a larger role in the economy. Diversification of the Saudi economy remains a strategic goal of the government to counter abrupt fluctuations in world oil prices which, in turn, have dictated the upturns and downturns in the economy. The private-sector contribution to GDP reached 37 percent in 1994.

The United States remains Saudi Arabia's leading trading and joint-venture partner. U.S. companies had the leading share of the Saudi import market, accounting for more than 20 percent of total Saudi imports in 1994. U.S. exports for 1995 were estimated to be $6.8 billion, up 13 percent from 1994.

Exporters find that the Saudi market is similar to that of most intermediate developing nations. Saudi law requires a local sponsor to conduct business in the kingdom and, generally, the sponsor is the agent, representative, or distributor the U.S. company has selected to promote and market its goods and services. Import tariffs are generally 12 percent but can be, in selected cases, raised to 20 percent to protect Saudi infant industries. Staple foods and animal feed are duty free.

There are no taxes, fees, or controls on remissions of earnings abroad, and the Saudi riyal is freely convertible with other currencies.

See Gulf Cooperation Council, Gulf nations, Iraqi-Saudi Non-Aggression Pact.

Saudi riyal: The currency of Saudi Arabia.

Schwab, Klaus: See World Economic Forum.

science and technology: See Secretariat (Middle East/North Africa Economic Summit).

Science and Technology Exchange (STE): a program of the Secretariat of the Middle East/North Africa Economic Summit; STE is to facilitate technology exchanges between countries in the Middle East/North Africa region as well as with other regions, through projects that contribute directly to economic development. STE will not attempt to duplicate generalized exchange mechanisms which already exist, but will target technologies associated with cooperative projects in the framework of the summit process. By selectively emphasizing technologies specifically identified with cooperative projects, the Secretariat hopes to provide a user-friendly mechanism of scientific and technical exchange which can show bottom-line results in the near term. In addition, STE will particularly focus on technologies and projects in the following high priority sectors: water, agriculture, energy, information technology, and tourism.

STE will promote scientific and technological exchange by servicing a clearinghouse function for:

- private or public firms and research institutes which have technologies to offer in the fields targeted;
- private firms or public enterprises in search of appropriate technologies for cooperative ventures;
- public or private research institutes attempting to build on technologies which are near the stage of commercialization; and
- scientists, based in both public and private organizations, who will have better information concerning the projects (e.g., demands for technology) proposed in the framework of the summit process.

Examples of cooperative projects could include the integrated use of wind energy to pump water destined for tourism and agricultural use, the marketing of new, low-cost information technology for public and private educational use, waste water reuse for high-valued agriculture, and so on.

In order to carry out its mandate, STE will engage in the following activities: the organization of conferences and seminars on targeted technologies as described above, the dissemination of printed and electronic media describing promising technologies and sources of those technologies, user-friendly database access to promising technologies (via the World Wide Web, for example), and promotion of funding for selected technologies with high potential for immediate financial and economic returns.

See Secretariat (Middle East/North Africa Economic Summits).

scientific databases: See Secretariat (Middle East/North Africa Economic Summits).

scientific exchanges: See Israel-Jordan Peace Treaty.

sea transport: See Amman Middle East/North Africa Economic Summit, transport, Transportation (Center of the Middle East/North Africa Community).

secondary boycott: See boycott.

Secretariat (Middle East/North Africa Economic Summits): Projects and programs of the Secretariat include:

A. The Regional Investment Promotion Program developed and structured with the assistance of the World Bank's Foreign Investment Advisory Service. It is based on two major principles:

1. the program uses a regional approach to investment promotion to incite member countries to complement their country-based

approach to investment promotion with a regional component. Moreover, it encourages cooperation in creating regional mechanisms aimed at fostering the attractiveness of the region as a whole, with the goal of increasing the total flow of foreign direct investments (FDI) in the region.
2. the program adopts a market-oriented approach to FDI promotion to design a marketing strategy that reflects priorities and concerns of potential investors, as opposed to investment promotion strategies based solely on governmental priorities. This approach is based on existing understanding of what serious foreign investors look for when selecting a location for their investment, then strives to foster the creation of an environment conducive to the flows of FDI, and finally promotes such an environment to potential investors.

The Regional Investment Promotion Program consists of three distinct, yet interrelated and mutually supportive components:

- image building campaigns to convey to the world the potential benefits that countries in the region and foreign investors may derive from the positive developments in the region;
- raising the awareness in country members of the Economic Summit about what needs to be done to take advantage of these potential benefits and how to do it, using the experiences of countries around the world which have been successful in attracting FDI; and
- offering expert services to government members of the Economic Summit to complement their efforts at better positioning themselves to take advantage of these potential benefits.

To support the conduct of the proposed activities of the Regional Investment Promotion Program, the Secretariat will establish the Regional Investment Promotion Resource Center. Its mission will be to develop and administer activities in support of the three-prong approach to the regional investment promotion effort described above, that is, image building, awareness raising, and expert services. The center will be a resource to existing government investment promotion agencies. The center will act as a focal point for gathering and distributing relevant information on the regional level, and will develop over time a source of comparative experiences of countries within the region. For example, it will use the MENAnet web site set up by the Secretariat to provide online information relevant to private investors.

B. Regional Cooperative Network in Science and Technology has been initiated by the Executive Secretariat, and has received the endorsement and support of several scientists and research institutions in and outside the region. The goal of this network is to facilitate technology exchanges between countries in the Middle East/North Africa region as well as with other regions, through projects that contribute directly to economic development. Initial fields identified for cooperation include water, energy, management science, and information technology. Regional scientific meetings, seminars, and workshops will be organized as part of this program which could include exchanges of scientists and professors between universities and research centers, the development of scientific databases, as well as distance learning activities.

C. Framework Agreement between the Executive Secretariat and the World Trade Centers Association (WTCA) signed to set the stage for joint programs between the two institutions in the area of investment and trade promotion, and facilitation of business partnerships. Specific actions will be undertaken within this context to effectively utilize the existing network of WTCs in the region (a total of about 20) as well as the WTC Information Network to promote the goals of the Economic Summit process. It is also planned, as part of the framework agreement, to jointly encourage the establishment of "trade points" within existing WTCs in the region.

D. The MENAnet service—this information server, with access on the World Wide Web, will provide online information on the summit process. This network will offer different types and levels of memberships, with a service package and obligations corresponding to each type. There will be various levels of authorization, access codes to authorized parties through system passwords and electronic keys. So as not to disadvantage those who cannot access the MENAnet, the Secretariat will provide them with services whereby faxes and hard copies can be scanned and put into digital form for integration in the MENAnet.

The types of membership will include:

- business;
- chambers of commerce and regional organizations;
- governments; and
- steering committee of the MENA Economic Summit.

See Amman Middle East/North Africa Economic Summit, Casablanca Declaration, Casablanca Middle East/North Africa Summit, Middle East/North Africa Series; Ferrab, Mostafa.

securities: See Amman Middle East/North Africa Economic Summit, Banking and Financial Services (Center of the Middle East/North Africa Community), Mediterranean Crans Montana Forum.

security: See Interim Agreement (on the West Bank and the Gaza Strip), Israel-Jordan Common Agenda, United States.

security and arms control: The Middle East arms market remains active. The many conflicts throughout the region—Arab-Israel, inter-Arab, domestic, religious, surrogate cold war—have led to spiralling arms proliferation. At the November 1991 Madrid Peace Conference a multilateral Regional Security and Arms Control Working Group (RSAC) was created to slow the race to regional military superiority. (Although the group has made no substantive breakthroughs there has been symbolic progress. For the first time 13 regional nations are conducting discussions on reducing the arms race).

See Camp David Accords, Gaza Strip and the Jericho Area Agreement (Article VIII), Israel-Jordan Peace Treaty, Madrid Conference.

Security Arrangements for Hebron: See Agreement on Security Arrangements for Hebron.

Security Council: The principal executive organ of the United Nations.

See Agreement on Security Arrangements for Hebron, Camp David Accords, Casablanca Declaration, Declaration of Principles, Egyptian-Israeli Peace Treaty, Gaza Strip and the Jericho Area Agreement, Hamas Covenant, London Agreement, PLO Phased Plan, Resolution 242, Resolution 338, Resolution 687, Stockholm PLO-American delegation statement, United Nations entries.

security seam: See fences.

Self-Rule Accord: See Interim Agreement.

SEMED: See Southeast Mediterranean Development Project.

separation: Following the terrorist bombing in Israel in the winter of 1996, separation between Israelis and Palestinians is being considered by many politicians in Israel as the best antidote to terrorism.

See border conflicts, fences, Interim Agreement.

separatism: See fences, free trade (theory).

services: See Treaty of the Middle East/North Africa Community (Article 10).

settlements (and the Oslo accords): There are 144 Jewish settlements in the West Bank and Gaza Strip, with an estimated population of 130,000. The Arab population of the territories is approximately 1.8 million (1 million in the West Bank, 800,000 in Gaza).

It is no coincidence that the Declaration of Principles makes hardly any mention of the settlements. In fact, the word "settlements" appears only three times in the DOP and its various appendices and minutes. And when "settlements" does appear in the DOP, it is only to push off the issue until the permanent status negotiations begin. Article V of the agreement states that settlements, as well as other issues will be dealt with when permanent status negotiations commence. The Oslo accords left the settlement issue for the final status negotiations in the hope that the peace process would create a positive momentum of its own before an issue as delicate as the settlements would have to be handled.

See Declaration of Principles, Har Homa, Moderate Territorial Compromise, Sharon Plan.

Shaahak, Amnon (General): See Agreement on Security Arrangements for Hebron.

Sha'ath, Nabil: See Agreement on Security Arrangements for Hebron.

Shai: The Haganah intelligence service established in October 1940, which later became the Israel Defense Forces Intelligence Branch. The organization was divided into regional sectors, a counterintelligence department, and the Arab Shai, which functioned on a national level. Shai supplied intelligence for Haganah operations such as the release of imprisoned illegal immigrants and the bombing of police stations. Secrecy was so strict within Shai that the service functioned as an underground movement within an underground movement.

See Hebrew Insurrection Movement.

Shaikh Hamad Bin Khalifa Al-Thani: An assertive ruler in the Gulf, who was backed by his family and received almost immediate international support on ousting his father Shaikh Khalifa Al-Thani in a bloodless coup in June 1995. Shaikh Hamad's rational for the coup was simple: the state had been unforgivably slow in the race to develop natural gas revenues and new international relations (where he favors a more independent policy apparently including eventual recognition of Israel). In a brief message at an emergency cabinet meeting he stated: "I am not happy with what has happened but it has to be done and I had to do it." As crown prince and defense minister since May 1977 (he is a royal graduate of Britain's Sandhurst Royal Military Academy) Shaikh Hamad developed a reputation for single-mindedness and hard work. He piloted the establishment of a welfare state, a job creation program for Qatar's 100,000 nationals, modernization of the armed forces and the bureaucracy, and presided over oil and gas developments.

See Qatar.

Shaikh Isa bin Sulman Hamad al-Khalifa: Born 1933, Shaikh Isa became emir in November 1961 on his father's death. He has overseen Bahrain's emergence as a financial services center in the Gulf, but faces rising political tensions, especially involving Bahrain's Shi'ite population.
See Bahrain.

Shaikh Zayed bin Sultan: Ruler of Abu Dhabi from 1966 who enjoys hunting and falconry. Assumed UAE presidency in December 1971, Shaikh Zayed oversees all major decisions in the emirates.
See United Arab Emirates.

Shalom Al Israel Synagogue: See Cairo Agreement.

Shamir, Yitzhak: Born in 1915, Shamir was chosen as Israel's Prime Minister twice, in 1983 and again in 1986. A member of the conservative party he is an opponent of any compromise of land for peace with the PLO.
Shamir participated in the October 1991 Middle East Peace Conference in Madrid at the same time argued for the expansion of Jewish settlements in the Occupied Arab Territories. In 1993 he stepped down as Likud party leader.
See Fatah.

Shamir Plan: See Fatah.

sharing power: See Interim Agreement.

Sharm el Sheikh: In March 1996 it was the location for the "Summit of the Peacemakers," held to cope with the increasing terrorism in the Middle East.
See Camp David Accords, Summit of the Peacemakers, terrorism.

Sharon, Ariel: See Sharon Plan.

Sharon Plan: A settlement plan suggested by Ariel Sharon in 1992; would have Israel annex some 50 percent of the territories. In the other half of the territories 11 cantons would be created. All told, there would be seven cantons in Judea and Samaria, and four in Gaza. The Palestinians would have autonomy in these regions and control Jericho. Two strips of land, one along the Jordan River and one along the 1967 border, would be security zones under Israeli control.
See settlements (and the Oslo accords).
Compare Alpher Plan, Beilin-Abu Mazen Plan, Palestinian National Authority Plan, Third-Way Plan.

Sheikh Hussein Bridge: See Japan.

shekel: The official currency of Israel, resurrected from Biblical times, that replaced the pound in 1980.

shipping: See Amman Middle East/North Africa Economic Summit, transport, Transportation (Center of the Middle East/ North Africa Community).

Shohat, Avraham: See private sector, Protocol on Economic Relations.

Shuhada Street: See Hebron (American Plan on Al-Shuhada Street).

Siegman, Henry: Director, U.S./Middle East Project for the Council on Foreign Relations.
See Council on Foreign Relations.

Sinai: This peninsula saw fighting between Egypt and Israel during the Suez War, the War of Attrition, and the October 1973 Arab-Israeli War. It was occupied by Israel between November 1956 and March 1957, and again from June 1967 to April 1982.
See Camp David Accords, Egyptian-Israeli Peace Treaty.

Sinai Campaign: See Nasser, Gamal Abd al-.

"Singapore of the Middle East": The past Arab refusal to trade with Israel added to the political and economic isolation of Israel, and stunted an economy many analysts believe and still believe has the potential to become the "Singapore of the Middle East."
See boycott.

single banking license: See banking, Banking and Financial Services (Center of the Middle East/North Africa Community).

Six-Day War: Commenced on June 5, 1967; Israel captures the West Bank and reunites the city of Jerusalem. Jordan continues to maintain social, religious, and economic ties with the West Bank and muslim religious institutions in Jerusalem. Travel for Palestinians and foreigners is permitted between the West Bank and Jordan through a number of border passages.
Synonymous with Arab-Israeli War (III) of 1967, June War.
See Nasser, Gamal Abd al-, Syria.

size: See Middle East/North Africa (MENA).

social insurance: See Protocol on Economic Relations (Article VII).

social rights: See Protocol on Economic Relations (Article VII), Social Rights (Center of the Middle East/North Africa Community).

Social Rights (Center of the Middle East/North Africa Community): The community can only endure when equality is the rule. Equality not only among the people of the community but also among member nations. No citizens of the community may be placed at a disadvantage or discriminated against because of his or her nationality. All member citizens are equal before the law. As far as the member nations are concerned, the principle of equality means that no community country has precedence over another and natural differences such as size, population, and differing organization must be dealt with only in accordance with the principle of equality.

Providing a charter of social rights at differing stages of the Middle East/North Africa Community may indeed be the most difficult aspect of regional integration. Considering the disparity of rights, personal, communal, employment, health, and so on, gaps are so great from one nation to another, that a wand promising equity may be interpreted as a threat to the nation(s) that have more.

Full implementation of the community treaty must ultimately include a program for adjusting the social needs of people in general, and specifically for workers to establish an equitable and consistent protection for all. Poorer nations will resist economic and wage equity in that, they will argue, inequity will provide opportunities of fuller employment with hopes of raising living standards. Social rights to be legislated will include the:

- freedom of movement across borders;
- employment and proper remuneration;
- overcoming poverty with appropriate living and working conditions.

As far as workers' social protection, the community must have:

- a principle of equal treatment for men and women in matters of social security;
- freedom of association and collective bargaining;
- vocational training to secure future employment;
- promotional possibilities for working men and women;
- equal treatment for working men and women;
- adequate health protection and safety at the workplace;
- protection of children and adolescents in the workplace;
- protection of the elderly at the time of retirement; and
- concrete measures of improving conditions for the disabled.

A major emotional and complex issue is the enforcement of religious ideology within nations and across borders. Religion must not be introduced or re-enforced across borders as dogma for member nations. The separation of church and state in its deliberations and preparation of legislation among individual nations will continue to determine their relationship within their borders. The community should encourage individual choice. It should prevent any one group from imposing its beliefs on other small groups. It should educate and legally bind individuals to accept freedom of religious choice, without discrimination or bias. The machinery and executives of the community should be driven by a combination of blindness to a sensitivity toward religious beliefs, thus setting itself as a model of full acceptance of religious tolerance.

See Centers of the Middle East/North Africa Community, Middle East/North Africa Community.

Sons of the Stones: Self-proclaimed Palestinians who rose up against Israel in the 1980s.

Southeast Mediterranean: The region extends along the coastline from Israel's southern coastal plain to the Suez Canal in Egypt. The region encompasses the southern shore of Israel, the Palestinian Authority, North Sinai, and the western end of the Suez Canal. It includes major cities such as Port Said, Ashdod, Ashkelon, and Gaza, as well as the Northern Sinai cities of El-Arish and Raffah.

See East Mediterranean Region, Southeast Mediterranean Region.

Southeast Mediterranean Development (SEMED) Project: A project for the coordinated development of the Southeast Mediterranean, including Israel's southern coastal strip from Ashdod, the Palestinian Authority area in the Gaza Strip, and extending to the Bardawil Lagoon in the North Sinai region of Egypt. Economic development of the sub-region focuses primarily on industrial, agricultural, tourism, and commercial development projects benefiting the parties comprising the sub-region.

This European Union-facilitated project intended that the SEMED initiative is of limited duration. Its central purpose is to focus investment on specific near-term sub-regional projects to enhance the area's attractiveness to private investors, advance economic integration between the three partners and encourage joint ventures, job creation and new enterprises in SEMED.

See TSU.

Southern Security Zone: See Cairo Agreement.

sovereignty: As defined by the United Nations and other international bodies "sovereignty" is the "absolute right of the ruling entity of the state to exercise power within it. Territory is a tangible attribute of statement and within that particular geographical area which it occupies, a state enjoys and exercises sovereignty."

Islamic constitutional law has defined an alternative model for political entities within the Islamic world. As far as the Islamic state and sovereignty are concerned, very different factors operate from those involved in international law: the basis of the Islamic state is ideological, not political, territorial, or ethnical and the primary purpose of government was to defend and protect the faith, not the state, with the whole of the Islamic world forming a single political unit, the *umma*. The crucial aspect of this type of constitutional structure was that it was concerned primarily with community and not with territory.

In North Africa the states are defined by its peoples. Traditional linguistic and cultural homogeneity—as illustrated by Muslims and by Arabs or Berberophones—extends across the postcolonial state boundaries. Many governments of North Africa assume that their statehood is legitimized as a discrete nation, which is being rejected by a growing Islamist movement that exist in all of them. Sovereignty is thus an ambiguous concept, with the territorial imperative being contested by the doctrinal communal vision.

See powerless state.

Spain: See Euro-Arab Management School.

Special Drawing Rights: See Free Trade Agreement (United States and Israel).

standards: See regional economic integration.

Standards (Center of the Middle East/North Africa Community): Setting common standards across community borders is a declaration of accepting the most sophisticated, advanced demands of any community. Such standards cut across all aspects of life, from the simple label on a jar of food, to the more advanced telecommunication or computer device. Yet, aspiring to this goal, increases the economies of scale and provides the population with increased choices, a lower frustration level, gives the purchaser a feeling of security and protection, and simplifies life a little.

Differing technical regulations and national standards in the member nations are very real obstacles to the creation of an effective and unified regional market. The effects are widespread, adding real costs and wasting valuable resources, restricting consumer choice and impeding the community's full competitive potential.

The existence of different technical regulations in individual community nations forces manufacturers to concentrate on national rather than community-wide markets. It is necessary to have production lines for different nations, and so the opportunities to reap economies of scale are reduced. As a result, costs are higher, often as much as 10 percent of total product costs.

To move toward the goals of removing technical barriers to trade a new approach to standardization is called for, which should be based on the following principles:

- a distinction will be drawn in future internal market initiatives between what is essential to standardize legislation and what may be left to be standardized by community bodies;
- legislative standardization will in the future be restricted to laying down health, safety, and other essential requirements;
- industrial standardization will be achieved by the elaboration of standards throughout the region. The result will be that a product manufactured in one community country in conformity with community legislation, as regards its essential safety requirement and with a standard in other respects, will be guaranteed automatic access to the markets of all other participating nations.

In many areas of the economy the lack of common standards and product regulations frustrates the creation of a unified market. Goods cannot move freely from one nation of the community to another, testing and certification procedures have to be repeated for each nation, and consumers are provided with varying levels of protection and information. Clearly the standardization of essential product regulations and procedures is necessary to create a truly common market, and to allow the Middle East/North Africa Community to become an effective force of competition in world markets. The emphasis must be on providing a high level of health and safety for all community citizens and to protect the environment.

See Centers of the Middle East/North Africa Community, gas, logistic trade centers, Middle East/North Africa Community, oil.

state: See powerless state.

State of Israel: See Israel.

State of Kuwait: See Kuwait.

State of Qatar: See Qatar.

state-owned enterprises: See privatization.

Statutory Road Accident Victims Compensation Fund: See Protocol on Economic Relations (Article VII).

STE: See Science and Technology Exchange.

Steering Committee: See Casablanca Middle East/North Africa Economic Summit, Middle East and North Africa Business Interaction Group, Middle East/North Africa Economic Strategy Group, Moscow Multilateral Middle East Conference, Secretariat (Middle East/North Africa Economic Summits).

Steering Committee of the Multilateral Negotiations: See Middle East/ North Africa Economic Strategy Group.

STIMENA: See Swiss Trade Initiative for the Middle East and North Africa.

stock exchanges: Perhaps investors attracted to MENA stock exchanges may become the catalyst for future economic expansion. By 1996, 11 Arab nations spread across MENA had stock exchanges, with a total market capitalization of $70 billion. Factoring in Israel and Turkey, that figure more than doubles to $143 billion.

By the end of 1996 the Palestine Securities Exchange (PSE) is planning to open in the West Bank town of Bataineh with initial market capitalization of about $500 million and expectations by the year 2000 of $2 billion.

Of all the Arab countries Saudi Arabia is by far the largest investment center. The Saudi bourse accounts for more than half of the market capitalization of the region, excluding Israel and Turkey, and has the most sophisticated trading system.

In addition, a number of MENA countries have moved to reform their stock markets, for example:

- In Morocco a new stock market law was adopted in 1994, the stock exchange was privatized, and an independent commission entrusted with its oversight. The new law, combined with a strong privatization program, has caused capitalization to almost double in relation to GDP and turnover to surge since 1993.
- In Tunisia the modernization of the legal and regulatory environment, which began in 1989, was significantly broadened in 1993. As

a result, the stock market has boomed, with capitalization increasing ninefold during 1990-1995.

- In Egypt the Capital Market Act (1993) revamped the legal framework for the securities market, allowing for foreign participation, increasing competition, prohibiting insider trading and other practices, and enforcing international accounting standards with more stringent regulation. The law helped revive the Cairo and Alexandria exchanges, which are among the oldest in the region and had languished for many years.

See investment (MENA), Palestinian Stock Exchange.

Stockholm PLO-American delegation statement: Signed on December 7, 1988, the text read:

The Palestinian National Council met in Algiers from November 12-15, 1988, and announced the declaration of independence which proclaimed the state of Palestine and issued a political statement.

The following explanation was given by the representatives of the PLO of certain important points in the Palestinian declaration of independence and the political statement adopted by the PNC in Algiers.

Affirming the principle incorporated in those U.N. resolutions which call for a two-state solution of Israel and Palestine, the PNC:

1. Agreed to enter into peace negotiations at an international conference under the auspices of the U.N. with the participation of the permanent members of the Security Council and the PLO as the sole legitimate representative of the Palestinian people, on equal footing with the other parties to the conflict; such an international conference is to be held on the basis of the U.N. Resolutions 242 and 338 and the right of the Palestinian people to self-determination, without external interference, as provided in the U.N. Charter, including the right to an independent state, which conference should resolve the Palestinian problem in all its aspects;
2. Established the independent state of Palestine and accepted the existence of Israel as a state in the region;
3. Declared its rejection and condemnation of terrorism in all its forms, including state terrorism;
4. Called for a solution to the Palestinian refugee problem in accordance with international law and practices and relevant U.N. resolutions (including the right of return or compensation).

stock markets: See Palestinian stock market, stock exchanges.

Strait of Tiran: See Camp David Accords, Egyptian-Israeli Peace Treaty.

Strategy Group: See Middle East/North Africa Economic Strategy Group.

submarines: Slang for long-term secret agents.

subsidies: See Free Trade Agreement (European Union and Israel).

subversion: See Syria, terrorism.

Sudan: Self-government was granted in 1953, with full independence in 1955. In 1958 there was a military coup, with many more to follow. Famine and civil war in the south continued into the 1990s.

See minorities, terrorism.

Suez Canal: Constructed between 1859 and 1869, the Six-Day War of 1967 led to its closing. The canal was again reopened by President Sadat on June 5, 1975.

A continuing drop in Suez Canal revenue and the volume of ship traffic appeared in 1996. The government slashed rates, began to deepen the 127-year-old waterway so it could handle heavier ships, and attempted programs to quash competitive means of moving oil and goods between the Middle East and Europe. Egypt agreed to slash tolls by 35 percent for the next 25 years to adjust for a 1995 traffic drop of 8.1 percent. Oil and oil products dropped by 30.4 percent to 41.3 tons in 1995.

See Camp David Accords, Egypt, Egyptian-Israeli Peace Treaty, Yom Kippur War.

Suez War: See Arab-Israel War (II).

Sufa-Gush Katif Road: See Cairo Agreement.

sulh: See PLO Phased Plan.

Sultanate of Oman: See Oman.

Summit of Peacemakers: Leaders of 27 nations, including sheiks, princes, and kings from 14 Arab nations from North Africa to the Persian Gulf, joined together on March 13, 1996 in Sharm El Sheik, Egypt in a sweeping condemnation of the February-March suicide bombings in Israel and vowed to cooperate in fighting terrorism and promoting peace throughout the Middle East. Syria and Lebanon, declined to attend.

Sponsored by the United States and Egypt, these nations pledged to work together to prevent terrorist organizations from recruiting members, raising funds or supplying arms on their soil, and to share training and equipment "to cooperate in cutting them off." President Clinton declared the following day that the United States would set aside more than $100 million to upgrade the fight against terrorism in the Middle East.

Summit participants at Sharm el Sheik issued the following statement:

> The Summit of Peacemakers has just concluded. This meeting took place at a time when the peace process confronts serious threats. The summit had three fundamental objectives: to enhance the peace process; to promote security; and to combat terror.
>
> Accordingly, the participants today expressed their full support for the Middle East peace process and their determination that this process continue in order to accomplish a just, lasting and comprehensive peace in the region; affirm their determination to promote security and stability; and to prevent the enemies of peace from achieving their ultimate objective of destroying the real opportunity for peace in the Middle East;
>
> And re-emphasize their strong condemnation of all acts of terror in all its abhorrent forms, whatever its motivation and whoever its perpetrator, including recent attacks in Israel, considering them alien to the moral and spiritual values shared by all peoples of the region; and reaffirm their intention to stand staunchly against all such acts and to urge all governments to join them in this condemnation and opposition.
>
> To that end, we decided to support the Israeli-Palestinian agreements, the continuation of the negotiating process, and to politically and economically reinforce it; to enhance the security situation for both, with special attention to the current and pressing economic needs of the Palestinians; to support the continuation of the negotiating process in order to achieve a comprehensive settlement; to work together to promote security and stability in the region by developing effective and practical means of cooperation and further assistance; to promote coordination of efforts to stop acts of terror on bilateral, regional and international levels, insuring instigators of such acts are brought to justice, supporting efforts by all parties to prevent their territories from being used for terrorist purposes and preventing terrorist organizations from engaging in recruitment, supplying arms or fundraising; to exert maximum efforts to identify and determine the source of the financing for these groups and to cooperate in cutting them off; and by providing training, equipment and other forms of support to those taking steps against groups using violence and terror to undermine peace, security or stability;
>
> Finally, to form a working group open to all summit participants to prepare recommendations on how best to implement the decisions contained in this statement through ongoing work, and to report to the participants within 30 days.

For Israel, the main significance of the meeting was that leaders of 13 Arab countries, including Saudi Arabia and other states that don't have official ties with Israel, attended a conference designed to show solidarity with the Jewish state. This marks the first time since its 1948 founding that Israel and the Arab countries have openly agreed to become partners in an issue related to Middle East security.

See Sharm el Sheikh, terrorism.

Summud: The Arab word for endurance or passive resistance and limited, Intifada-like demonstrations.

Sweden: See Free Trade Agreement (European Free Trade Association and Israel).

Swiss Trade Initiative for the Middle East and North Africa (STIMENA): Following a commitment made by the Swiss government at the Casablanca Economic Summit in October 1994, the former Director-General of GATT, Arthur Dunkel, was entrusted with the task of giving effect to the Swiss Trade Initiative for the Middle East and North Africa. Its secretariat, located in Geneva, has been operational since July 1, 1995.

STIMENA is based on the conviction that sustained regional trade, economic cooperation, and attractive investment opportunities will improve the living conditions in the Middle East and, thus, contribute to the peace process. STIMENA's objective is to explore with the main parties concerned the feasibility and content of establishing a framework for working toward this goal and assist them in implementing it.

Since taking up the work, the STIMENA secretariat has:

- worked in close cooperation with the responsible authorities of the four core parties to the peace process, that is, Egypt, Israel, Jordan, and the Palestinian Authority. In view of STIMENA's focus on trade relations, the trade subcommittee of the Regional Economic Development Working Group REDWG has been a natural partner and contact point for this cooperation;
- maintained close contact with the core parties's major trading partners (especially the EU and the United States) and academic and other organizations working on trade-related issues in the region;
- worked on an analysis of trade relations among the core parties and with the key third parties (the EU and the United States). In the context of a roundtable discussion in Geneva in December 1995, the core parties had requested such a study, which should identify areas of inconsistency between their trade policies and practices and make recommendations for harmonization. The analysis and the recommendations should serve as the starting point for a common work program of the core parties;
- been requested to present its views on topics in its field of work through speeches or participation in workshops, seminars, and other meetings.

Although financed by the Swiss government, the STIMENA secretariat works independently and is solely responsible for its substantive contributions. STIMENA's approach to the tasks at hand is problem-oriented; its aims at pragmatic solutions in a step-by-step fashion and through close cooperation with the responsible regional partners.

SYP: The international standard code for the currency of Syrian Arab Republic—the Syrian pound.

Syria: Officially the Arabic Republic of Syria; declared independence from France on January 1, 1944. Syria has been at war with Israel since Israel's founding and provoked the Six-Day War of 1967 and opposed the Camp David Accords.

Syria finally agreed to sit down with Israel at the historic Madrid peace conference in October 1991. This marked the first time that Syria engaged in direct, bilateral negotiations with Israel. In fact, Syria neither publicly supported nor opposed the September 1993 Israel-PLO accord.

Since 1990 Syria has experienced real economic growth. Foreign exchange constraints have eased due to a renewal of external foreign exchange flows, both official and private, and the government has taken liberalization measures to open up the economy. Until recently, the government pursued policies aimed at expanding the public sector, with tight controls imposed on private sector activity.

In 1994 the government's general budget was about US$3.4 billion, up 17 percent from 1993 and 55 percent from 1992, with most increases reflected in the capital or project component of expenditures.

During 1990-1993, after years of stagnation, the economy registered average annual growth rates of 7-8 percent. Since 1991 private-sector investment also has picked up noticeably, as businessmen have taken advantage of incentives offered under a new investment law. Liberalization measures implemented in recent years now permit private exporters to retain foreign exchange from exports, 75 percent for industrial products and 100 percent for agricultural commodities, and to finance permitted imports for inputs and a number of basic commodities using foreign exchange gained from exports.

See Agreement of Understanding between Israel and Lebanon; Asad, Hafez (President), Lebanon, Pentagon Plan, terrorism.

Syrian pound: The currency of the Syrian Arab Republic.

T

Taba Agreement: See Interim Agreement.

Taba Declaration: Just moments across the border from Eilat, Israel lies Taba, a desert town preparing to welcome tourists and other visitors crossing into the Sinai. There on February 7-8, 1995 a meeting was held (as proposed at the Casablanca Middle East/North Africa Economic Summit) hosted by the government of Egypt and brought together five trade leaders—Egyptian, Israeli, Jordanian, Palestinian, and United States—to further cooperation on private-sector development and trade in the region. The signatories agreed:

- They will assemble a group of government experts to identify and study ways to remove the barriers to a free and open market for trade and investment in the region and in relation to the United States, a major trading partner. The group will coordinate closely with the efforts underway in the monitoring committee of the regional economic development working group and it will report to the trade leaders within six months.
- They will establish a group of government and private-sector experts to develop a Middle East information highway designed to benefit all participants. This information network would be designed to collect and disseminate trade and investment opportunities in the region.
- They welcome the convening of the upcoming Amman Economic Summit during October 1995 and call on the international public and private community to fully support this important initiative which would promote trade and investment in the region.
- They will cooperate in the organization of training seminars, business missions, and other business promotion events in support of regional private-sector development.

- They will continue and increase their efforts to further private-public partnership for regional economic development to ensure the best possible environment for unfettered business investment and trade.
- In support of private-sector development in the region, the parties endorse the efforts being made to study the possibility of establishing a regional business council.
- They will meet again within 12 months to review progress on the achievement of the objectives in this declaration.

See Amman Middle East/North Africa Economic Summit, Cairo Agreement, Casablanca Middle East/North Africa Economic Summit.

Taba-Eilat-Aqaba Macro Area: See TEAM-Area.

Taif Agreement: See chronology (May 1991), Lebanon.

Talal, King: See Jordan.

Task Force of the Council on Foreign Relations on the U.S. Middle East Peace Policy and Peace Process: See Council on Foreign Relations.

taxation: See informal economy, Protocol on Economic Relations (Articles V and VI).

TEAM Area: The Taba-Eilat-Aqaba Macro Area. A working group involving Israel, Jordan, and Egypt for the coordinated development of the Gulf of Aqaba sub-region which has been supported by the European Union. TEAM Area has been defined as:

- in Egypt: the town of Taba and the triangle defined by the inland road to Ras el Naqeb and Nuweiba, including the Nuweiba seaport and Ras el Naqeb Airport;
- in Jordan: the administrative border of the Aqaba region;
- in Israel: the city of Eilat, including 17 kilometers from the Jordanian and Egyptian border.

According to the TEAM Area concept, Egypt, Israel, and Jordan will continue to pursue their development objectives within their national territories. Coordinated development of the region will complement national planning and is deemed imperative, given the scarcity of resources, ecological fragility, and logistic importance of the region. Vast opportunities for private-sector investment in this region exist, primarily in tourism and transport services. Rough estimates of new investments in the TEAM AREA sub-region during the coming 15 years comes to $5 billion.

See TSU.

technical regulations: See Standards (Center of the Middle East/North Africa Community).

technical standards: See oil, standards.

Technical Support Unit: See TSU.

telecommunications: Cooperation in telecommunications throughout MENA will strengthen ties between neighboring countries and their citizens, and will create a sounder base for cross-border business activities, foreign investment as well as tourism and trade. In the long run, advanced telecommunications systems could assist MENA nations in improving their competitiveness in world markets.

Main lines and teledensity rates in 1992 in MENA were:

Country	*Main Lines (thousands)*	*Main Lines/100 inhabitants*
Algeria	962.2	3.65
Bahrain	112.5	21.23
Egypt	2173.6	3.94
Israel	1845.5	35.56
Jordan	278.3	6.49
Kuwait	345.6	17.54
Libya	235.8	4.84
Morocco	653.9	2.48
Oman	130.1	7.93
Qatar	104.6	23.25
Saudi Arabia	1568.4	9.86
Syria	513.4	3.96
UAE	540.5	32.36
Yemen	143.4	1.10

See European Union, Telecommunications (Center of the Middle East/North Africa Community.

Telecommunications (Center of the Middle East/North Africa Community): The marriage of communications and information will serve as a major springboard for economic cooperation in nations of the community. The region today remains far behind in the communications revolution with significant opportunities for the development of joint, advanced systems across borders and throughout the community. Services of all types can be expanded including, public telephone services, cellular telephone services and paging systems, leased lines and private networks, data and information services, satellite communication and teleports, radio and television broadcasts, international telecom services, and rural telephony.

The relatively short distances between Israel and its neighbors will encourage them to consider working as a single network, with all the advantages of shared services and backup and significantly lower rates.

The community must create convenient incentive conditions as well as suitable regulations and agreements by member states. To achieve this, nations of the region will have to dismantle the obstacles which constrain the cross- border supply of these new information-communication services. It also has to solve problems caused by the differing technical standards of the equipment used.

Community states will cooperate with neighboring nations with infrastructure, in such a manner as to facilitate mutual backup and support of the communications systems. Likewise, there must be a differentiation based on relative advantage of the respective member nations and the creation of economies of scale to optimize investments within the infrastructure.

Cooperation in telecommunications will hasten to strengthen cross-border business, foreign investment, as well as greater tourism and trade. Over time, advanced communication systems will be a major force with the community nations to improve their competitiveness in world markets.

See Amman Middle East/North Africa Summit, Cairo Middle East/North Africa Economic Conference, Centers of the Middle East/North Africa Community, Israel-Jordan Peace Treaty, Middle East/North Africa Community, telecommunications, television.

teledensity rates: See telecommunications.

telephone services: See Telecommunication (Center of the Middle East/North Africa Community).

television: Satellite transponders and state-of-the-art television production studios are spreading throughout MENA. For example, the Emirates Dubai Television likes to call itself the second-largest television network by coverage throughout the world.

Competition for regional viewers is generating substantial investments in programming, with advertisers beginning to shift more to regional satellite networks and less to local stations.

Arab television will not only evolve as the most powerful form of regional communications, but also become more export-oriented to reach followers around the world. At the same time, advertisers increasingly devote revenue to these expanding programs. For example, in 1995 television advertising expenditure in the Middle East was:

Country	*Amount ($ millions)*
Lebanon	$245.1
Pan-Arab	114.9
Egypt	69.6
Saudi Arabia	46.8
UAE	33.7
Kuwait	14.6
Bahrain	8.6
Oman	7.2
Jordan	6.4
Syria	5.3
Qatar	4.0
Yemen	3.1

See Telecommunications.

Temporary International Presence in Hebron (TIPH): On May 31, 1994 Israel and the PLO agreed to establish a three-month TIP in Hebron. The issue was raised by the Palestinians after the Hebron massacre and the TIPH agreement marks the resumption of Israeli-Palestinian talks. The TIPH was deployed on May 5 and left Hebron on August 9.

See Agreement on Security Arrangements for Hebron (1994), Agreement on the Temporary International Presence in the City of Hebron (1997), Gaza Strip and the Jericho Area Agreement (Article XXI).

Terrab, Mostafa: Secretary-General, Executive Secretariat of the Middle East/North Africa Economic Summits based in Rabat, Morocco.

See capital markets development, Casablanca Declaration, Casablanca Middle East/North Africa Economic Summit, Middle East/North Africa Community, Secretariat (Middle East/North Africa Economic Summits).

territorial issues: See Israel-Jordan Common Agenda.

terrorism: On April 30, 1996 President Clinton and Prime Minister Shimon Peres of Israel signed an antiterrorism accord. The signing ceremony in Washington formalized antiterrorism steps by providing a legal basis for antiterrorism cooperation, including the exchange of intelligence about terrorists and rocket attacks.

The U.S. State Department's "Patterns of Global Terrorism—1996" established that a handful of states, predominantly Middle Eastern, continue to endorse international terrorism. Of the seven "state sponsors of terrorism" five are Middle Eastern: Iran—identified as the "primary state sponsor," Iraq, Libya, Sudan, and Syria (Cuba and North Korea complete the list). A sixth Middle Eastern state—Syrian-dominated Lebanon—while not designated a "state sponsor" is identified as the base for numerous terrorist organizations.

Iran: The "premier state sponsor of terrorism" provides "direction and support to terrorist groups...Iran continues to assassinate dissidents abroad and also provide support to other terrorist groups that oppose Israel and the Middle East Peace Process." Iran is "involved in the planning and execution of terrorist acts by its own agents and by surrogates such as Lebanese Hizballah and continues to fund and train known terrorist groups."

Iraq: "Iraq's ability to carry out terrorism abroad has been curbed by UN sanctions." Within Iraq, however, the "regime continues to murder dissidents."

Libya: "Terrorism by Libyans has been sharply reduced by UN sanctions"...Libya still evades its obligation to hand over those indicted for airline bombings.

Sudan: While "Sudan was not directly involved in any acts of international terrorism in 1996," it remains a "sanctuary and training center for several international terrorist groups."

In August 1997 the Palestine Liberation Organization reached a final settlement with Leon Klinghoffer's two daughters who had sued the PLO for causing their father's death. On October 7, 1985, four Palestinian gunmen hijacked an Italian cruise ship—the *Achille Lauro*. When the terrorists received no response to their demands they executed Klinghoffer, who had suffered two earlier strokes. He and his wheelchair were then shoved overboard. The undisclosed payment indicates that the PLO can be brought to account in their acknowledging responsibility by settling this suit.

See Agreement of Understanding between Israel and Lebanon, Arafat; Arafat, Yasir (Letter to Y. Rabin recognizing Israel's right to exist in peace, September 9, 1993); Arafat, Yasir (Press Statement of December 14, 1988), Black September, Fatah Revolutionary Council, fences, Force 17, Hamas, HizbAllah, Interim Agreement (on the West Bank and the Gaza Strip), Iran and Libya Sanctions Act, Israel-Jordan Peace Treaty, Jihad, Letters of Mutual Recognition, London Agreement, Palestine Liberation Front, Palestine National Congress, Pentagon Plan, Popular Front for the Liberation of Palestine, Popular Front for the Liberation of Palestine-General Command, Sharm el Sheikh, Stockholm PLO-American delegation statement, Summit of Peacemakers, United States; vision, the.

tertiary boycott: See boycott.

testing: See Standards (Center of the Middle East/North Africa Community).

textiles: See Amman Middle East/North Africa Economic Summit, manufacturing, Multi-Fiber Arrangement.

Third Way: See Third Way Plan.

Third Way Plan: A border plan between Israel and Palestine; proposed by the Third Way, a political party that is part of Israel's ruling coalition. The map is believed to be the closest to the position of the Netanyahu government, which is believed to advocate a 50-50 split of the territory. Israel annexes most Jewish settlements and two security columns, one running along the 1967 border, the other on the border with Jordan. Israel would also control several corridors, including main highways, in the West Bank.

Compare Alpher Plan, Beilin-Abu Mazen Plan, Palestinian National Authority Plan, Sharon Plan.

TIP: See temporary international presence.

TIPH: Temporary International Presence in the City of Hebron.

See Agreement on Security Arrangements for Hebron (1994), Agreement on the Temporary International Presence in the City of Hebron (1997).

TND: The international standard code for the currency of Tunisia—the Tunisian dinar.

Tomb of the Patriarchs: In Hebrew, Makhpelah. Since the 1994 massacre of 29 Muslims inside the tomb, by Baruch Goldstein, the building has been divided between Muslims and Jews, with the largest section given to the Muslims which contains the spacious hall devoted to Isaac and Rebecca.

For 10 days a year, on Jewish holidays, the entire building is given over to Jews; likewise, for 10 days a year the whole structure becomes a mosque. For the rest of the year Muslims and Jews are forbidden to enter each other's zones—though the waaf, the Muslim religious authority that administers the building, maintains an office in the courtyard on the Jewish side, the Israelis control security.

See Agreement on Security Arrangements for Hebron, Hebron, Interim Agreement (the West Bank and the Gaza Strip).

tourism: Although tourism has been an important source of foreign currency for many MENA nations, in many respects large-scale tourism remains underdeveloped.

The Middle East nations have led the way to attracting overseas tourists. For example, in 1994 Israel recorded 2.17 million inbound visitors, nearly five times the arrivals in 1970, who spent about $2.35 billion. Egypt's receipts from tourism totalled $1.52 billion, while Jordan's receipts totalled $836 million.

The expansion of hotel development is slow compared to the number of tourists. At the end of 1994 there were 35,000 hotel rooms in Israel, with average occupancy of 67 percent. In 1993 there were 56,000 hotel rooms in Egypt with an occupancy rate of 60 percent and in Jordan 7,700 rooms with an average occupancy of 47 percent.

In 1994, 15 European Union states and 13 Mediterranean countries set up a framework for joint projects in various fields; the first launched in 1997 includes three in the field of tourism:

- Museums without Frontiers sets out in maps, guidebooks, and databases a series of cultural routes concentrating on Islam's influence on Israel, Jordan, Egypt, Turkey, Tunisia, Morocco, the Palestinian Authority, Spain, and Austria. There are no actual museums involved, just sites, buildings, and social changes influenced by these cultures.
- Network of Ancient Theaters for the Performing Arts is a database of ancient theater venues currently in use. Fourteen countries, including France, Cyprus, Malta, Greece, Jordan, and Portugal, are involved. Among the Israeli venues—the Roman amphitheater at Beit She'an.
- Network of Major Popular and Traditional Mediterranean Festivals promotes modern cultural events based on age-old traditions. The Spaniards list the famed running of the bulls at Paplona as one of their entries, Italy picked the horse race in Siena's famed Piazza del Campo, and Israel designated festivities around the Shavuot holiday.

See Amman Middle East/North Africa Economic Summit, Cairo Middle East/North Africa Economic Conference, Israel-Jordan Peace Treaty, parks development, Protocol on Economic Relations (Article X), Tourism (Center of the Middle East/North Africa Community).

Tourism (Center of the Middle East/North Africa Community): At the outset tourism should be encouraged throughout the region as it has the potential to attract both vacation seekers and religious tourists, with monies usually delivered upon signing. Appropriate investment in infrastructures and joint and/or combined projects are likely to give an enormous boost to the existing tourist movement, with all its economic implications for the economies of the community region.

Projects, as part of signed trade agreements, should include creating the infrastructure for tourism—airports, ports, marine, and overland transport—and the development of joint tourist areas, free tourist agreements offering regional tourist packages to religious sites, archaeological sites, and vacation seekers.

Tourism provides the sharing of culture, an introduction to varying lifestyles of the past, along with a demonstration of the genius and creativity of citizens in all participating community nations. The community, by establishing the appropriate mechanism for expanding cross-border tourism, will witness an explosive increase in revenue received from the traveling population.

In the area of cultural development among community members, services can deal with historic, artistic, and religious treasures and moments in a nation's history.

In addition, the entertainment fields of broadcasting and audio-visual programming can also be cross-fertilized and enriched with community participants.

See aviation, Centers of the Middle East/North Africa Community, Middle East/North Africa Community, tourism.

Track II: A long list of academic institutions and think tanks from the United States and Western Europe regularly sponsor "workshops" during which nothing that is said is binding, and the documents that are at times written are not considered agreements. These meetings proved to be especially valuable during stagnant periods of the Middle East Peace Process, and indeed recommendations from these groups are often fed back to governments for consideration and at times, implementation.

trade: See Industrial Cooperation and Trade (Center of the Middle East/North Africa Community), input sharing, Middle East/North Africa (trade), trade reform (MENA).

trade bridge: Utilizing Israel to gain tariff-free access to the United States, the EU, and other regional communities for goods manufactured in Israel.

trade center: See Industrial Cooperation and Trade (Center of the Middle East/North Africa Community).

trade diversion: See Industrial Cooperation and Trade (Center of the Middle East/North Africa Community), input sharing.

trade expansion: See Industrial Cooperation and Trade (Center of the Middle East/North Africa Community).

trade liberalization: See trade reform (MENA).

trademark: See intellectual property rights.

Trade Points: See Secretariat (Middle East/North Africa Economic Summits).

trade reform (MENA): The International Monetary Fund summarizes the major reasons for the MENA region's disappointing performance on trade reform:

- Large revenues deriving from oil and other natural resources, as well as transfers in the form of aid, allowed many MENA countries to postpone needed reforms, including trade liberalization.
- A number of restrictive economic factors—excessive government intervention, distorted prices, lack of transparency in the regulatory environment, and nonconvertibility of the local currency—were maintained, discouraging private investment in the tradable sectors and hampering the conduct of trade.
- Excessive protectionism and distorted incentives were allowed to survive, leading to production structures inconsistent in some cases with the comparative advantages of the individual countries. This resulted in turn in limited complementarity in the production and trade structures among the countries in the region, and hence little intra-regional trade.
- Lack of necessary political will to bear the short-term costs involved in correcting the production structure through the liberalization of trade.
- Formal agreements aimed at trade liberalization at the regional level have resorted to the inclusion of extensive provisions for exemptions and have lacked a clear time frame for implementation, thus making them virtually ineffectual.
- The high level of political tension in the region has hampered trade.

See Middle East/North Africa (trade).

trade structure: See trade reform (MENA).

transfer of authority: See Gaza Strip and the Jericho Area Agreement.

transfer of technology: See income gap.

transitional arrangement: See Declaration of Principles, interim arrangement.

Transjordan: The name given in July 1922 to the region east of the Jordan River that earlier has been southern Syria. Originally it became the autonomous Emirate of Transjordan and in February 1928 entered into a treaty with Britain.

Transjordan acquired independence in May 1946 and the king renamed his realm the Hashemite Kingdom of Jordan.

See Jordan.

transmission grids: See Energy and Environment (Center of the Middle East/North Africa Community).

transport: The evolution of cross-border transportation systems is one of the preconditions for building an economically developed MENA. Furthermore, accessibility and transport ties between nations of the region can be improved by making optimal use of its strategic geographic location, as a crossroads between Asia, Africa, and Europe.

Road links between population and activity centers, through joint, mutually beneficial projects, would serve trade, tourism, and industry as well as other economic sectors throughout MENA. Opportunities for new types of economic enterprises will ensue and the efficiency of current transport systems will improve, lowering production and distribution costs for local and imported goods.

Efficient and cost effective railroads were once in place throughout the region. In particular, an extensive railway system was built in the Middle East at the beginning of the 1900s, connecting Turkey, Syria, Iraq, Jordan, Lebanon, Palestine, Egypt, and Saudi Arabia. Regional conflicts and lack of economic development caused the dismemberment and collapse of this system in both physical and economic terms. As rail transport can be a more economical method of transporting goods over land, rehabilitation of the railway system in the region could have a major impact on intra-regional and international trade.

See aviation, Council on Foreign Relations, Middle East Regional Transportation Study, shipping, Transportation (Center of the Middle East/North Africa Community), Transport Corridors in Middle East.

Transportation (Center of the Middle East/North Africa Community): The development of a Community transportation system (land transport, marine transport, and air transport) is a primary precondition for sustaining a newly integrated and economically developing region. In addition, accessibility and transport ties between community nations can be upgraded by making optimal use of its strategic geographic location, as a crossroads between Asia, Africa, and Europe. The road link between population and activity centers, through joint mutually beneficial projects, would serve trade, tourism, industry, as well as other economic sectors in all community nations. Opportunities for new forms of economic enterprises will evolve and the efficiency of existing transport systems will improve, lowering production and distribution costs for local and imported goods.

The transport market is characterized at the international level by quotas and other restrictions. The costs to importers and exporters that arise from insufficient competition in this market impede the growth and integration of the community's economy. To remedy this unsatisfactory situation and given community suppliers and users of transport services as a genuine common market, the community must adopt a two-phase approach applicable to the main transport sectors—road, sea, and air.

In the first phase the objective will be to liberalize transport services between governments of the community. In the second phase the objective will be to liberalize transport within nations of the community by opening up the national markets to nonresident carriers. Any adoption of such measures will permit the elimination of the border control of the current bilateral quotas in transport and the possibility for a transport carrier to operate in any member nation, either occasionally or on a permanent basis.

As the wheels, wings, and rudders of transport move throughout the community, rates will be cut, time will be shortened, and efficiencies will encourage the further expansion of facilities to move goods and people. In a spread of more than 2,500 miles from east to west, the community will benefit from a more flexible, cost effective, improved transportation network.

See Amman Middle East/North Africa Economic Summit, aviation, Centers of the Middle East/North Africa Community, Israel-Jordan Peace Treaty, Middle East/North Africa Community, transport.

Transport Corridors in the Middle East: In the beginning of 1995 regional discussion on projects for transportation infrastructures and policies reached a standstill, due to the lack of an impartial evaluation methodology to facilitate identification of priority projects for the enhancement of regional transport. Thus, it was decided to work separately on fast track projects (for short-term realization), on a regional transport study (which would help build a frame for the long-term scheme of regional transport), and on a corridor methodology, which would link the long-term schemes and the short-term approach.

Cofinanced by the European Union and the French government a prime objective is the development of methodology which will enhance the development of a regional transport system and policy, which is a prerequisite for regional integration and economic development.

See Middle East Regional Transportation Study.

Treaty of Brotherhood, Cooperation and Coordination between Syria and Lebanon: See Lebanon.

Treaty of MENAC: See Treaty of the Middle East/North Africa Community.

Treaty of Peace (Egypt and Israel): See Casablanca Declaration, Egyptian-Israeli Peace Treaty.

Treaty of Peace (Jordan and Israel): See Casablanca Declaration, Israel-Jordan Peace Treaty.

Treaty of Peace between the State of Israel and the Hashemite Kingdom of Jordan: See Israel-Jordan Peace Treaty.

Treaty of the Middle East/North Africa Community (proposed):

PREAMBLE

Whereas, liberation of the people can best be achieved by creating an economic interdependence of citizens from neighboring nations through liberation of the market economy;

Whereas, the market system can do more than any other system yet known to improve human welfare;

Whereas, all must endeavor to attain both the humanistic and social goals and productive power of an economic system based on individual effort and competition;

Whereas, the need exists to reconcile economic growth with moral concern for fellow men and women and human cost;

Whereas, there is a demand for a competitive market economy with social-development programs designed to provide economic opportunities for all;

Whereas, market-distorting prices and protectionist controls, often adopted to help the less fortunate, have negative economic and social consequences;

Whereas, the responsibility of the state is to develop human resources for both economic and social ends, especially in the fields of health, education, and the strengthening of community and family, but not with permanent dependency;

Whereas, the need exists to improve productivity, but also to reduce the cost of labor, including benefits and a virtually guaranteed lifetime employment;

Whereas, concentrated production impedes the free flow of goods and services;

Whereas, a relatively strong and stable currency and banking system is required;

Whereas, foreign investors must be protected against arbitrary confiscation of their assets;

Whereas, the region's system must shift from a mercantilist approach, where the state has been captured by special-interest groups from bureaucrats and business and labor people, to a capitalist system;

Whereas, solidarity should be developed within the context of a free, competitive, and flexible economy, with a less interventionist and regulatory state, creating instead the policy environment for economic success, while channeling budgetary resources to assist the less fortunate;

Whereas, the need exists for an institutional structure to inspire solidarity and charity in individual economic decision making;

Whereas, there must be a ban in the export of hazardous wastes, a reduction in air and water pollution, and maintenance of other environmental standards, all paralleling the sharing of water and the growth of industry;

Whereas, long-term security implies a political bond able to minimize separate nationalistic attitudes, thus enabling all to speak and act as one;

Be It Resolved, these goals can be achieved by cooperative efforts and genuine aspirations of inspired leaders and citizens in the development of a Middle East Community, in a framework that protects private contracts, provides for currency convertibility, encourages the development of free trade in a private enterprise system, and protects the human rights of all.

TREATY

Article 1—Objectives of the Community

The community shall have as its task, by establishing a common market and progressively bringing together the economic policies of member nations, to promote throughout the Middle East/North Africa Community a harmonious development of economic activities, a continuous and balanced expansion, an increase in stability, an accelerated raising of the standard of living, and closer relations among the countries belonging to it.

Article 2—Activities of the Community

To achieve the objectives of the Middle East/North Africa Community, activities shall include, as provided in this treaty:

1. the elimination of custom duties and of quantitative restrictions on the import and export of goods among member nations, and of all other measures having equivalent effect;
2. the establishment of a common customs tariff and of a common commercial policy toward nations outside the Middle East/North Africa Community;
3. the abolition of obstacles to freedom of movement for persons, goods, services, and capital among the member nations;
4. the adoption of common and coordinated policies in the spheres of agriculture, transport, water, energy, health, education, monetary, tourism, research and development and standards, and so on;
5. the institution of a system ensuring that competition in the common market is not distorted;
6. the application of procedures by which the economic policies of member nations can be coordinated and disequilibria in their internal and external balance of payments remedied;
7. the development of a common set of laws of member nations to the extent required for the proper functioning of the common market;
8. the creation and funding of a Middle East/North Africa Community Social Fund in order to improve employment opportunities for workers and to contribute to the raising of their standard of living;
9. the establishment of a Middle East/North Africa Community Development Bank to facilitate the economic expansion of the community by opening up fresh resources;
10. the association of overseas nations and territories in order to increase trade and to promote jointly economic and social development.

Article 3—Nondiscrimination

Within the scope of application of this treaty, and without prejudice to any special provisions contained therein, any discrimination against people, goods, and/or services on the grounds of nationality shall be prohibited.

Article 4—Internal Market

The Middle East/North Africa Community shall adopt measures with the aim of fostering and eventually establishing the internal market which shall comprise an area without internal economic borders in which the free movement of goods, persons, services, and capital is ensured in accordance with the provision of this treaty.

Article 5—Free Movement of Goods

The Middle East/North Africa Community shall be based upon a customs union which shall cover all trade in goods and which shall evolve the prohibition between member nations of customs duties on imports and exports and of all charges having equivalent effect among member nations, and the adoption of a common customs tariff in their relations with countries that are not members.

Article 6—The Elimination of Quantitative Restrictions

Quantitative restrictions on imports and all measures having equivalent effect shall, without prejudice to the following provisions, be prohibited among member nations.

Article 7—Exception

The provisions shall not preclude prohibitions or restrictions on imports, exports, or goods in transit justified on grounds of public morality, public policy, or public security; the protection of health and life of humans, animals, or plants; the protection of national treasures possessing artistic, historic, or archeological value; or the protection of industrial or commercial property. Such prohibitions or restrictions shall not, however, constitute a means of arbitrary discrimination or a disguised restriction on trade among member nations.

Article 8—Free Movement of Persons

1. Freedom of movement for workers shall be secured within the Middle East/North Africa Community by the end of the transition period at the latest.
2. Such freedom of movement shall entail the abolition of any discrimination in terms of employment, remuneration, and other conditions of work based on nationality among workers of the member nations.

Article 9—Right of Establishment

Companies or firms formed in accordance with the laws of the Middle East/North Africa Community and having their registered office, central administration, or principle place of business within the community shall be treated in the same way as natural persons of member countries.

Article 10—Services

Restrictions on freedom to provide services within the Middle East/North Africa Community shall be progressively abolished during the transitional period. Nationals of member nations who are established in a nation of the community will be free to render services in any member country.

The council may, acting unanimously on a proposal from its commission, extend the provision to nationals of a third country who provide services and who are established within the Middle East/North Africa Community.

Article 11—Rules of Competition

1. The following shall be prohibited as incompatible with the common market: all agreements among undertakings, decisions by associations of undertakings, and concerted practices which may affect trade among member nations and which have as their object the prevention, restriction, or distortion of competition within the common market, and in particular those which:
   a. directly or indirectly fix purchase or selling prices or any other trading conditions;
   b. limit or control production markets, technical development, or investment;
   c. share markets or sources of supply;
   d. apply dissimilar conditions to equivalent transactions with other trading parties, thereby placing them at a competitive disadvantage;
   e. making the conclusion of contracts subject to acceptance by the other parties of supplementary obligations which, by their nature or according to commercial usage, have no connection with the subject of such contracts.
2. Any agreements or decisions prohibited pursuant to this article shall be automatically void.
3. The provisions of paragraph 1 may, however, be declared inapplicable in the case of:
   a. any agreement or category of agreements between undertakings;
   b. any decision or category of decisions by associations of undertakings;
   c. any concerted practice or category of concerted practices which contributes to improving the production or distribution of goods or to promoting technical or economic progress, while

allowing consumers a fair share of the resulting benefit which does not:

i. impose on the undertakings concerned restrictions which are not indispensable to the attainment of these objectives;
ii. afford such undertakings the possibility of competition in respect to a substantial part of the products in question.

Article 12—Abuses

Any abuse by one or more undertakings of a dominant position within the common market or in a substantial part of it shall be prohibited as incompatible with the common market insofar as it may affect trade among member nations. Such abuse may, in particular, consist in:

1. directly or indirectly imposing unfair purchase or selling prices or other unfair trading conditions;
2. limiting production, markets, or technical development to the prejudice of consumers;
3. applying dissimilar conditions to equivalent transactions with other trading parties, thereby placing them at a competitive disadvantage;
4. making the conclusion of contracts subject to acceptance by the other parties of supplementary obligations which, by their nature or according to commercial usage, have no connection with the subject of such contracts.

Article 13—Undertakings

1. In the case of public undertakings and undertakings to which member nations grant special or exclusive rights, member nations shall neither enact nor maintain in force any measure contrary to the rules contained in this treaty.
2. Undertakings entrusted with the operation of services of general economic interest or having the character of a revenue-producing monopoly shall be subject to the rules contained in this treaty, in particular to the rules on competition, insofar as the application of such rules does not obstruct the performance, in law or fact, of the particular tasks assigned to them. The development of trade must not be affected to such an extent as would be contrary to the interests of the Middle East/North Africa Community.
3. The Secretariat shall ensure the application of the provisions of this article and shall, where necessary, address appropriate directives or decisions to member nations.

Article 14—Aid

1. Save as otherwise provided in this treaty, any aid granted by a member nation or through national resources in any form whatsoever

which distorts or threatens to distort competition by favoring certain undertakings or the production of certain goods shall, insofar as it affects trade among member nations, be incompatible with the common market.

2. The Secretariat shall, in cooperation with member nations, keep under constant review all systems of aid existing in those nations. It shall propose to the latter any appropriate measures required by the progressive development or by the functioning of the common market.

Article 15—Approximation of Laws

1. The council shall, acting by a qualified majority of the proposal from the Secretariat in cooperation with the Community Parliament—its legislative body—and the Economic and Social Committees, adopt the measures for the approximation of the provisions laid down by law, regulation, or administrative action in member nations which have as their object the establishment and functioning of the unified market.
2. Paragraph 1 shall not apply to fiscal provisions, to those relating to the free movement of persons, or to those relating to the rights and interests of employed persons. In such matters, unanimity shall be required.
3. The Secretariat, in its proposals laid down in paragraph 1, concerning health, safety, environmental, and consumer protection, will take as a base a high level of protection.
4. If, after the adoption of a standardization measure by the council acting by a qualified majority, a member nation deems it necessary to apply national provisions, on grounds of major need, or relating to protection of the environment or the working environment, it shall notify the Secretariat of these provisions. The Secretariat shall confirm the provisions involved after having been verified that they are not a means of arbitrary discrimination or a disguised restriction on trade between member nations. By way of derogation from procedures, the Secretariat or any member nation may bring the matter directly before the Court of Justice if it considers that another Middle East/North Africa Community country is making improper use of the powers provided for in this article. The standardization measures referred to above shall, in appropriate cases, include a safeguard clause authorizing the member nation to take, for one or more of the noneconomic reasons, provisional measures subject to a Middle East/North Africa Community control procedure.

Article 16—Enforcement Provision

Decisions of the council or of the Secretariat which impose a pecuniary obligation on persons rather than nations shall be enforceable. Enforcement shall be governed by the rules of civil procedure in force in the country in the territory of which it is carried out. The order for its enforcement shall be appended to the decision, without formality other than verification of the authenticity of the decision, by the national authority which the government of each Middle East/North Africa Community country shall designate for this purpose and shall make known to the Secretariat and to the Court of Justice. When the formalities have been completed on application by the party concerned, the latter may proceed to enforcement in accordance with the national law, by bringing the matter directly before the competent authority. Enforcement may be suspended only by a decision of the Court of Justice. However, the courts of the country concerned shall have jurisdiction over complaints that enforcement is being carried out in an irregular manner.

* * * * *

There should be no threat to the future of the nation-state with the creation of a Middle East/North Africa Community Treaty. Countries are today more sophisticated than ever in their formation and sense of political and social purpose. Under the treaty, the only initial, significant shift will be in the surrendering of some sovereignty of a nation's economic independence; other social and political concerns may follow. In the end the nation-state should survive and prosper. No other organizational structure can better deal with the governments of large societies within their boundaries. It has been and it shall remain the primary objective of global trade, defense, and internationalism, to have nations negotiate, unite, and disunite based on needs of citizens and the politics of the controlling parties. It works and should be perpetuated.

This treaty will create a needed level of solidarity in the definition of freedom, for the ruthless exercise of freedom by some is always at the expense of others. For this reason, if the MENAC Treaty is to endure, it must always recognize the solidarity of its members as a fundamental principle, and share both the advantages, that is, prosperity, and the burdens equally and justly among its members. This treaty will fulfill the needs of a broad membership of the contiguous nations touching the boundaries of Israel, and beyond. These concepts described can be applied to two or more nations and even expanded away from the traditionally defined lines of geography of the Middle East, such as other Arab and Muslim nations in Africa, the Persian Gulf, and Asia. (The treaty can be amended calling for a name change.) Obviously, the larger the membership base the greater the need and financial feasibility to create the organizational institutions of the court, Secretariat, and so on.

In its infancy, alternative institutions will have to be established, but a long-term commitment growing out of stage two should involve a full-fledged superstructure, that bows to the subsidiarity rights of nations to minimize the growth and power of empire building and empire control. Indeed, the foundation and support for the Middle East/North Africa Community rests with its treaty, this one or another preferred one, negotiated by founding members. By design it should be general and brief, allowing the exceptional issue to be dealt with later, thereby not restricting future opportunities by minimizing flexibility or attending to a larger audience of situation and need.

The treaty must command more attention than any other structural or legislative detail, for challenges, which are sure to come, must provide the various pressure groups and the Court of Justice with ample clarity for interpretation. Amendments will flow, as they should, but the treaty, the Middle East/North Africa Community Constitution, should remain as the guiding principle for continuity, determination, and to provide a heritage worthy of its efforts. It should strengthen the commitment to regional security, peace, a rising standard of living, equal opportunity for all, and territory where people living in artificially created borders learn that neighbors enrich community life.

See Middle East/North Africa Series.

Treaty on the Nonproliferation of Nuclear Weapons: See United Nations (nuclear-weapon-free zone in the region of the Middle East).

TRIDE: See Trilateral Industrial Development Program.

Trilateral Industrial Development Program (TRIDE): Announced by Israel, Jordan, and the United States in 1996; a new fund that will offer applicants grants of up to 50 percent of the direct costs of bringing new products to market. The TRIDE project will begin with the three countries jointly investing $1 million in the effort to encourage joint ventures involving private companies from all three countries.

Trilateral Israel-Jordan-U.S. Economic Committee: In addition to bilateral talks between Israel and Jordan, this committee was established at the October 1, 1993 White House meeting between President Clinton, Crown Prince Hassan, and Foreign Minister Peres to discuss economic cooperation and development. This forum first convened in Washington on November 30, 1993. Specific areas for cooperation include:

Trade, Finance and Banking

- Trade based on most-favored-nation agreement;
- Trade liberalization in phases;
- Regional trade agreement;
- Opening of Jordanian bank subsidiaries in West Bank.

Cooperative Projects

- Jordan Rift Valley master plan;
- Red Sea-Dead Sea canal;
- Jordan Rift Valley binational park;
- Regional tourism master plan;
- Free movement for third country tourists;
- Red Sea Riviera;
- Israel-Jordan road.

Civil Aviation

- Flyover rights;
- Air corridors;
- Communications;
- Security.

See Washington Declaration.

TRL: The international standard code for the currency of Turkey—the Turkish lira.

Truman, Harry S: See chronology (May 15, 1948).

TSU: Technical Support United; an European Union financed unit, based in the region, with offices in Aqaba and Gaza to service both the TEAM and SEMED projects.
See SEMED, TEAM Area.

Tunisia (Tunis): Habib ibn Ali Bourguiba was elected President of this independent republic on July 25, 1957. Tunisia has been used as the home base of the Palestinian Liberation Organization and also was a founder of the Maghreb Union in 1989.

Tunisia has sound macroeconomic policies and has successfully completed an International Monetary Fund and World Bank-sponsored restructuring program (1987-1994). Thanks to ongoing privatization and market liberalization programs, the economy has grown at an average rate of 4.5 percent since 1993. Progress has been made in reforming the banking and financial sectors. The local stock market, while still closed to foreign investors, is becoming more active.

The private sector is playing an ever-increasing role in the economy. Textiles and tourism are two of its biggest revenue earners. Tunisia is committed to a free trade regime and export-led growth. Most goods can be imported without restriction. Significantly, Tunisia and the European Union initialed a free trade accord in 1995. The pact enters into force in 1997 and phase-in over 12 years. The accord will allow the free trade of goods and services between Tunisia and its larger partner.

Tunisia welcomes foreign investment. Although residual bureaucratic bottlenecks can still hamper business, Tunisia has a comprehensive investment code, offering a variety of incentives to foreign investors.

See Bourguiba, Habib ibn Ali; Maghreb Agreement, Moscow Multilateral Middle East Conference, Union du Maghreb Arabe.

Tunisian dinar: The currency of Tunisia.

Turkey: Seventy-five percent of its 64 million people live in cities; more than half of its people are under 25 years of age. An Islamic, but secular nation, Turkey's inflation is at 60-90 percent. In 1995 her overall growth was 7.6 percent with manufacturing output up by 13.2 percent.

Private consumption per person is $3,617, cars per 1,000 people is 43, television sets per 1,000 people is 269, telephone lines per 1000 people is 201, with an infant mortality per 1,000 live births at a high 52.6 percent.

Turkey is the only member of the Islamic Conference Organization to belong to NATO and to have a customs union with the European Union.

In 1996 Israel and Turkey had signed agreements calling for everything from free trade between them to joint air-force training, naval visits, and intelligence sharing.

On July 8, 1996, breaking with Turkey's 73-year record of secularism, the Parliament gave narrow approval to a coalition government led by an Islamic party that has campaigned to restore the country's Muslim identity. The coalition, a marriage between the Islamic Welfare Party and a center-right party won a confidence vote, just three more than the majority required. Under the terms of the coalition protocol, the post of prime minister will rotate after a year.

The new Prime Minister Necmettin Erbakan dropped his past references to replace the Turkish lira with an Islamic dinar and stopped urging the creation of an Islamic common market and an Islamic United Nations and introducing Islamic law alongside Turkey's secular code.

On June 18, 1997 Prime Minister Erbakan, the first head of an Islamic party to lead Turkey, yielded to pressure from the military and resigned after nearly a year in power. The secular leader, Mesut Yilmaz, was named Prime Minister by month's end.

See League of Nations Mandate for Palestine, Middle East Free Trade Agreement.

Turkish Empire: See League of Nations Mandate for Palestine.

Turkish lira: The currency of Turkey, divided into 100 kurus (piastres).

TVT: Technical and Vocational Training.

U

UAE: See United Arab Emirates.

UAE dirham: The currency of the United Arab Emirates.

ultranational development: See New Middle East, The.

UMA: See Union du Maghreb Arabe.

umma: In Arabic, a single nation. See sovereignty.

unconventional weapons: Nonproliferation of weapons is failing throughout MENA. Properly, one of the highest priorities of the MENA Community will be to form a unit to deal with the spread of doomsday weapons. Proliferation of biological, chemical, and nuclear weapons remains the main strategic threat for the region.

Egypt has resumed efforts to acquire long-range ballistic missiles and is stepping up research in chemical and biological weapons. Israel is developing missiles that can reach as far as the former Soviet Union, an achievement that might alter the regional balance of power. Saudi Arabia and the United Arab Emirates have acquired ballistic missiles capable of carrying conventional and unconventional warheads. Syria is conducting tests of an extended-range Scud-C missile, and its chemical-weapons program is the most advanced in the Arab world.

undertakings: See Treaty of the Middle East/North Africa Community (Article 13).

UNDOF: See United Nations Disengagement Observer Force.

unemployment (West Bank and Gaza Strip): The Palestinian Central Bureau of Statistics in February 1996 released unemployment figures for the first time. In the West Bank and the Gaza Strip unemployment is 18.3 percent. Another 20 percent of the labor force are underemployed, while 61.6 percent have jobs.

By mid-July 1996 the Palestinian Statistics Office reported that of the 2,527,000 citizens (Gaza 962,000 and the West Bank 1,565,000) among them:

- 53 percent are over 15 years of age;
- 43 percent are under 15 years of age;
- 57 percent work full time;
- 14.4 percent work part time;
- 23.5 percent work in Israel and the territories;
- 28.6 percent are without work;
- 58.8 percent are outside the employment circuit: 15.9 percent because of old age; 21.6 percent for study or training; 52.8 percent are housewives; 9.7 percent for other reasons

Totally, in Gaza and the West Bank, 29.9 percent of the men are unemployed and 20.8 percent of the women.

See fences.

Union du Maghreb Arabe (UMA): Since October 1964, the date of the creation at Tunis of the Comite Permanent Consultatif Maghrebin, the unification of these five nations has been targeted. Maghreb is an Arabic word that means the west, including all countries from Egypt to Yemen. It is usually not confused with Morocco even though Morocco is the English transcription of Maghreb. North Africa includes only Morocco, Algeria, and Tunisia. The treaty of the UMA was signed in February 1989 at Marrakech in Morocco.

United Arab Emirates (UAE): Britain conceded the UAE independence in 1971. Seven Emirates created an independent federation between December 1971 and February 1972. The Sultan of Abu Dhabi became the first UAE president.

The United Arab Emirates remains the most dynamic, and perhaps only expanding economy in the Gulf region, despite the continued weakness in world oil prices. U.S. exports to the UAE dropped by about l0 percent in 1994, with first quarter 1995 showing that U.S. exports were up more than 25 percent.

The UAE's economic growth is being fueled by two main factors: an ongoing construction boom in both Abu Dhabi and Duai, and expansion of the greater Dubai economic zone as a regional business center and trading hub.

The UAE remains a highly competitive market dominated by government procurement, where the emphasis is on being lowest bidder. The UAE is the most attractive, growing business center between the Mediterranean and India, and has direct trade links with a region encompassing almost one quarter of the world's population.

See Gulf Cooperation Council, Gulf nations.

United Nations: See Camp David Accords, Egyptian-Israeli Peace Treaty; Gulf crisis, Hamas Covenant, Hussein, Saddam (President); Israel (Declaration of Establishment), Israel-Jordan Peace Treaty, Japan, League of Nations Mandate, Lebanon, London Agreement, Resolution 242, Resolution 338, Resolution 425, Security Council, United Nations (Middle East Peace Process), terrorism, United Nations (nuclear-weapon-free zone in the Middle East), United Nations (Partition Plan on Palestine), Washington Declaration, World Bank.

United Nations (Middle East Peace Process): Resolution adopted by the General Assembly on December 12, 1995:

The General Assembly,

Recalling its resolution 49/88 of December 16, 1994 and Economic and Social resolution 1995/52 of July 28, 1995,

Stressing that the achievement of a comprehensive, just, and lasting settlement of the Middle East conflict will constitute a significant contribution to the strengthening international peace and security,

Recalling the convening of the Peace Conference on the Middle East at Madrid on October 30, 1991, on the basis of Security Council resolution 242 (1967) of November 22, 1967 and 338 (1973) of October 22, 1973, and the subsequent bilateral negotiations, as well as the meetings of the multilateral working groups, and noting with satisfaction the broad international support for the peace process,

Noting the continuing positive participation of the United Nations as a full extra-regional participant in the work of the multilateral working groups,

Bearing in mind the Declaration of Principles on Interim Self-Government Arrangements, signed by the government of the State of Israel and the Palestine Liberation Organization, the representative of the Palestinian people, in Washington on September 13, 1993, and the subsequent Agreement on the Gaza Strip and the Jericho Area signed by the government of the State of Israel and the Palestine Liberation Organization at Cairo on May 4, 1994, their August 29, 1994 Agreement on the

Preparatory Transfer of Powers and Responsibilities, the Protocol on Further Transfer of Powers and Responsibilities signed by the government of Israel and the Palestine Liberation Organization at Cairo on August 27, 1995, and the Interim Agreement on the West Bank and Gaza Strip, signed by the government of Israel and the Palestine Liberation Organization in Washington on September 28, 1995,

Also bearing in mind the Agreement between Israel and Jordan on the Common Agenda, signed in Washington on September 14, 1993, the Washington Declaration, signed by Jordan and Israel on July 25, 1994, and the Treaty of Peace between the State of Israel and the Hashemite Kingdom of Jordan, of October 26, 1994,

Welcoming the Declaration of the Middle East/North Africa Economic Summit held at Casablanca from October 30 to November 1, 1994, as well as the Declaration of the Middle East/North Africa Economic Summit held in Amman from October 29-31, 1995,

1. Welcomes the peace process started at Madrid, and supports the subsequent bilateral negotiations;
2. Stresses the importance of, and need for, achieving a comprehensive, just, and lasting peace in the Middle East;
3. Expresses its full support for the achievements of the peace process thus far, in particular the Declaration of Principles on Interim Self-Government Arrangements, signed by the government of the State of Israel and the Palestine Liberation Organization, the representative of the Palestinian People, the subsequent Agreement on the Gaza Strip and the Jericho Area, signed by the government of the State of Israel and the Palestine Liberation Organization, their August 29, 1994 Agreement on the Preparatory Transfer of Powers and Responsibilities, the Protocol on Further Transfer of Powers and Responsibilities signed by the government of Israel and the Palestine Liberation Organization at Cairo on August 27, 1995, the Interim Agreement on the West Bank and Gaza Strip, signed by the government of Israel and the Palestine Liberation Organization in Washington on September 28, 1995, the Agreement between Israel and Jordan on the Common Agenda, the Washington Declaration, signed by Jordan and Israel on July 25, 1994, and the Treaty of Peace between the State of Israel and the Hashemite Kingdom of Jordan, of October 26, 1994, which constitute important steps in achieving a comprehensive, just, and lasting peace in the Middle East, and urges all parties to implement the agreements reached;
4. Stresses the need to achieve rapid progress on the other tracks of the Arab-Israeli negotiations within the peace process;

5. Welcomes the results of the Conference to Support Middle East Peace, convened in Washington on October 1, 1993, including the establishment of the Ad Hoc Liaison Committee, and the subsequent work of the World Bank Consultative Group, welcomes also the appointment by the Secretary-General of the "United Nations Special Coordinator in the Occupied Territories," and urges Member States to expedite economic, financial, and technical assistance to the Palestinian people during the interim period;
6. Calls upon all member states to extend economic, financial, and technical assistance to parties in the region and to render support for the peace process;
7. Considers that an active United Nations role in the Middle East peace process and in assisting in the implementation of the Declaration of Principles can make a positive contribution;
8. Encourages regional development and cooperation in areas where work has begun within the framework of the Madrid Conference.

79th plenary meeting
December 4, 1995

See United Nations (nuclear-weapon-free zone in the region of the Middle East).

United Nations (nuclear-weapon-free zone in the region of the Middle East): Resolution adopted by the United Nations General Assembly on January 6, 1996.

ESTABLISHMENT OF A NUCLEAR-WEAPON-FREE ZONE IN THE REGION OF THE MIDDLE EAST

The General Assembly

Recalling its resolution 3263 (XXIX) of 9 December 1974, 3474 (XXX) of 11 December 1975, 31/71 of 10 December 1976, 32/82 of 12 December 1977, 33/64 of 14 December 1978, 34/77 of 11 December 1979, 35/147 of 12 December 1980, 36/87 A and B of 9 December 1981, 37/75 of 9 December 1982, 38/64 of 15 December 1983, 39/54 of 12 December 1984, 40/82 of 12 December 1985, 41/48 of 3 December 1986, 42/28 of 30 November 1987, 43/65 of 7 December 1988, 44/108 of 15 December 1989, 45/52 of 4 December 1990, 46/30 of 6 December 1991, 47/48 of 9 December 1992, 48/71 of 16 December 1993 and 49/71 of 15 December 1994 on the establishment of a nuclear-weapon-free zone in the region of the Middle East,

Recalling also the recommendations for the establishment of such a zone in the Middle East consistent with paragraphs 60 to 63, and in particular paragraph 63 (d), of the Final Document of the Tenth Special Session of the General Assembly,

Emphasizing the basic provisions of the abovementioned resolutions, which call upon all parties directly concerned to consider taking the practical and urgent steps required for the implementation of the proposal to establish a nuclear-weapon-free zone in the region of the Middle East and, pending and during the establishment of such a zone, to declare solemnly that they will refrain, on a reciprocal basis, from producing, acquiring, or in any other way possessing nuclear weapons and nuclear explosive devices and from permitting the stationing of nuclear weapons on their territory by any third party, to agree to place all their nuclear facilities under International Atomic Energy Agency safeguards and to declare their support for the establishment of the zone and to deposit such deliberations with the Security Council for consideration, as appropriate,

Reaffirming the inalienable right of all states to acquire and develop nuclear energy for peaceful purposes,

Emphasizing the need for appropriate measures on the question of the prohibition of military attacks on nuclear facilities,

Bearing in mind the consensus reached by the General Assembly at its thirty-fifth session that the establishment of a nuclear-weapon-free zone in the region of the Middle East would greatly enhance international peace and security,

Desirous of building on that consensus so that substantial progress can be made toward establishing a nuclear-weapon-free zone in the region of the Middle East,

Welcoming all initiatives leading to general and complete disarmament, including in the region of the Middle East, and in particular on the establishment therein of a zone free of weapons of mass destruction, including nuclear weapons,

Noting the peace negotiations in the Middle East, which should be of a comprehensive nature and represent an appropriate framework for the peaceful settlement of contentious issues in the region,

Recognizing the importance of credible regional security, including the establishment of a mutually verifiable nuclear-weapon-free zone,

Emphasizing the essential role of the United Nations in the establishment of a nuclear-weapon-free zone in the region of the Middle East,

Having examined the report of the Secretary-General on the implementation of resolution 49/71,

1. Urges all parties directly concerned to consider seriously taking the practical and urgent steps required for the implementation of the

proposal to establish a nuclear-weapon-free zone in the region of the Middle East in accordance with the relevant resolutions of the General Assembly, and, as a means of promoting this objective, invites the countries concerned to adhere to the Treaty on the Non-proliferation of Nuclear Weapons;

2. Calls upon all countries of the region that have not done so, pending the establishment of the zone, to agree to place all their nuclear activities under International Atomic Energy Agency safeguards;
3. Takes note of resolution GC (39)/RES/24, adopted on September 22, 1995 by the General Conference of the International Atomic Energy Agency at its thirty-ninth regular session, concerning the application of Agency safeguards in the Middle East;
4. Notes the importance of the ongoing bilateral Middle East peace negotiations and the activities of the multilateral working group on arms control and regional security in promoting mutual confidence and security in the Middle East, including the establishment of a nuclear-weapon-free zone;
5. Invites all countries of the region, pending the establishment of a nuclear-weapon-free zone in the region of the Middle East, to declare their support for establishing such a zone, consistent with paragraph 63 (d) of the Final Document of the Tenth Special Session of the General Assembly, and to deposit those declarations with the Security Council;
6. Also invites those countries, pending the establishment of the zone, not to develop, produce, test, or otherwise acquire nuclear weapons or permit the stationing on their territories, or territories under their control, of nuclear weapons or nuclear explosive devices;
7. Invites the nuclear-weapon states and all other states to render their assistance in the establishment of the zone and at the same time to refrain from any action that runs counter to both the letter and the spirit of the present resolution;
8. Takes note of the report of the Secretary-General;
9. Invites all parties to consider the appropriate means that may contribute toward the goal of general and complete disarmament and the establishment of a zone free of weapons of mass destruction in the region of the Middle East;
10. Requests the Secretary-General to continue to pursue consultations with the states of the region and other concerned states, in accordance with paragraph 7 of resolution 46/30 and taking into account the evolving situation in the region, and to seek from those states their views on the measures outlined in chapters III and IV of the study annexed to his report or other relevant measures, in order to

move toward the establishment of a nuclear-weapon-free zone in the region of the Middle East;

11. Also requests the Secretary-General to submit to the General Assembly at its fifty-first session a report on the implementation of the present resolution;
12. Decides to include in the provisional agenda of its fifty-first session the item entitled "Establishment of a nuclear-weapon-free zone in the region of the Middle East."

90th plenary meeting
December 12, 1995

See United Nations (Middle East Peace Process).

United Nations (Partition Plan on Palestine): The United Nations Special Committee on Palestine proposed a Majority Plan of Partition with Economic Union. It was passed by the General Assembly on November 29, 1947, by over two-thirds majority. It reads:

1. The basic premise underlying the partition proposal is that the claims to Palestine of the Arabs and Jews, both possessing validity, are irreconcilable, and that among all of the solutions advanced, partition will provide the most realistic and practicable settlement, and is the most likely to afford a workable basis for meeting in part the claims and national aspirations of both parties.
2. It is a fact that both of these peoples have their historic roots in Palestine, and that both make vital contributions to the economic and cultural life of the country. The partition solution takes these considerations fully into account.
3. The basic conflict in Palestine is a clash of two intense nationalisms. Regardless of the historical origins of the conflict, the rights and wrongs of the promises and counterpromises, and the international intervention incident to the Mandate, there are now in Palestine some 650,000 Jews and some 1,200,000 Arabs who are dissimilar in their ways of living and, for the time being, separated by political interests which render difficult full and effective political cooperation among them, whether voluntary or induced by constitutional arrangements.
4. Only by means of partition can these conflicting national aspirations find substantial expression and qualify both peoples to take their places as independent nations in the international community in the United Nations.
5. The partition solution provides that finality which is a most urgent need in the solution. Every other proposed solution would tend to induce the two parties to seek modification in their favor by means

of persistent pressure. The grant of independence to both states, however, would remove the basis for such efforts.

6. Partition is based on a realistic appraisal of the actual Arab-Jewish relations in Palestine. Full political cooperation would be indispensable to the effective functioning of any single-state scheme, such as the federal-state proposal, except in those cases which frankly envisage either an Arab or a Jewish dominated state.
7. Partition is the only means available by which political and economic responsibility can be placed squarely on both Arabs and Jews, with the prospective result that, confronted with the responsibility for bearing fully the consequences of their own actions, a new and important element of political amelioration would be introduced. In the proposed federal-state solution, this factor would be lacking.
8. Jewish immigration is the central issue in Palestine today and is the one factor, above all others, that rules out the necessary cooperation between the Arab and Jewish communities in a single state. The creation of a Jewish state under a partition scheme is the only hope of removing this issue from the arena of conflict.
9. It is recognized that partition has been strongly opposed by Arabs, but it is felt that the opposition would be lessened by a solution which definitively fixed the extent of territory to be allotted to the Jews with its implicit limitation on immigration. The fact that the solution carries the sanction of the United Nations involves a finality which should allay Arab fears of further expansion of the Jewish state.
10. In view of the limited area and resources of Palestine, it is essential that, to the extent feasible, and consistent with the creation of two independent states, the economic unity of the country should be preserved. The partition proposal, therefore, is a qualified partition, subject to such measures and limitations as are considered essential to the future economic and social well-being of both states. Since the economic self-interest of each state would be vitally involved, it is believed that the minimum measure of economic unity is possible, where that of political unity is not.
11. Such economic unity requires the creation of an economic association by means of a treaty between the two states. The essential objectives of this association would be a common customs system, a common currency, and the maintenance of a country-wide system of transport and communications.
12. The maintenance of existing standards of social services in all parts of Palestine depends partly upon the preservation of economic unity, and this is a main consideration underlying the provisions for an economic union as part of the partition scheme. Partition, however, nec-

essarily changes to some extent the fiscal situation in such a manner that, at any rate during the early years of its existence, a partitioned Arab state in Palestine would have some difficulty in raising sufficient revenues to keep up its present standards of public services.

One of the aims of the economic union, therefore, is to distribute surplus revenue to support such standards. It is recommended that the division of the surplus revenue, after certain charges and percentages of surplus to be paid to the city of Jerusalem are met, should be in equal proportions to the two states. This is an arbitrary proportion but it is considered that it would be acceptable, that it has the merit of simplicity and that, being fixed in this manner, it would be less likely to become a matter of controversy. Provisions are suggested whereby this formula is to be reviewed.

13. This division of customs revenue is justified on three grounds:
    (1) The Jews will have the more economically developed part of the country embracing practically the whole of the citrus-producing area which includes a large number of Arab producers;
    (2) the Jewish state would, through the customs union, be guaranteed a larger free trade area for the sale of the products of its industry;
    (3) it would be to the disadvantage of the Jewish state if the Arab state should be in a financially precarious and poor economic condition.
14. As the Arab state will not be in a position to undertake considerable development expenditure, sympathetic consideration should be given to its claims for assistance from international institutions in the way of loans for expansion of education, public health and other vital social services of a non-self-supporting nature.
15. International financial assistance would also be required for any comprehensive irrigation schemes in the interest of both states, and it is to be hoped that constructive work by the Joint Economic Board will be made possible by means of international loans on favorable terms.

See Israel (Declaration of Establishment).

United Nations (Resolutions): See Gulf crisis, Resolutions.

United Nations Disengagement Observer Force (UNDOF): After the disengagement agreement between Israeli and Syrian forces on the Golan Heights on May 31, 1974, a U.N. Disengagement Observer Force was posted to establish an area of separation and verify troop levels with more than 1,000 personnel coming from three European nations and Canada.
See Japan.

United Nations Special Commission on Iraq: The Secretary-General appointed a Special Commission in April 1991, to carry out on-site inspection of Iraq's biological, chemical, and missile facilities, and to cooperate with the International Atomic Energy Commission regarding on-site inspection of Iraq's nuclear capabilities and to establish a monitoring system to verify Iraq's continued compliance of its undertaking in these areas.

See Iraq.

United Nations Special Coordinator in the Occupied Territories: See United Nations (Middle East Peace Process).

United States: The fact is that there has not been a single successful agreement in the Arab-Israeli conflict without United States involvement. Just prior to the end of World War II, President Roosevelt while attending a meeting in Yalta emphasized the need to maintain U.S. access to oil, with stability in the region. Stability and peace remain the primary focus of all presidential administrations.

Today, there are numerous factions split over the issue of U.S. support. Included are many in the United States who in the past were either supportive or silent when it came to U.S. support of Israel and the peace process.

The essential question for the United States is whether or not it is prepared to be a full partner in the process. The parties to the peace will of course have to be the ones whom make the tough decisions, but it remains clear that the process will falter without real leadership from the United States. Therefore, it will take a good deal of convincing to persuade the American people that an American bridge to peace is not only necessary, but also in the vital interests of the United States.

On August 12, 1997 the U.S. government announced that the Israeli and Palestinian authorities had agreed to report all they had learned about the July 30, 1997 suicide bombing in Jerusalem to a three-way panel whose American representative would be the Central Intelligence Agency (CIA) station chief in Tel Aviv.

This mechanism allows the United States to serve as referee and arbiter to decide whether the Palestinians are doing enough to cooperate with Israeli demands to crack down on suspected Islamic militants. The United States will play a continuing role to ensure that security cooperation continues beyond the investigation into the bombing of July 30, 1997.

See Agreement of Understanding between Israel and Lebanon, chronology (July 30, 1997), Council on Foreign Relations, Har Homa, Middle East Peace and Stability Fund, United States Trade with the Middle East.

United States of America (Concurrent Resolution): Dated December 17, 1945 and passed by the Senate, and December 19, 1945 and passed by the House of Representatives of the 79th Congress, this Concurrent Resolution reads:

Whereas the Sixty-seventh Congress of the United States on June 30, 1922, unanimously resolved "That the United States of America favors the establishment in Palestine of a national home for the Jewish people, it being clearly understood that nothing shall be done which may prejudice the civil and religious rights of Christians and all other non-Jewish communities in Palestine, and that the holy places and religious buildings and sites in Palestine shall be adequately protected," and

Whereas the ruthless persecution of the Jewish people in Europe has clearly demonstrated the need for a Jewish homeland as a haven for the large number who have become homeless as a result of this persecution; and

Whereas these urgent necessities are evidenced by the President's request for the immediate right of entry into Palestine of one hundred thousand additional Jewish refugees; and

Whereas the influx of Jewish immigration into Palestine is resulting in its improvement in agricultural, financial, hygienic, and general economic conditions; and

Whereas the President and the British Prime Minister have agreed upon the appointment of a "Joint Anglo-American Committee of Enquiry" to examine conditions in Palestine as they bear upon the problem of Jewish immigration and the Jewish situation in Europe and have requested a report within one hundred and twenty days;

Therefore Be it

Resolved by the Senate (the House of Representatives concurring),

That the interest shown by the President in the solution of this problem is hereby commended and that the United States shall use its good offices with the mandatory power to the end that Palestine shall be opened for free entry of Jews into that country to the maximum of its agricultural and economic potentialities, and that there shall be full opportunity for colonization and development, so that they may freely proceed with the upbuilding of Palestine as the Jewish national home and, in association with all elements of the population, establish Palestine as a democratic commonwealth in which all men, regardless of race or creed, shall have equal rights.

See Pentagon Plan.

United States (Department of Commerce): In 1996 a Middle East/North Africa Business Home Page was established to provide information on private-sector trade and investment in the Middle East and North

Africa. It represents a substantial step forward in the Commerce Department's ability to provide U.S. companies with the best data and resources for achieving their business goals in the region. The page offers: access to government information and services from project design to financing; lists of commercial projects in the region, complete with details on value, status, and contact information; country-specific information on trade opportunities, resources, and assessments; material on Middle East/North Africa Economic Summits; a bulletin board for companies to post public messages, such as offers and requests for goods or services or listings for joint ventures, and so on; a password-protected listing of a sample of firms interested in doing business in the region; and general resources related to trade and commerce in the region, including trade law, chambers of commerce, and trade publications. Located at the Internet address of "http://www.ita.doc.gov/region/amman."

See United States Trade with the Middle East.

United States-Israel Free Trade Area Agreement: See Free Trade Agreement (United States and Israel).

United States Middle East Peace Policy and Peace Process: See Council on Foreign Relations.

United States Trade with the Middle East: In 1995 U.S. trade with nations of the Middle East, in millions of dollars, was:

	Imports		*Exports*		*Surplus (Deficit)*
	$Mil	*%*	*$Mil*	*%*	*$Mil*
Total	638,030	100.0	480,987	100.0	(157,043)
Industrial	364,261	57.1	276,146	57.4	(88,115)
Countries					
Middle East	16,432	2.5	18,634	3.9	2,202
Bahrain	129	633	0.1	504	
Egypt	678	0.1	2,750	0.5	3,428
Israel	4,980	0.8	4,593	0.9	(357)
Kuwait	2,096	0.3	993	0.2	(1,103)
Oman	387		280		(127)
Qatar	81		164		83
Saudi Arabia	6,889	1.1	6,082	1.3	(807)
UAE	564		218		(346)
Syria	95		218		
Jordan	25		334		

Upper Gulf of Aqaba Oil Spill Contingency Project: Egypt, Israel, and Jordan, supported by the European Union, agreed to initiate an operational oil spill contingency project for the Northern part of the Gulf of Aqaba. The project was launched at a meeting of the Multilateral Working Group on the Environment in Cairo in late 1993.

The project is designed to protect the marine environment of the Upper Gulf of Aqaba, through the establishment of individual oil spill combat centers in Egypt, Israel, and Jordan. The objective is the development of a regional oil spill contingency partnership aimed at joint utilization of all available oil spill combat resources in the area for combating larger spills. An added objective, in relation to the Middle East Peace Process, is for the project to act as a catalyst for development of regional cooperation between Egypt, Israel, and Jordan.

It is envisaged that the individual centers in the Gulf of Aqaba will be integrated, not only into more comprehensive national oil spill contingency organizations in Egypt, Israel, and Jordan respectively, but also into regional plans covering the eastern Mediterranean and the Red Sea areas.

See Japan.

Uruguay Round: The International Monetary Fund claims that "The Uruguay Round incorporates progress in removing nontariff measures facing MENA exporters, especially in agriculture, textiles, and clothing. However, elements of the round could have adverse consequences for MENA, particularly in the short term. For example, the most-favored nation (MFN) tariff cuts incorporated in the round will reduce preference margins that some MENA countries have enjoyed with the European Union, under the Mediterranean Agreements, as well as under other preferential trade agreements.

A number of studies conclude that there will be some important initial losses for MENA. However, these losses can be mitigated to the extent that MENA countries themselves undertake important domestic policy reforms, leading eventually to the development of more efficient productive capacity as well as a more diversified economic trade structure.

See General Agreement on Tariffs and Trade.

USAID: See Hebron.

U.S.-Egyptian Partnership for Economic Growth and Development: See Egypt.

U.S.-Israel antiterrorism accord: See terrorism.

U.S.-Israel Free Trade Area Agreement: See Free Trade Agreement (United States and Israel).

U.S.-Middle East Project (of the Council on Foreign Relations): See Council on Foreign Relations.

U.S.-Qatar Business Council: A nonprofit, private-sector organization to promote commercial relations between the United States and Qatar was formed in Washington during the summer 1996. The council is intended to provide practical guidance to American firms interested in operating in the Persian Gulf country.

See Qatar.

U.S.-West Bank and Gaza Strip Free Trade Agreement: See Free Trade Agreement (United States and the West Bank and the Gaza Strip).

usury: See Amman Middle East/North Africa Economic Summit, Banking and Financial Services (Center of the Middle East/North Africa Community), Islamic banking.

V

Value Added Tax (VAT): See Preparatory Transfer of Powers and Responsibilities, Protocol on Economic Relations (Article VI).

Vanguard of the Popular War of Liberation: See Saiqa.

VAT: See Value Added Tax.

Vatican: See chronology (December 31, 1993), Fundamental Agreement.

Versailles Arab-Jewish Agreement: See Arab-Jewish Agreement at Versailles.

veterinary services: See Regional Animal Health Cooperation Program.

violence: See Interim Agreement (on the West Bank and the Gaza Strip), London Agreement, terrorism.

vision, the: The population of the Middle East/North Africa Community in 1995 was approximately 280 million people, with a gross domestic product of about $600 billion, larger than that of Canada or Spain, and about two-thirds the GDP of the United Kingdom.

This community is an idea whose time has arrived, the only way to secure papered peace treaties among former enemies. Tradeoffs of interdependence binds people and governments together, where hopefully all parties benefit from the relationships.

The community will come only after considerable argument, an abundance of heated debate, and then, some failures. The community will not evolve easily, as mistrusts and counteraccusations of imperialism, colonization, loss of sovereignty, and so on will be reheard time and again.

A successful launching of the community will more than adequately communicate to the rest of the Arab region that the idea for a common market can be realistic and productive. It will not be easy to release the energies from differing nations. Many of the countries of the community are still dominated by state bureaucrats who have only a cursory grasp of modern economics and market realities, and little political will to institute the changes necessary to realize the enormous benefits awaiting those willing to open up their economies and liberalize trade.

There are some who argue that a regional community in the Middle East and North Africa will be fraught with danger and accomplish little other than stirring up further troubles. These thinkers promote a vertical form of relationship, based on inequality and domination. Arab nations they claim, must develop their own societies without fears of being dependent on a far superior Israel, while Israel must be free from the fear of terrorists of contiguous nations.

In a golden age over a millennium ago, the region was the commercial and cultural crossroads of the world. A more peaceful, more prosperous, and more integrated Middle East and North Africa can once again take its position among other movers of the global economy, making this vision a reality.

Risk-taking will be followed by reform; it will yield the spirit and the means for a community; it must inspire a determination to have an uplifted region of healthier, happier, and wealthier people.

First must come the series of free trade accords, leading to a common market of the Middle East. Gathered evidence will outweigh all the questioning and criticism. In a short time nations of the region, now expanded to include countries from the expanded circle and North Africa could be entwined into a colossus of supreme management and harmony, paving the way in the new millennium to a centrifugal force of unparalleled advance, ensuring a continuing peace.

For the community to have a chance of success involves a significant degree of mutual trust and influence between member nations. Once accomplished, the gains will be economic, political, and decisively human.

See Middle East Community, Middle East/North Africa Community.

vocational training: See Education (Center of the Middle East/North Africa Community), Social Rights (Center of the Middle East/North Africa Community).

Volcker, Paul: Chair, Middle East/North Africa Economic Strategy Group and former Chair, Federal Reserve Board of the United States.

See Amman Economic Summit, Casablanca Middle East/North Africa Economic Summit, Middle East/North Africa Economic Strategy Group.

voluntary boycott: See boycott.

voucher schemes: A system that calls for the preparation of vouchers for transactions involving payments and for the recording of such vouchers in a special book of original entry.

See privatization.

W

Wall Mountain: See Har Homa.

War of Independence: Synonymous with Arab-Israeli War.

War of Liberation: See Arab-Israeli War.

Washington Declaration: Signed by Israel's Prime Minister Rabin and King Hussein, with President Clinton serving as a witness, on July 25, 1994 in Washington, DC. Its major achievements are a series of agreements and concrete steps. It reads:

A. After generations of hostility, blood, and tears and in the wake of years of pain and wars, His Majesty King Hussein and Prime Minister Yitzhak Rabin are determined to bring an end to bloodshed and sorrow. It is in this spirit that His Majesty King Hussein of the Hashemite Kingdom of Jordan and Prime Minister and Minister of Defense, Mr. Yitzhak Rabin of Israel, met in Washington today at the invitation of President William J. Clinton of the United States of America. This initiative of President William J. Clinton constitutes a historic landmark in the United States' untiring efforts in promoting peace and stability in the Middle East. The personal involvement of the President has made it possible to realize agreement on the content of this historic declaration.

The signing of this declaration bears testimony to the President's vision and devotion to the cause of peace.

B. In their meeting, His Majesty King Hussein and Prime Minister Yitzhak Rabin have jointly reaffirmed the five underlying principles of their understanding on an Agreed Common Agenda designed to reach the goal of a just, lasting, and comprehensive peace between the Arab states and the Palestinians, with Israel.

1. Jordan and Israel aim at the achievement of just, lasting, and comprehensive peace between Israel and its neighbors and at the conclusion of a Treaty of Peace between both countries.
2. The two countries will vigorously continue their negotiations to arrive at a state of peace, based on Security Council Resolutions 242 and 338 in all their aspects, and founded on freedom, equality, and justice.
3. Israel respects the present special role of the Hashemite Kingdom of Jordan in Muslim Holy shrines in Jerusalem. When negotiations on the permanent status will take place, Israel will give high priority to the Jordanian historic role in these shrines. In addition the two sides have agreed to act together to promote interfaith relations among the three monotheistic religions.
4. The two countries recognize their right and obligation to live in peace with each other as well as with all states within secure and recognized boundaries. The two states affirmed their respect for and acknowledgment of the sovereignty, territorial integrity, and political independence of every state in the area.
5. The two countries desire to develop good neighborly relations of cooperation between them to ensure lasting security and to avoid threats and the use of force between them.

C. The long conflict between the two states is now coming to an end. In this spirit the state of belligerency between Jordan and Israel has been terminated.

D. Following this declaration and in keeping with the Agreed Common Agenda, both countries will refrain from actions or activities by either side that may adversely affect the security of the other or may prejudice the final outcome of negotiations. Neither side will threaten the other by use of force, weapons, or any other means, against each other, and both sides will thwart threats to security resulting from all kinds of terrorism.

E. His Majesty King Hussein and Prime Minister Yitzhak Rabin took note of the progress made in the bilateral negotiations within the Jordan-Israel track last week on the steps decided to implement the sub-agenda on borders, territorial matters, security, water, energy, environment, and the Jordan Rift Valley.

In this framework, mindful of items of the Agreed Common Agenda (border and territorial matters) they noted that the boundary sub-commission has reached agreement in July 1994 in fulfillment of part of the role entrusted to it in the sub-agenda. They also noted that the sub-commission for water, environment, and energy agreed to mutually recognize, as the role of their negotiations, the rightful allocations of the two sides in Jordan River and Yarmouk River waters and to fully respect and comply with the negotiated rightful allocations, in accordance with agreed acceptable principles with mutually acceptable quality. Similarly, His Majesty King Hussein and Prime Minister Yitzhak Rabin expressed their deep satisfaction and pride in the work of the trilateral commission in its meeting held in Jordan on Wednesday, July 20, 1994, hosted by the Jordanian Prime Minister, Dr. Abdessalam al-Majali, and attended by Secretary of State Warren Christopher and Foreign Minister Shimon Peres. They voice their pleasure at the association and commitment of the United States in this endeavor.

F. His Majesty King Hussein and Prime Minister Yitzhak Rabin believe that steps must be taken both to overcome psychological barriers and to break with the legacy of war. By working with optimism toward the dividends of peace for all the people in the region, Jordan and Israel are determined to shoulder their responsibilities towards the human dimension of peace making. They recognize imbalances and disparities are a root cause of extremism which thrives on poverty and unemployment and the degradation of human dignity. In this spirit, His Majesty King Hussein and Prime Minister Yitzhak Rabin have today approved a series of steps to symbolize the new era which is now at hand:

1. Direct telephone links will be opened between Jordan and Israel.
2. The electricity grids of Jordan and Israel will be linked as part of a regional concept.
3. Two new border crossings will be opened between Jordan and Israel—one at the southern tip of Aqaba-Eilat and the other at a mutually agreed point in the north.
4. In principle, free access will be given to third country tourists traveling between Jordan and Israel.
5. Negotiations will be accelerated on opening an international air corridor between both countries.
6. The police forces of Jordan and Israel will cooperate in combating crime, with emphasis on smuggling and particularly drug smuggling. The United States will be invited to participate in this joint endeavor.
7. Negotiations on economic matters will continue in order to prepare for future bilateral cooperation including the abolition of all economic boycotts.

All these steps are being implemented within the framework of regional infrastructural development plans and in conjunction with the Jordan-Israel bilaterals on boundaries, security, water, and related issues and without prejudice to the final outcome of the negotiations on the items included in the Agreed Common Agenda between Jordan and Israel.

G. His Majesty King Hussein and Prime Minister Yitzhak Rabin have agreed to meet periodically or whenever they feel necessary to review the progress of the negotiations and express their firm intention to shepherd and direct the process in its entirety.

H. In conclusion, His Majesty King Hussein and Prime Minister Yitzhak Rabin wish to express once again their profound thanks and appreciations to President William J. Clinton and his Administration for their untiring efforts in furthering the cause of peace, justice, and prosperity for all the peoples of the region. They wish to thank the President personally for his warm welcome and hospitality. In recognition of their appreciation to the President, His Majesty King Hussein and Prime Minister Yitzhak Rabin have asked President William J. Clinton to sign this document as a witness and as a host to their meeting.

See Ein Avrona, Israel-Jordan Common Agenda, Israel-Jordan Peace Treaty, Trilateral Israel-Jordan-U.S. Economic Committee.

watan: Arab nationalists refer to Palestine as watan, the homeland or fatherland.

Watani: See PLO Charter.

Wataniyya: See PLO Charter.

water: The region constitutes one of the largest arid regions in the world. Almost three-quarters of the land from Morocco through Iraq is desert, lacking the water resources to support urban and sedentary agrarian populations. For example, the proportion of land use in the Middle East is:

Desert	74%
Cropland	4%
Pasture	20%
Forest	2%

Throughout the parched Middle East, water is the natural resource many consider more valuable than oil. While Americans have access to 10,000 cubic meters (2.6 millon gallons) of water per person annually, Egyptians rely on only 1,100, Israelis 460, and Jordanians only 260 cubic meters per person each year. In Israel, water demand in 1994 within small and medium-sized towns shot up by an average of over 32 percent in the past four years and by more than 14 percent in big cities.

Projected demand for water in billion cubic meters is:

	Year		
	1985	*2000*	*2030*
Egypt	59.500	72.400	112.800
Israel & PA	1.926	2.113	3.600
Jordan	0.499	1.075	1.700
Lebanon	0.859	1.448	3.106
Saudi Arabia	3.530	6.521	13.365
Syria	6.883	8.498	14.915

In 1979 then-Egyptian Foreign Minister Boutros Boutros-Ghali commented, that the next regional war would be fought over water, not politics. Since that time, high population increases, over-drilling leading to polluted aquifers and groundwaters, lack of sufficient rainfall, and wasteful industrial and agricultural programs, have only exacerbated the need to ensure a plentiful water supply across the region.

At the Madrid Peace Conference, a Water Working Group was formed as part of the multilateral track.

According to Franklin M. Fischer, an economist at the Massachusetts Institute of Technology water policy doesn't have to be divisive. His water project found that the actual value of the water now in dispute is at most $110 million a year, and even by the year 2020 will not exceed $500 million. He urges the creation of a single multinational authority to control disputed water resulting in all groups having incentives to treat water as a commodity with a price that reflects scarcity. For example, if Jordan priced water to discourage waste, it would be able to spend less for water from the authority. On the other hand, each participating nation would retain the option of subsidizing internal water use for reasons of social policy or expediency. Professor Franklin believes that water really is not exceptionally scarce in the region, and is thus not very valuable at the equivalent of a competitive market price.

The Middle East water supply is shorter than most places, and getting shorter all the time. Lebanon is the only tiny part that has more than enough. The region faces drought every four years. Rivers and renewable underground sources now supply it with about half the water it could use; by the time the twenty-first century is a quarter way through it may require four times as much as it can get from natural resources.

Farming, which swallows at least 80 percent of the region's water is most affected.

On May 27, 1997 King Hussein marked the beginning of the second stage of the Israel-Jordan water agreement, in accordance with which Israel had begun to supply Jordan with 3,000 cubic meters of water per hour.

See Agriculture (Center of the Middle East/North Africa Community), Amman Middle East/North Africa Economic Summit,Cairo Middle East/ North Africa Economic Conference, desertification, Interim Agreement (on the West Bank and the Gaza Strip), Israel-Jordan Common Agenda, Israel-Jordan Peace Treaty, Middle East Desalination Research Center, Moscow Multilateral Middle East Conference, Water & Desertification (Center of the Middle East/North Africa Community), Water Data Banks Project.

Water & Desertification (Center of the Middle East/North Africa Community): In most community nations, water is a scarce commodity and the situation is deteriorating. This crisis is due mainly to natural factors—climate and geography, minister precipitation, heating, and so forth. However, the shortage of water is also exacerbated by human factors—a tremendous growth in population within the region, wasteful utilization, the destruction of water sources, and so on.

At the Beijing October 1993 meeting nations agreed to establish a regional database to facilitate water sharing, and to create a training program for regional water personnel. A primary task is to preserve the quality of the existing aquifer, which is already deteriorating because of the massive increase in wastewater generated by the growing urban population. The options include either importing water or opting for the desalination of sea water.

A water authority is needed to be initiated by the community. Not only to find and secure water, but the center and its units must protect the quality and usage of existing waters. Legislation must be directed at protecting the high quality of water supplies, the use of available potential resources, and at elevating the forecasting of its consumer and industrial demand. Directives are needed to set qualitative criteria for surface waters intended for drinking or for other forms of human consumption.

Water activity will include the development of new water sources, erection of dams, prevention of pollution and salinity of existing sources, rain intensification, purification of runoff water, desalination of seawater, joint development of economical irrigation techniques, and the community should be instrumental in negotiating equitable water lines separating participating nations.

In addition, the degradation of natural resources with climatic changes and human intervention exacerbate the problem of dryland regions in the community. Accompanying the intense sun radiation and poor quality of water resources with highly saline or brackish water, the challenge for the community is to turn the ecological disadvantages of the drylands to economic advantage. Methods for preventing desertification are needed.

See Agriculture (Center of the Middle East/North Africa Community), Centers of the Middle East/North Africa Community, desertification, environment, Middle East North Africa Community, Moscow Multilateral Middle East Conference, water.

Water Data Banks Project: The concept of a Regional Water Data Bank to improve the availability of data for Palestinian, Jordanian, and Israeli water managers, planners, and operators was adopted at the fourth Water Working Group meeting in Beijing.

The main activities in this European Union-financed project are the development of a Palestinian Water Data Unit, the updating of hydrological publications including a hydrological bulletin, the identifying and inputting of historical data, network review and evaluation, and the development of standards for hydrological, hydrogeological, hydrometeorological, and water quality standards.

The first phase of the project started in January 1996.

Water Working Group: See Moscow Multilateral Middle East Conference, water.

WEF: See World Economic Forum.

weapon-free zone: See unconventional weapons, United Nations (nuclear-weapon-free zone in the region of the Middle East).

Weizmann, Chaim: The first president of the State of Israel, Zionist leader, and scientist. Born in Russia in 1874, he was raised amid an atmosphere of love for Israel. Weizmann was one of the first Zionist visionaries who supported Theodore Herzl. Thanks to his prominence and scientific contributions to the British military effort during World War I, the British government issued the Balfour Declaration which expressed sympathy with the Zionist movement. Weizmann headed the World Zionist Organization for 20 years. After the establishment of the State of Israel, he was elected its first president in recognition of his critical role in laying the ground work for a Jewish state. He died in 1952.

See Arab-Jewish Agreement at Versailles, chronology (January 3, 1919).

West Bank: An area of Palestine west of the River Jordan; allocated by U.N. Partition Plan of 1947 as a separate Arab nation. The Camp David Accords between Israel and Egypt in September 1978 included an agreement for autonomy for the Palestinians in the West Bank and Gaza.

In July 1988 Jordan cut its ties with the territory.

See Agreement on Security Arrangements for Hebron, Casablanca Declaration, Declaration of Principles, Free Trade Agreement (United States and the West Bank and the Gaza Strip), Interim Agreement (on the West Bank and the Gaza Strip), Intifada, Moscow Multilateral Middle East Conference, Palestinian, Paris Economic Agreement, settlements, unemployment.

Western Sahara: See Ouguiya.

West Jerusalem: Captured by Israeli forces in the 1948-1949 Arab-Israeli War and retained by Israel. In December 1949 Israel relocated its capital to West Jerusalem, though it was not recognized by most world governments.

Compare East Jerusalem.

Westoxication: an Iranian term for all the evils from the Western way of life.

White House: See Clinton, William (President).

withdrawal: See redeployment.

Working Group on Economic Development: See Economic Development.

Working Group on the Environment: See Environment.

World Bank: A specialized agency of the United Nations to provide loans to governments of U.N. members for the purpose of economic development. As early as April 1993 the World Bank envisaged a 4-5 percent growth in the economies throughout the Middle East—the highest growth rate of the last 20 years.

On May 2, 1994, two days before the White House signing of the Palestinian-Israel Peace Accord, the World Bank unveiled its $1.2 billion, three-year economic development plan for the West Bank and Gaza, with financial pledges from 40 donor nations to assist in the implementation of the economic arrangements agreed upon between Israel and the Palestinians.

In its 1996 "Getting Connected Report-Private Participation in Infrastructure in the Middle East and North Africa" the World Bank urged that private participation in infrastructure—including power, telecommunications, transportation, and water and sanitation—brings the promise of better, faster, and cheaper ways of providing the services that link economies and improves lives. "Most countries in the Middle East and North Africa are in the early stages of preparing or implementing their first transactions involving private participation in infrastructure. These transactions present demanding choices and challenges for the region's governments." The World Banks believes that private participation in infrastructure in MENA will bring:

- better infrastructure for plugging into the world economy;
- opportunities to upgrade infrastructure are large;
- the quality of service is as much of an issue as the quantity;
- fast-growing populations and rapid urbanization will require massive infrastructure investments that are beyond the public sector's capacity to finance;
- new roles for governments to play;
- better connectors in which private participation can bring firms and households will be able to reap the benefits of an increasingly integrated world economy.

Synonymous with International Bank for Reconstruction.

See Amman Middle East/North Africa Economic Summit, Cairo Middle East/North Africa Economic Conference, chronology (May 2, 1994), industrial zones, Middle East/North Africa, Moscow Multilateral Middle East Conference, Paris Economic Agreement, Secretariat (Middle East/ North Africa Economic Summits).

World Economic Forum (WEF): The World Economic Forum based in Geneva, Switzerland is the foremost international membership organization integrating leaders from business, government and academia into a partnership committed to improving the state of the world. It provides a platform for interaction that is direct, efficient, global, personalized, and on the highest level.

Founder and President is Klaus Schwab, who encourages the foundation to serve the global public interest. It is independent, impartial, and not-for-profit, tied to no political, partisan, or regional interests. It is self-financed through annual membership fees and cost contributions from participants in specific events.

Organization and preparation for the Middle East/North Africa Economic Summits have rested with Gregory Blatt, WEF Director.

At the Amman Economic Summit in October 1995 Dr. Schwab said "Yet, recognizing that peace is good for business is not sufficient to attract the foreign investment needed to rebuild and restructure the Middle East and North Africa. Governments must also demonstrate that they realize that business is good for peace. They must do so quickly if they wish to compete with other regions of the world. To achieve this, governments must ensure political stability and create an environment conducive to business in which foreign investors feel both welcome and secure. This requires seemingly simple demands, which in practice could involve radical change. It requires from governments enlightened leadership and vision."

See Amman Middle East/North Africa Economic Summit, Cairo Middle East/North Africa Economic Conference, Casablanca Declaration, Casablanca Middle East/North Africa Economic Summit, Energizing Business Actors, Middle East and North Africa Business Interaction Group.

World Trade Centers Association (WTCA): See Energizing Business Actors, Secretariat (Middle East/North Africa Economic Summits).

World Trade Organization (WTO): The General Agreement on Tariffs and Trade (GATT) was replaced by the WTO on January 1, 1995. The Middle East and North Africa Community will be responsible for bringing non-WTO nations into the new world trade institution. Only a handful of regional nations—Kuwait, Bahrain, Egypt, Morocco, Tunisia, and Israel are members. Saudi Arabia, the United Arab Emirates, and Jordan are in the process of applying, with others showing interest in membership. The MEC/MENAC must seek to have its members liberalize trade and open their markets to compete against nationally protected industries.

For years, most Middle East and North Africa nations had seen little value in GATT or WTO membership. State-dominated regimes like Syria and Iraq had little interest in liberalizing trade or opening markets to compete against nationally protected industries. Oil-rich states, like those of the Arab Gulf, had a number of reasons for disinterest: heavy dependence on basic imports meant already low tariffs, high revenues from oil gave little incentive to diversify. Small economies, like Jordan and Yemen, refrained because of fears over outside competition.

Nations of the region, with a legacy of oil-rich states and state-run regimes, have a long and difficult road of economic adjustment ahead of them. The possible benefits, however, would seem to far outweigh the costs involved in joining with the world economy of the 1990s.

As of 1997, 10 of the 23 MENA countries have become members of the WTO and four countries are in the process of accession.

See General Agreement on Tariffs and Trade, Uruguay Round.

World Wide Web (WWW): See Energizing Business Actors, MENAnet, Regional Investment Promotion Program, Science and Technology Exchange, Secretariat (Middle East/North Africa Economic Summits).

World Zionist Organization: Founded during the First Zionist Congress at Basel in 1897 by Theodore Herzl, the organization's mission was defined in the Basel Program as "establishing a homeland in the Land of Israel for the Jewish people." From then on the World Zionist Organization became the primary Jewish organization working for an independent state for the Jewish people. Affiliation with the Zionist movement was open to anyone who accepted the Basel Program and paid membership dues—called purchasing the Zionist shekel. The following activities were determined as the means of achieving the goals of the World Zionist Organization: settlement in the Land of Israel, unifying the people of Israel through local organizations, and increasing Jewish national recognition and sentiment. Following the establishment of the State of Israel, the World Zionist Organization's mission was redefined as cooperating with the government to encourage aliyah, immigrant absorption and settlement.

See Herzl, Theodore.

WTCA: World Trade Centers Association. See Secretariat (Middle East/North Africa Economic Summits).

WTC Information Network: See Secretariat (Middle East/North Africa Economic Summits).

WTO: See World Trade Organization.

WWW: See World Wide Web.

XYZ

Yarmouk River: See Israel-Jordan Peace Treaty.

Yeltsin, Boris: See Amman Middle East/North Africa Economic Summit, Casablanca Declaration, Casablanca Middle East/North Africa Economic Summit, Middle East/ North Africa Economic Summit.

Yemen: Officially the Republic of Yemen, comprising the states of Yemen Arab Republic and the People's Democratic Republic of Yemen. The unified state was proclaimed in May 1990.

Conditions in Yemen are changing rapidly, mostly for the better. A populous country strategically located at the southern tip of the Arabian Peninsula, Yemen is pursuing the transition to democracy and a free market economy. The political and security problems which led to the 1994 civil war have lessened and the Yemeni government is committed to improving the business climate. A series of economic reforms announced in March 1995 should have positive impact in the medium to long term. However, Yemen remains a challenging market with real risks.

Both the long-term and the short-term outlook for Yemen are uncertain. Major road blocks to doing business include rampant corruption and red tape, import restrictions, a weak legal system, poor shipping links, a poorly functioning financial system, multiple exchange rates, near-hyper inflation, lack of financing, and poor dispute resolution mechanisms.

Top government reform priorities include cutting expenditures (to reduce inflation), privatizing state-owned enterprises, and reducing red tape and corruption. Positive changes include implementation of a new labor law, reduction in subsidies, and reform of currency exchange regulations, interest rates, and customs procedures.

In 1995, with the assistance from the International Monetary Fund and the World Bank, Yemen embarked on structural adjustments which, in 1997, have led to significant results. Inflation is down to 9 percent. Growth in the non-oil sector is 6 percent and the budget deficit has been almost eliminated.

Yemini: See minorities.

Yeminite Jews: See Operation Magic Carpet.

Yilmaz, Mesut: See Turkey.

Yom Kippur War: Commenced with surprise Arab attacks on Israeli forces on the Day of Atonement (Yom Kippur) on October 6, 1973. A cease-fire was called on October 24, with disengagement along the Suez Canal and the formation of a U.N. peace force on the Golan Heights.

Synonymous with October War.

See chronology (1973).

youth: See Interns for Peace, Japan.

Zine el-Abdine Ben Ali: See Ben Ali.

Zionism: The movement advocating the return of Jews to their homeland of Palestine and the resumption of sovereign Jewish life there. Not until 1914 did the British government mention, probably for the first time, that Zionism was an attractive idea, to keep both Russia and France away.

The Zionist movement enunciated its goals—a national home for the Jewish people in the Land of Israel. In 1997 Zionism celebrated its one hundredth birthday.

See Balfour Declaration, boycott, Fatah, Hamas, Hamas Covenant, Israel (Declaration of Establishment), League of Nations Mandate, PLO Charter, World Zionist Organization.

Zionist Organization: See Arab-Jewish Agreement at Versailles.

Zionist shekel: See World Zionist Organization.

Zofar Area: See Israel-Jordan Peace Treaty.

Zuckerman, Mortimer B.: See Council on Foreign Relations.

BIBLIOGRAPHY

1997

Eliyahu Kanovsky, "The Middle East Economies: The Impact of Domestic and International Politics," Bar-Ilan University, Mideast Security and Policy Studies, Israel, February, 33 pages.

1996

"Building on Progress: Reform and Growth in the Middle East and North Africa," Middle Eastern Department, The International Monetary Fund, Washington, DC, November, 69 pages.

"Getting Connected: Private Participation in Infrastructure in the Middle East and North Africa," The World Bank, Washington, DC, November, 28 pages.

Ishac Diwan, "External Debt in the MENA Region-1994-1995," Cairo: Economic Research Forum.

"Programs for Regional Cooperation-1997," The Government of Israel, November, unpaginated order.

Willem Van Eeghen, *Poverty in the Middle East and North Africa*, Washington, DC, World Bank.

"Economic Trends in the MENA Region," The Economic Research Forum for the Arab Countries, Iran and Turkey, Cairo, 107 pages.

"Middle East North Africa Capital Markets-Investor's Guide," Middle East & North Africa Economic Summit, Rabat, 142 pages.

Jerry M. Rosenberg,*The Peace Dividend: Creating a Middle East/North Africa Community*, Rabat, Morocco, Middle East/North Africa Economic Summit Secretariat, The MENA Series, 273 pages.

Peter L. Sud, "Infrastructure Investment in the Middle East," Keynote speech to a conference sponsored by the Middle East Economic Digest, Bahrain, February.

Percy Barnevik, "Foreign Investment and Industrial Development in the Middle East and North Africa: Opportunities and Constraints," Forum, ERF Newsletter, Volume 3, No. 2, June.
Joel Peters, "Pathways to Peace: The Multilateral Arab-Israeli Peace Talks," The Royal Institute of International Affairs, London, 110 pages.
"Middle East Regional Co-operation," The European Commission, Brussels, November, 16 pages.

1995

Benjamin R. Barber, *Jihad vs. McWorld*, New York, Times Books, Random House, 381 pages.
"Building a Prosperous Peace: Amman '95, Middle East and North Africa Economic Summit." October 29-31, International Press Office, The Royal Hashemite Court, Amman, Jordan.
Heba Handoussa, "Potential for Arab Cooperation for Industrial Development," United Nations Industrial Development Organization.
"Middle East Economic Strategy Group," October 30, Paul A. Volcker, Chair, Council on Foreign Relations, New York, 9 pages.
R. Shaban, R. Assaad, and S. Al-Qudsi, "The Challenge of Unemployment in the Arab Region," International Labor Review, Vol. 134, No. 1.
Claiming the Future: Choosing Prosperity in the Middle East and North Africa. The World Bank, Washington, DC, 119 pages.
David Kimche, "Sowing-the Seeds of Economic Cooperation," Institute of the World Jewish Congress, Jerusalem, 23 pages.
"Development Options for Cooperation: The Middle East/ East Mediterranean Region," Government of Israel, Ministry of region, Version IV, August, unpaged.
"The Jordan-Israel Peace Treaty," Jordan Media Group, Publication No. 181, Third printing, August, unpaged.
"The Casablanca Report," The Council on Foreign Relations and The World Economic Forum, 65 pages.
Albert Martinez, "Private Sector Development in the Middle East and North Africa," World Bank, Europe and Central Asia-Middle East and North Africa Technical Department, Washington, DC.
Jerry M. Rosenberg, *Encyclopedia of the North American Free Trade Agreement, the New American Community, and Latin-American Trade*, Greenwood Press, Westport, CT, 562 pages.
World Bank, "Will Arab Workers Prosper or Be Left Out in the Twenty-First Century?" Regional Perspective on World Development Report, Washington, DC.

1994

Mohamed Ben El Hassan Alaoui, "La Cooperation entre l'Union Europeenne et les pays du Maghreb," Nathan, 237 pages.

Jeremy Berkoff, *A Strategy for Managing Water in the Middle East and North Africa*, A world Bank Directions in Development book, Washington, DC.

"Beyond the White House Lawn: Current Perspectives on the Arab-Israeli Peace Process," Anti-Defamation League, New York, 117 pages.

M. Karsheans, "Structural Adjustment and Employment in the Middle East and North Africa," Department of Economics, School of Oriental and African Studies, University of London, November.

Clive H. Schofield and Richard N. Schofield (Ed), *The Middle East and North Africa*, Routledge, London, 204 pages.

Jerry M. Rosenberg, *Dictionary of International Trade*, John Wiley & Sons, Inc., New York, 314 pages.

Daniel Hillel, *Rivers of Eden: The Struggle for Water and the Quest for Peace in the Middle East*, Oxford, New York, 355 pages.

Stanley Fisher, Leonard J. Hausman, Anna D. Karasik, and Thomas C. Schelling, *Securing Peace in the Middle East*, The MIT Press, Cambridge, MA, 166 pages.

1993

Yehuda Raveh, "Free Trade from Egypt, Jordan, Syria & other Middle Eastern States," Middle East Edition, Raveh & Co., Kesselman, Tel-Aviv, 99 pages.

"Occupied Palestinian Territory," Palestinian Academic Society for the Study of International Affairs, PASSIA, 129 pages.

Shimon Peres, *The New Middle East*, Henry Holt, New York, 224 pages.

C. Pissarides, "Labor Markets in the Middle East and North Africa," World Bank, Middle East and North Africa Region, Discussion Paper 5, Washington, DC.

Stanley Fischer, Dani Rodrik, and Elias Tuma (Ed) *The Economics of Middle East Peace*, The MIT Press, Cambridge, 370 pages.

1992

Jerry M. Rosenberg, *The New American Community: A Response to the European and Asian Economic Challenge*, Praeger, Westport, CT, 187 pages.

1991

Jerry M. Rosenberg, *The New Europe: An A to Z Compendium on the European Community*, The Bureau of National Affairs, Washington, DC, 206 pages.

Magazines

Middle East Insight, various issues.
The Jerusalem Report, various issues.
MEED, Middle East Business Weekly, various issues.
Worldlink, various issues.
Civil Society, various issues.
Al-Ahram, various issues.

ABOUT THE AUTHOR

Jerry M. Rosenberg is Professor of International Business, Department Chairman, and Director of the Center for Middle East/North Africa Business Studies at Rutgers University, Graduate School of Management, Newark, New Jersey. He was also a visiting professor at the University of British Colombia (Canada), Middlesex Polytechnic University (England), John Cabot University (Italy), and has lectured at the London School of Economics and Political Sciences, the University of Hawaii, Ben-Gurion University, and Hebrew University (Israel).

The U.S. Library of Congress lists 25 of his books, including seven new editions. Many of these works have been translated into several languages, including Spanish, Chinese, Slovak, Russian, and Japanese. Some of his most recent books are: *The New Europe*, *The New American Community*, and *Dictionary of International Trade*.

An expert leader at the fall 1994 Casablanca Economic Summit and the 1995 Amman Economic Summit, Dr. Rosenberg's *The Peace Dividend: Creating a Middle East/North Africa Community* was initially published by the Secretariat of the Middle East/North Africa Summit (Morocco) in the fall 1996 and distributed at the Cairo Middle East/North Africa Economic Conference. The book dealt with a model of institutional structure and governance for the region. Based on his 15 years of research on regional economic integration and alliances dealing with the European Union, NAFTA, and Latin-American trade, he has made numerous trips throughout North Africa and the Middle East to collect pertinent data and interview many of the leading visionaries in creating a new prosperous and peaceful area.

Professor Rosenberg is married, father of two daughters and two granddaughters.